Sustaining North American Salmon: Perspectives Across Regions and Disciplines

Sustaining North American Salmon: Perspectives Across Regions and Disciplines

Edited by

Kristine D. Lynch
Michael L. Jones
William W. Taylor

Michigan State University
Department of Fisheries and Wildlife
13 Natural Resources Building
E. Lansing, Michigan 48824-1222

American Fisheries Society
Bethesda, Maryland
2002

Suggested citation formats follow.

Entire book

Lynch, K. D., M. L. Jones, and W. W. Taylor, editors. 2002. Sustaining North American salmon: perspectives across regions and disciplines. American Fisheries Society, Bethesda, Maryland.

Chapter within the book

Whoriskey, F. G., and J. Glebe. 2002. The Atlantic salmon recreational angling industry: economic benefits. Pages 77–92 *in* K. D. Lynch, M. L. Jones, and W. W. Taylor, editors. Sustaining North American salmon: perspectives across regions and disciplines. American Fisheries Society, Bethesda, Maryland.

Printed in the United States of America on acid-free paper.

Library of Congress Control Number 2002101549
ISBN 1-888569-25-5

American Fisheries Society
5410 Grosvenor Lane, Suite 110
Bethesda, Maryland 20814-2199

Contents

Contributors vii
Preface xi
Symbols and Abbreviations xvii

Part I Atlantic Salmon

From Game Fish to Tame Fish: Atlantic Salmon in North America, 1798 to 1998 3
John F. Kocik and Russell W. Brown

Sustaining Atlantic Salmon in the Northwest Atlantic: Considerations from a Life History Perspective 33
Jeffrey A. Hutchings

Atlantic Salmon Fisheries in Eastern North America: The Prince and the Pauper 61
Kenneth F. Beland and Alex T. Bielak

The Atlantic Salmon Recreational Angling Industry: Economic Benefits 77
F. G. Whoriskey and J. Glebe

Canadian and U.S. Atlantic Salmon Institutions and Politics: Where are the Fish and Who Cares? 93
Stephen G. Rideout and John A. Ritter

Emerging Science and Management Issues with Atlantic Salmon 117
Kevin D. Friedland

Part II Great Lakes Salmon

History of the Great Lakes Salmon Fishery: A Michigan Perspective 139
Howard A. Tanner and Wayne H. Tody

Ecological Factors Affecting the Sustainability of Chinook and Coho Salmon Populations in the Great Lakes, Especially Lake Michigan 155
Michael J. Hansen and Mark E. Holey

The Many Faces of Salmon: Implications of Stakeholder Diversity in the Great Lakes 181
Barbara A. Knuth

Uses of Resource Economics in Managing Great Lakes Fisheries: Michigan Examples 195
Frank Lupi and Douglas B. Jester

Managing Great Lakes Fisheries Under Multiple and Diverse Authorities 221
Margaret Ross Dochoda and Michael L. Jones

The Future of Pacific Salmon in the Great Lakes 243
C. I. Goddard

Part III Pacific Salmon

History of the Salmon Fisheries of the Pacific Northwest Coast 261
Terry Glavin

Ecological Perspectives on Pacific Salmon: Can We Sustain Biodiversity and Fisheries? 277
E. Eric Knudsen

Salmon, Society, and Politics: Moving Toward Ecosystem-Based Management for Pacific Salmon 321
Donald D. MacDonald, Megan Hanacek, and Laura Genn

The Pacific Salmon Dispute: Rationalizing a Dysfunctional Joint Venture 341
Robert McKelvey and Kathleen A. Miller

Challenges and Opportunities for the Future of Pacific Salmon Management 369
Jim Martin

Summary

Sustaining Salmon Fisheries in Canada and the United States: Recommendations for Policy and Management 379
K. D. Lynch, M. L. Jones, and W. W. Taylor

Contributors

Kenneth F. Beland (Chapter 3): Maine Atlantic Salmon Commission, 650 State Street, Bangor, Maine 04401, USA.

Alex T. Bielak (Chapter 3): New Brunswick Cooperative Fish and Wildlife Research Unit, Biology Department and Faculty of Forestry and Environmental Management, University of New Brunswick, New Brunswick, Canada; Honorary Research Associate, Columbia Rivers Institute, 77 Flanders Drive, Waterdown, Ontario L0R 2H7, Canada.

Russell W. Brown (Chapter 1): National Oceanic and Atmospheric Administration, NOAA-Fisheries, Northeast Fisheries Science Center, 166 Water Street, Woods Hole, Massachusetts 02543-1097, USA.

Margaret Ross Dochoda (Chapter 11): Great Lakes Fishery Commission, 2100 Commonwealth Boulevard, Suite 100, Ann Arbor, Michigan 48105, USA.

Kevin D. Friedland (Chapter 6): UMass/NOAA Cooperative Marine Research and Education, Blaisdell House, University of Massachusetts, Amherst, Massachusetts 01003, USA.

Laura Genn (Chapter 15): Sustainable Fisheries Foundation, 2376 Yellow Point Road, Nanaimo, British Columbia V9X 1W5, Canada.

Terry Glavin (Chapter 13): Pacific Fisheries Resource Conservation Council, Box C-25, Fernhill Road, Mayne Island, British Columbia V0N 2J0, Canada.

J. Glebe (Chapter 4): University of New Brunswick, Joy Kidd Residence, P. O. Box 473, Fredericton, New Brunswick E3B 5A3, Canada.

C. I. Goddard (Chapter 12): Great Lakes Fishery Commission, 2100 Commonwealth Boulevard, Suite 209, Ann Arbor, Michigan 48109, USA.

Megan Hanacek (Chapter 15): Sustainable Fisheries Foundation, 2376 Yellow Point Road, Nanaimo, British Columbia V9X 1W5, Canada.

Michael J. Hansen (Chapter 8): University of Wisconsin-Stevens Point, College of Natural Resources, 1900 Franklin Street, Stevens Point, Wisconsin 54481, USA.

Mark E. Holey (Chapter 8): U.S. Fish and Wildlife Service, Green Bay Fishery Resources Office, 1015 Challenger Court, Green Bay, Wisconsin 54311, USA.

Jeffrey A. Hutchings (Chapter 2): Department of Biology, Dalhousie University, Halifax, Nova Scotia B3H 4J1, Canada.

Douglas B. Jester (Chapter 10): 350 Whitehills Drive, East Lansing, Michigan 48823, USA.

Michael L. Jones (Chapter 11): Michigan State University, Department of Fisheries and Wildlife, 13 Natural Resources Building, East Lansing, Michigan 48824, USA.

E. Eric Knudsen (Chapter 14): USGS, Biological Resources Division, Alaska Biological Science Center, 1011 East Tudor Road, Anchorage, Alaska 99503, USA.

Barbara A. Knuth (Chapter 9): Human Dimensions Research Unit, Cornell University, Department of Natural Resources, 122A Fernow Hall, Ithaca, New York 14853, USA.

John F. Kocik (Chapter 1): National Oceanic and Atmospheric Administration, NOAA-Fisheries, Northeast Fisheries Science Center, 166 Water Street, Woods Hole, Massachusetts 02543-1097, USA.

Frank Lupi (Chapter 10): Department of Agricultural Economics and Department of Fisheries and Wildlife, Michigan State University, East Lansing, Michigan 48824, USA.

Donald D. MacDonald (Chapter 15): Sustainable Fisheries Foundation, 2376 Yellow Point Road, Nanaimo, British Columbia V9X 1W5, Canada.

Jim Martin (Chapter 17): Conservation Director, Pure Fishing, 1900 18th Street, Spirit Lake, Iowa 51360, USA.

Robert McKelvey (Chapter 16): University of Montana, 1740 Madeline Avenue, Missoula, Montana 59801, USA.

Kathleen A. Miller (Chapter 16): Environmental and Societal Impacts Group, National Center for Atmospheric Research, P. O. Box 3000, Boulder, Colorado 80307, USA.

Stephen G. Rideout (Chapter 5): USGS, S.O. Conte Anadromous Fish Research Center, P. O. Box 796, Turners Falls, Massachusetts 01376, USA.

John A. Ritter (Chapter 5): 4191 Highway #2, Wellington, Nova Scotia B2T 1A3, Canada.

Howard A. Tanner (Chapter 7): 5755 Green Road, Haslett, Michigan 48840, USA.

Wayne H. Tody (Chapter 7): 942 Snow Apple Drive, Traverse City, Michigan 49684, USA.

F. G. Whoriskey (Chapter 4): Atlantic Salmon Federation, P. O. Box 5200, St. Andrews, New Brunswick E5B 3S8, Canada.

Preface

Salmon are among the most economically and culturally significant fishes in North America, yet they are also among the most challenging species to manage. In large part, this challenge results from the fact that, in all regions of North America where they are found, salmon's migration route crosses state, provincial, and national boundaries. Salmon produced in streams or hatcheries from one jurisdiction are often harvested in another. Changes to stocks and the habitats upon which they depend have forced managers to implement costly measures to protect and restore salmon, raising difficult social and political issues regarding who pays and who reaps the benefits. This book, based on a symposium held at the 1998 American Fisheries Society meeting in Hartford, Connecticut, documents an attempt to examine these issues in new ways—by contrasting salmon management experiences in different regions of North America and by viewing the problem from the perspective of an array of relevant disciplines.

Salmon management, like all resource management, is not simply a matter of defining the ecological limits for sustainable use of the species and then prescribing management actions that are in keeping with these limits. Yet, this problem alone has proven to be an enormous challenge, even though salmon are among the most thoroughly researched fish species on the planet. Because of the immense significance of salmon to people, wise salmon management requires an equally firm appreciation of the social, economic, and political influences on the relationship between people and salmon. In developing the symposium upon which this book is based, we first adopted the view that a thorough examination of "sustaining salmon" necessitates consideration of each of these disciplines, not to mention the relationships among them.

Regionally-important salmon fisheries exist in three areas of North America: the Pacific Northwest, the Great Lakes, and the Atlantic Coast. Each area has experienced a starkly different history, and managers in each region face an array of challenges that are sometimes similar and sometimes distinct. For each of these regions, we have compiled multidisciplinary assessments of the salmon fisheries and the issues affecting their sustainability. Our hope was that insight might be gained by formally contrasting experiences with these fisheries, to see whether lessons learned in one region might be applied to others.

This book is organized into three main sections, one for each region. Each section contains a series of chapters covering the history, ecology, economics, sociology, politics and institutions, and future of salmon management. The general topics of the comparable chapters for each region are consistent across regions, but the details vary considerably. Our original intent was to develop a very similar set of chapters for each region, but we soon realized that we would gain more by allowing the authors of each chapter to make their own judgments of how to tackle their topics. The result is a rather eclectic set of contributions whose character provides fascinating, if not completely unbiased, insights into the problems and issues faced within each region and discipline. Although the content of some chapters may occasionally overlap with other chapters when discussing important common issues, we have elected to retain these redundancies in compiling this book so that each chapter can serve as a coherent, informative, and independent contribution to salmon policy and management.

Part One: North Atlantic Salmon Fisheries

Salmon fisheries in the North Atlantic Ocean have had a long history of commercial harvest, but this harvest proved to be unsustainable. The combined effects of overharvesting and habitat loss have all but eliminated a commercial wild salmon fishery, and remnant populations now mainly support a specialized (and sometimes exclusive) recreational fishery. While managers attempt to rebuild Atlantic salmon stocks to self-sustaining populations, hatcheries and fish farms supplement salmon production. This section opens with a history of North Atlantic salmon fisheries by John F. Kocik and Russell W. Brown, "From Game Fish to Tame Fish: Atlantic Salmon in North America, 1798 to 1998." Next, Jeffrey A. Hutchings provides an ecological perspective in "Sustaining Atlantic Salmon in the Northwest Atlantic: Considerations from a Life History Perspective." Hutchings describes the remarkable diversity of Atlantic salmon life histories and explains how this diversity may be key to the management of the species.

Three chapters follow that explore many of the social aspects of the North Atlantic salmon fishery. Kenneth F. Beland and Alex T. Bielak, in "Atlantic Salmon Fisheries in Eastern North America: The Prince and the Pauper," present a vivid illustration of today's recreational salmon fishery, highlighting how different social conditions influence various approaches to salmon conservation and management. Frederick G. Whoriskey and J. Glebe discuss the economic forces that influence salmon sustainability in "The Atlantic Salmon Recreational Angling Industry: Economic Benefits." The binational aspects of Atlantic salmon management are touched upon in these chapters, but they receive more detailed attention in the contribution by Stephen G. Rideout and John A. Ritter, "Canadian and U.S. Atlantic Salmon Institutions and Politics: Where are the Fish and Who Cares?" Finally, Kevin D. Friedland looks to the future in "Emerging Science and Management Issues with Atlantic Salmon," which summarizes and synthesizes how ecological and social systems are affecting the long-term outlook for salmon in the North Atlantic Ocean. In particular, Friedland notes how the fate of Atlantic salmon populations may be closely tied to large-scale environmental changes over which humans have relatively little control.

Part Two: Great Lakes Salmon Fisheries

Salmon are not native to the Great Lakes, yet they now support a thriving recreational fishery throughout the basin. While salmon in the Great Lakes are not facing immediate threats of collapse, effective management of this fishery presents challenges to those who manage salmon habitat, hatcheries, harvests, and the coordination of stocking efforts. One perspective on the history of the Great Lakes salmon fishery is recounted in "History of the Great Lakes Salmon Fishery: A Michigan Perspective." This chapter was written by two managers who were instrumental in bringing the species to the Great Lakes in the 1960s, Howard A. Tanner and Wayne H. Tody. Michael J. Hanson and Mark E. Holey evaluate the performance and impacts of introduced Pacific salmon in "Ecological Factors Affecting the Sustainability of Chinook and Coho Salmon Populations in the Great Lakes, Especially Lake Michigan." They highlight the challenges of managing an artificially-sustained fishery within the ecological constraints created by a natural but disturbed food web.

Several social aspects of the Great Lakes fishery are discussed in the next three chapters on stakeholders, economics, and politics surrounding the Great Lakes salmon fishery. In "The Many Faces of Salmon: Implications of Stakeholder Diversity in the Great Lakes," Barbara A. Knuth describes the complex social setting of U.S. and Canadian salmon fishery stakeholders; this chapter highlights the importance of values in determining management goals, approaches, and conflicts. Other kinds of values are discussed in Frank Lupi and Douglas B. Jester's chapter, "Uses of Resource Economics in Managing Great Lakes Fisheries: Michigan Examples" They explain how resource economics has been and continues to be a significant tool in developing the region's fisheries. In "Managing Great Lakes Fisheries Under Multiple and Diverse Authorities," Margaret R. Dochoda and Michael L. Jones discuss the complex institutional arrangements that have evolved to promote high levels of coordination and collaboration among the region's state/provincial, national, and tribal aquatic resource managers. The outlook for the Great Lakes salmon fishery is discussed in "The Future of Pacific Salmon in the Great Lakes" by Chris I. Goddard. He explores how long-term ecosystem productivity may vary as the Great Lakes ecosystem changes, yet the social forces that affect the goals, priorities, and approaches to salmon management will still have to be coordinated throughout the basin.

Part Three: North Pacific Salmon Fisheries

As in the North Atlantic, stocks of salmon in the Pacific region of North America face serious declines due to overfishing and habitat loss. In contrast to their Atlantic counterparts, however, many Pacific salmon stocks remain the object of highly valued commercial fisheries. Pacific salmon constitute a key ingredient of the cultural as well as economic heritage of the region. For these reasons and many others, the stakes in the Pacific Northwest are very high. We have tried to convey a sense of the magnitude and scope of these issues in Part 3 of this volume.

In "History of the Salmon Fisheries of the Pacific Northwest Coast," Terry Glavin lays the foundation for understanding the complex issues that have long surrounded salmon fisheries in the North Pacific. Glavin uses this historical perspective to argue for a fundamental shift in salmon management in the region. E. Eric Knudsen provides a comprehensive overview of how ecosystem impacts can affect salmon sustainability in "Ecological Perspectives on Pacific Salmon: Can We Sustain Biodiversity and Fisheries?" He offers many suggestions for action to restore and protect healthy salmon populations. As in other regions, the ecological issues that lie at the heart of sustainability cross political boundaries and therefore require interjurisdictional, cooperative management approaches. Complex and often contentious interjurisdictional relationships have led to various forms of management agreements, some of which are discussed in "The Pacific Salmon Dispute: Rationalizing a Dysfunctional Joint Venture" by Robert McKelvey and Kathleen A. Miller. Using economic and game theory, they present a compelling analysis of the strengths and weaknesses of the 1985 Pacific Salmon Treaty and its more recent replacement. Donald D. MacDonald, Megan Hanacek, and Laura Genn discuss other social factors that affect society's relationship with salmon and their ecosystems in their chapter, "Salmon, Society, and Politics: Transitioning Toward Ecosystem-Based Management for Pacific Salmon."

The Pacific salmon section concludes with an essay by Jim Martin, "Challenges and Opportunities for the Future of Pacific Salmon Management," in which he reflects on the symposium as a whole and on his career as a senior fishery manager in the Pacific Northwest. He uses his unique insight and perspectives to further encourage holistic, collaborative management frameworks as ways to increase the likelihood of achieving sustainable management systems.

In the final chapter, we attempt to draw together common and contrasting themes from the three regions covered by the book. This comparative analysis leads into an assessment of the factors that appear to either hinder or promote sustainability of salmon fisheries in North America. Building on this synthesis, we present a set of recommendations covering three topics: policy and management, communications, and research. Our ultimate goal for this book and particularly the final chapter is to assist fisheries professionals in jointly pursuing and implementing policy and research agendas that benefit from a consideration of past practices. In turn, this should help lead to greater levels of coordination and collaboration in interjurisdictional approaches to common salmon management issues. We hope that this book will make important contributions to the fisheries profession and will help promote sustainable salmon management throughout North America and beyond.

This book would not be possible without the generosity, support, and assistance of many others. John F. Kocik contributed a great deal of time and effort in helping to organize and host the original symposium on this topic. A Canadian Embassy symposium grant and the Canadian Consulate General in Detroit provided financial and professional support for the symposium. Funding of this publication was also provided by Michigan State University's Office of the Provost, the College of Agriculture and Natural Resources, the Department of Fisheries and Wildlife, and the Canadian Studies Centre. These sources funded the travel and other meeting expenses so that speakers could attend the meeting, and they contributed to book publishing fees. Sue Robertson performed the technical editing, and the dedicated publications staff at the American Fisheries Society helped guide and oversee the entire publishing process.

Finally, we would like to thank all the fisheries professionals who contributed chapters and anonymous peer reviews (three per chapter) to this book. The chapters in this book represent their collective knowledge and insight, and we appreciate their dedication and concern for promoting awareness about the sustainability of North American salmon fisheries.

Symbols and Abbreviations

The following symbols and abbreviations may be found in this book without definition. Also undefined are standard mathematical and statistical symbols given in most dictionaries.

A	ampere
AC	alternating current
Bq	becquerel
C	coulomb
°C	degrees Celsius
cal	calorie
cd	candela
cm	centimeter
Co.	Company
Corp.	Corporation
cov	covariance
DC	direct current; District of Columbia
D	dextro (as a prefix)
d	day
d	dextrorotatory
df	degrees of freedom
dL	deciliter
E	east
E	expected value
e	base of natural logarithm (2.71828...)
e.g.	(exempli gratia) for example
eq	equivalent
et al.	(et alii) and others
etc.	et cetera
eV	electron volt
F	filial generation; Farad
°F	degrees Fahrenheit
fc	footcandle (0.0929 lx)
ft	foot (30.5 cm)
ft^3/s	cubic feet per second (0.0283 m^3/s)
g	gram
G	giga (10^9, as a prefix)
gal	gallon (3.79 L)
Gy	gray
h	hour
ha	hectare (2.47 acres)
hp	horsepower (746 W)
Hz	hertz
in	inch (2.54 cm)
Inc.	Incorporated
i.e.	(id est) that is
IU	international unit
J	joule
K	Kelvin (degrees above absolute zero)
k	kilo (10^3, as a prefix)
kg	kilogram
km	kilometer
l	levorotatory
L	levo (as a prefix)
L	liter (0.264 gal, 1.06 qt)
lb	pound (0.454 kg, 454g)
lm	lumen
log	logarithm
Ltd.	Limited

M	mega (10^6, as a prefix); molar (as a suffix or by itself)
m	meter (as a suffix or by itself); milli (10^{23}, as a prefix)
mi	mile (1.61 km)
min	minute
mol	mole
N	normal (for chemistry); north (for geography); newton
N	sample size
NS	not significant
n	ploidy; nanno (10^{29}, as a prefix)
o	ortho (as a chemical prefix)
oz	ounce (28.4 g)
P	probability
p	para (as a chemical prefix)
p	pico (10^{212}, as a prefix)
Pa	pascal
pH	negative log of hydrogen ion activity
ppm	parts per million
qt	quart (0.946 L)
R	multiple correlation or regression coefficient
r	simple correlation or regression coefficient
rad	radian
S	siemens (for electrical conductance); south (for geography)
SD	standard deviation
SE	standard error
s	second
T	tesla
tris	tris(hydroxymethyl) aminomethane (a buffer)
UK	United Kingdom
U.S.	United States (adjective)
USA	United States of America (noun)
V	volt
V, Var	variance (population)
var	variance (sample)
W	watt (for power); west (for geography)
Wb	weber
yd	yard (0.914 m, 91.4 cm)
α	probability of type I error (false rejection of null hypothesis)
β	probability of type II error (false acceptance of null hypothesis)
Ω	ohm
μ	micro (10^{26}, as a prefix)
′	minute (angular)
″	second (angular)
°	degree (temperature as a prefix, angular as a suffix)
%	per cent (per hundred)
‰	per mille (per thousand)

Part I Atlantic Salmon

Sustaining North American Salmon: Perspectives Across Regions and Disciplines, 3–32

Chapter 1

From Game Fish to Tame Fish: Atlantic Salmon in North America, 1798 to 1998

John F. Kocik and Russell W. Brown

Introduction

Atlantic salmon *Salmo salar* have an extensive pan-North Atlantic distribution that historically ranged from Portugal northward to the Arctic Circle in Europe across the Atlantic Ocean, from Ungava Bay to Long Island Sound in North America (Scott and Crossman 1973). Within North America, anadromous Atlantic salmon were native to nearly every coastal river north of the Hudson River (Atkins 1874; Kendall 1935; Scott and Crossman 1973). Although this represents a longitudinal range of approximately 2,100 km, the North American coastline length inhabited by the species exceeds 16,000 km (Dunfield 1985).

Though Atlantic salmon were likely never as numerous as the species complex of North Pacific salmonines (Carlson 1993), historical accounts suggest that they were a notable portion of the anadromous fish catch in eastern North America prior to Euro-American settlement (Stolte 1981; Dunfield 1985). A large body of literature details the importance of salmon to the peoples of the Pacific Northwest from aboriginal to modern times. Although there is less documentation, Atlantic salmon were important to aboriginal peoples in the Atlantic provinces dating back to at least 4000 B.C. (Dunfield 1985). Though native peoples relied on salmon to varying degrees as a resource, they apparently did not overexploit the resource. Given the distribution of American shad *Alosa sapidissima* and other alosids in North America, it is likely that, as latitude decreased, Atlantic salmon became a less prevalent component of the anadromous fish community and the alosid component increased. Many European settlers utilized Atlantic salmon in their homelands, so they naturally took advantage of this familiar resource (Beland and Bielak, Chapter 3).

In this chapter, our goal is to quantitatively review the history of the Atlantic salmon resource in North America. This analysis is presented within the framework of a review of the distribution and abundance of the North American stock complex. We utilize available data to document the trends in distribution and abundance of Atlantic salmon. Our reconstruction of history, incorporated within the broader context of this book on salmon across North America, is intended to facilitate comparisons related to the success and pitfalls of salmonine conservation. Though declines in abundance have not been uniform, overall patterns reflect a marked decline in abundance and a range contraction at the southern edge of their distribution in recent decades. Review of conservation efforts within North America should indicate some critical pitfalls to the conservation and restoration of Atlantic salmon populations.

This chapter provides descriptions of the North American range of Atlantic salmon, the diverse ecosystems they inhabit, and the resultant substructure of these populations; a description of the population trends of wild Atlantic salmon populations for the past 200 years; an analysis of the effect of hatcheries and fisheries upon the North American stock complexes; and an assessment of the status of Atlantic salmon held in private aquaculture facilities. Within this discussion, we compare the productivity of the natural environment with production occurring in aquaculture and mariculture environments.

Are There Discrete Populations of Atlantic Salmon in North America?

Fisheries managers have increasingly adopted a holistic approach to management activities that acknowledges the importance of local adaptations in fish populations. This approach was formalized by the Stock Concept International Symposium in 1980 (Berst and Simon 1981). Throughout their range, Atlantic salmon are naturally structured into genetically differentiated and reproductively isolated populations within and among coastal drainages and are referred to as stocks (Saunders 1981; Thorpe and Mitchell 1981; Ståhl 1987; Bermingham et al. 1991; King et al. 1993; Nielsen 1998; King et al. 1999). Mechanisms acting to differentiate stocks are reinforced by the strong homing tendencies of Atlantic salmon, and a decrease in straying to nonnatal rivers as distance between rivers increases is thought to be common (Johannsson et al. 1998). Both Canadian and U.S. managers accept river stocks as the standard for the smallest management unit of Atlantic salmon populations (Moring et al. 1995; Atkinson et al. 1998).

Recently, there has been an interest in describing unique population units for conservation purposes (Nielsen and Powers 1995; Nielsen 1998). The target population units for these efforts range in scope from substocks to stocks through the species level, depending upon the legal mandates or overall purpose of identifying these units. From a practical viewpoint, fisheries managers and most conservationists agree that individual stocks are important components of a species (Berst and Simon 1981). However, many managers recognize that an intermediate unit exists that is often the most practical one for conservation because it preserves biodiversity at the appropriate spatial and temporal scale (Callicott 1995) and forms spatially meaningful management units (Atkinson et al. 1998).

Within the United States, efforts to critically assess the discreteness of salmon stocks have been fostered by efforts to define stocks of Pacific salmon *Oncorhynchus* spp. regulated under the Endangered Species Act (ESA; Waples 1991). This effort led to the development of an objective method, the evolutionarily significant unit (ESU) approach, to define that portion of the species distribution considered to be unique and in need of protection. The ESU approach was primarily used by the National Marine Fisheries Service (NMFS) to assess Pacific salmon populations under consideration for ESA protection. In 1996, the NMFS and the U.S. Fish and Wildlife Service (FWS) extended the ESU concept to define appropriate units for vertebrate conservation. This policy consists of three elements to be considered in a decision regarding the status of a possible distinct population segment (DPS) as endangered or threatened under the ESA: the discreteness of the population segment in relation to the remainder of the species or subspecies to which it belongs, the significance of the population segment to the species or subspecies to which it belongs, and the conservation status of the population segment in relation to ESA listing standards. The first two criteria determine the appropriate units of conservation; the third is used to determine the health of the DPS delineated in the first two criteria.

Stock status reports for Atlantic salmon in Canadian waters document approximately 550 rivers containing Atlantic salmon. These rivers are divided among 34 management units—salmon fishing areas (SFAs)—that are functionally similar to DPSs. That is, the SFAs represent a partitioning that is primarily along biological and ecological lines but is also influenced by political boundaries and management authorities. Canadian Atlantic salmon resources in New Brunswick, Prince Edward Island, Nova Scotia, and Newfoundland and Labrador are managed by the Department of Fisheries and Oceans (DFO), while salmon populations in Quèbec are managed by the Ministère de l'Environnement et de la Faune. Atlantic salmon populations under DFO management are grouped into 23 SFAs, while populations managed by Quèbec are grouped into 11 areas designated by Q1–11. Population trends within Canadian SFAs are generally evaluated by monitoring index rivers within each SFA.

Biodiversity of Atlantic Salmon in North America

Under provisions of the U.S. Endangered Species Act and DPS guidelines, Colligan et al. (1999) examined genetic, life history, biogeographical, and environmental information in the process of evaluating Atlantic salmon populations throughout their U.S. range. Their analysis was complicated by the fact that elimination of stocks from 96% of their historical habitat by human activities has resulted in almost a 2-degree northward latitudinal shift in the southern extent of their historic distribution within the United States. Additionally, Colligan et al. (1999) considered historical stocking activities and their potential influence on genetic structure. Given these limitations, Colligan et al. (1999) used a zoogeographical approach, which integrated Atlantic salmon ecology and historical distribution, to hypothesize that at least three ancestral DPSs occurred in the United States: Long Island Sound, Central New England, and the Gulf of Maine.

The Long Island Sound DPS was spatially isolated from the other Atlantic salmon rivers in New England because all the rivers in this DPS empty into a relatively shallow and unique saltwater environment. All other Atlantic salmon rivers in the United States pass through river-dominated estuaries, characterized by relatively narrow, short, cylindrical channels, before entering the Gulf of Maine. The Central New England DPS extended northward from the Merrimack River, New Hampshire/Massachusetts border, to the Royal River, Maine. Though the habitat differences provide less distinction between the Central New England and Gulf of Maine DPSs, important ecological differences between these ecoregions are sufficient to expect distinctions in local adaptations—including growth rates, river residence time, and smolt migration timing between the two areas (Colligan et al. 1999). The native populations of the Long Island Sound and Central New England DPSs were extirpated.

The Gulf of Maine DPS, as currently defined, includes all coastal watersheds with native populations of Atlantic salmon from the Androscoggin River north to the mouth of the St. Croix River at the U.S.-Canadian border (Colligan et al. 1999). At present, remnant native populations have been documented in eight rivers: Cove Brook and the Sheepscot, Ducktrap, Narraguagus, Pleasant, Machias, East Machias, and Dennys. A key trait of all U.S. populations is that a high percentage (typically >80% of the population) spend two winters in the ocean (2SW) and are highly migratory during the marine phase. This heritable trait is common to the Gulf of Maine DPS (Baum 1997), and it is distinct from adjacent Canadian populations in the Bay of Fundy that are dominated by fish that typically spend one winter at sea (1SW). Functionally, the northern extent of this DPS likely included the St. Croix River and southwestern New Brunswick coastal tributaries northward to the St. John River. Since 1SW Atlantic salmon dominated (greater than 50%) the St. John River population, this river may have been transitional between the highly migratory Gulf of Maine DPS and the nonmigratory inner Bay of Fundy stocks. Though Maine and southern New Brunswick stocks are likely ecologically related, the rivers in Canada, including the mainstem St. Croix River, were eliminated from inclusion in the Gulf of Maine DPS because of the U.S. Endangered Species Act inclusion criteria related to international boundaries (Colligan et al. 1999). In addition to naturally spawning populations in these rivers, active enhancement operations are ongoing in six of these systems—Sheepscot, Narraguagus, Pleasant, Machias, East Machias, and Dennys rivers). For management purposes, captive broodstock and offspring held in hatchery systems are considered part of the DPS.

Within the U.S. range, larger rivers such as the Connecticut, Merrimack, Androscoggin, Kennebec, and Penobscot contained unique stocks or substocks that contributed to historical DPS structure and perhaps represented unique DPSs. With the possible exception of the Penobscot basin, all of the native large river populations in the United States have been extirpated. There have been active restoration programs in the Connecticut and Merrimack rivers for more than 25 years, and newer programs have been implemented in smaller streams in Rhode Island and New Hampshire, as well as in the Saco and Kennebec rivers in Maine (USASAC 1999). Most of these restoration programs were initiated with broodstock and offspring from the Penobscot River. Some of these programs still rely on smolt stocking of Penobscot River origin smolts, but several systems have developed sufficient sea-run returns and captive broodstock programs to meet enhancement requirements using within-basin production of offspring. In the past, some European strains were experimentally used in enhancement programs, but survival was substantially lower for European strains than for U.S. donor stocks (Orciari et al. 1987; Orciari et al. 1994). Because of the failure of out-of-basin stocking to restore self-sustaining populations, restoration programs are now increasingly relying on river-specific broodstock for restoration, with the postulate that local adaptations for these southernmost rivers will arise from the donor stocks through natural selection (Baum et al. 1995). U.S. managers consider these populations to be restoration populations and not specifically regulated as part of the remnant native populations within the Gulf of Maine DPS.

Canadian Atlantic salmon stocks occur in watersheds characterized by a broad range of terrestrial ecoregions (Bailey 1998). The southernmost Canadian stocks—in the Bay of Fundy, Nova Scotia, and parts of Quèbec—

reside in the warm continental division with watersheds characterized by mixed deciduous and coniferous forests. The remaining Quèbec stocks, Gulf of St. Lawrence stocks, and southern Newfoundland stocks reside in the subarctic division; watersheds are dominated by taiga and boreal forest. The northern Newfoundland stocks occur in the subarctic mountain division, characterized by mountainous taiga habitat. The Labrador stocks, the northernmost Atlantic salmon stocks in Canada, occur in the subarctic mountain division, at the interface of woodlands and tundra regions. Differences in terrestrial habitat within watersheds result in differing overall levels of productivity, which lead to divergence in life history patterns among Canadian stocks (Hutchings, Chapter 2). In addition to stark differences in watersheds, the range in latitude occupied by Atlantic salmon in Canada results in substantial differences in climate regimes and growing season.

Canadian Atlantic salmon stocks exhibit a broader range of life history strategies than U.S. stocks (Hutchings, Chapter 2). In-river residence periods of juvenile Atlantic salmon in Canadian rivers range from two to seven years, which is considerably longer than the residence times of U.S. stocks (i.e., one to three years). Canadian stocks also exhibit wider variation in marine residence times and migratory paths in marine systems. Stocks residing in the inner Bay of Fundy are characterized by a seawater phase with shorter migration distances, and most adults (70–90%) return to spawn as grilse after only one winter at sea. Other stocks, including those in the Gulf of St. Lawrence, generally have marine phases exceeding one year, undertaking longer migrations to the waters off Newfoundland and Greenland. The prevalence of iteroparity (single fish spawning in multiple years) is significantly higher in some Canadian stocks (e.g., Newfoundland) than in more southerly Canadian and U.S. stocks.

Review of Distribution and Abundance

Before European contact, Atlantic salmon were likely distributed throughout a watershed area in North America exceeding 600,000 km^2 (Dunfield 1985). Estimates of annual North American adult returns of Atlantic salmon during aboriginal times, prior to significant human disturbance associated with European settlement, ranged from 25,000 to 70,000 t (Rostlund 1953; Dunfield 1985).

Abundance of U.S. Populations

Estimates of the total historic Atlantic salmon adult population returning to U.S. rivers range from 300,000 (Stolte 1981) to 500,000 (Beland 1984). Estimates for the stocks south of Maine are speculative because there is little documentation relative to freshwater production regimes. However, better quantification of historic production in Maine is possible based on Beland's (1984) estimate of historic Atlantic salmon spawning and nursery habitat in Maine rivers. If we use a production estimate of three smolts per 100 m^2 (Baum 1997), which is likely conservative, Maine rivers may have produced in excess of 14 million smolts annually. With an average 1% marine survival rate, which is also conservative (Bley and Moring 1988), annual returns to Maine rivers may have exceeded 140,000. Because of the potential juvenile rearing habitat (Stolte 1981, 1994; Beland 1984; Meyers 1994; Baum 1997), the largest historical salmon runs in New England were likely in the Connecticut, Merrimack, Androscoggin, Kennebec, and Penobscot rivers.

By the early to mid-1800s, the Atlantic salmon runs in New England had been severely depleted, greatly reducing the species' distribution in the southern half of its range. The earliest impacts were from overfishing, water quality degradation, and barriers to migration caused by waste disposal and waterpower development (gristmills) associated with the Industrial Revolution (Stolte 1981; Meyers 1994; Baum 1997; Colligan et al. 1999). Recognition of fish passage issues occurred as early as the late 1600s, and there are records of local regulations mandating the free passage of fish through human-made barriers as early as 1709 (Stolte 1981).

Restoration efforts were initiated in the mid-1800s, but they had little success because of the continued development of dams, the inefficiency of early fishways, and the primitive state of knowledge regarding fish culture practices (Baum 1997). The U.S. Fish Commission built three salmonine egg collecting and hatchery facilities in the early 1870s. These early hatcheries were for Penobscot River Atlantic salmon, landlocked salmon in Grand Lake Stream, Maine, and Chinook salmon *Oncorhynchus tshawytscha* from the Sacramento River, California (Parsons 1973). Fish were sometimes transferred between stations and seem to have been used interchangeably if logistics were advantageous, as evidenced by the fact that many early Great Lakes salmon stocking efforts utilized Chinook salmon as well as Atlantic salmon (Parsons 1973; Tanner and Tody, Chapter 7).

Natural Atlantic salmon runs disappeared from southern New England rivers by 1865. There was a brief period in the late 19th century when limited adult returns were established in the Merrimack and Connecticut rivers through artificial propagation, but these runs were not self-sustaining, and the salmon were extirpated by the end of the 19th century (Stolte 1981; Meyers 1994; Stolte and McKeon 1996). By most historic accounts, anadromous Atlantic salmon stocking efforts prior to the 1970s were unsuccessful in reestablishing spawning populations and likely acted as marginally successful put-grow-and-take programs (MacCrimmon and Gots 1979; Baum 1997).

Atlantic salmon in the large rivers, from the mainstem Kennebec River and southward, became extirpated during this same time (Atkins 1874; Kendall 1894; Kendall 1935). Thus, by the end of the 19th century, four of the five largest salmon populations in New England were extirpated, shifting the southern extent of the species' range approximately 2 degrees north in latitude and 4 degrees east in longitude. Within the United States, less than 25,000 ha (52%) of the historic Atlantic salmon juvenile nursery habitat (all in Maine) are now freely accessible to naturally spawning adults (Baum et al. 1995).

The Penobscot River continued to support a substantial natural population during the late 1800s, with a reported commercial catch of more than 90 t in 1880 (Baum 1997). In subsequent years, an on-going artificial propagation program appeared to have limited success in slowing the decline of this population and maintaining commercial landings. As late as 1931, landings still exceeded 34 t and, in the face of this exploitation, the population continued to decline. It was not until 1947, once annual commercial landings declined to only 40 Atlantic salmon, that the commercial fishery was closed (Baum 1997). Simultaneously, the abundance of Atlantic salmon in the coastal rivers of the Gulf of Maine DPS also continued to decline. By the mid-20th century, the total adult runs of Atlantic salmon to U.S. rivers had declined from tens of thousands of fish in the early part of the previous century to a marginally stable, artificially enhanced population of 1,500–2,500 fish, mostly in rivers in eastern Maine.

At present, only eight recognized rivers in the Gulf of Maine DPS range still contain functioning natural populations, albeit at dramatically reduced abundance levels (Baum 1997; Colligan et al. 1999; King et al. 1999). The nucleus of this DPS is in the remnant populations located in the Dennys, East Machias, Machias, Narraguagus, Ducktrap, and Sheepscot rivers, which are all coastal rivers with significant amounts of historic river habitat remaining (Baum 1997). The Penobscot River is not included in the Gulf of Maine DPS because of on-going investigations into the genetic structure of these populations and the relative contributions of source populations stocked in the basin (Colligan et al. 1999). Conservation spawning target levels based on available spawning habitat have been established for the three major restoration programs of New England. In the late 1990s, adult returns in these three river systems were approximately 10%, less than 2%, and less than 1% of spawner escapement goals for the Penobscot, Merrimack, and Connecticut rivers, respectively (USASAC 1999).

Abundance of Canadian Populations

Many Canadian stocks of Atlantic salmon have declined because of the same causes (e.g., overfishing, habitat degradation, dams) as U.S. populations. In general, however, Canadian stocks are at higher levels than their southern counterparts. Estimates of two-sea-winter (2SW) spawners as a proportion of the estimated escapement requirement indicate that 1998 2SW spawner returns in Newfoundland, Quèbec, the Gulf of St. Lawrence, and Scotia-Fundy were 198%, 28%, 37%, and 16% of escapement goals, respectively (ICES-NASWG 1999). By comparison, overall returns of U.S. 2SW returns averaged only 5% of the estimated escapement requirement (ICES-NASWG 1999).

Colonization and industrialization of watersheds occurred later in most regions of Canada than in New England. Large increases in population and industrialization occurred during the 1700s in the northeastern United States, but these developments generally did not occur in Canada until the early to mid-1800s. In New Brunswick, the human population was estimated to be only 40,000 in 1815 but increased fivefold over the next 30 years (Dunfield 1985). In addition to delayed industrialization, early fisheries in coastal areas of Canada focused on abundant nearshore stocks of cod and other groundfish.

As industrialization proceeded, Canadian Atlantic salmon populations encountered similar fish passage issues as U.S. stocks because of the development of gristmill and flood control dams. Fish passage issues were also recognized in Canada, but, because of delayed industrial development, measures regulating fish passage were implemented almost a century later than those in the U.S. New Brunswick and Nova Scotia both implemented regulations on fish passage in 1786, and New Brunswick regulations mandated fish passage on all dams in the colony by 1810 (Dunfield 1985). Hydroelectric development has been less intense in Atlantic Canada than in U.S. river systems. Although hydroelectric development is responsible for localized impacts on Canadian rivers (Ruggles and Watt 1975), overall this type of development has had less impact on Canadian Atlantic salmon stocks than on U.S. stocks.

Modern Continental Collapse

Since 1980, North American stocks of Atlantic salmon have experienced a region-wide decline in abundance. The estimated abundance of nonmaturing 1SW Atlantic salmon (destined to return as 2+ SW adults) of North American origin has declined by 85%, from 800,000 fish in the mid-1970s to fewer than 110,000 fish in 1998 (ICES-NASWG 1999; Figure 1). Several studies have shown that Atlantic salmon stocks over this wide geographic range exhibit synchronous marine survival rates and that the winter period may be the critical stage for postsmolt survival and maturation (Reddin and Shearer 1987; Scarnecchia 1989; Friedland et al. 1993; Friedland et al. 1998). Two-parameter regression models using spawner abundance and area of suitable sea surface temperature in February (surrogate for winter thermal habitat) have been successful at forecasting the abundance of 1SW fish (ICES-NASWG 1999). Two multiyear periods of abnormally low sea surface temperatures since 1980 resulted in reduced availability of marine thermal habitat. Though most North American stocks experienced significant declines in abundance during these periods, those stocks at previously low

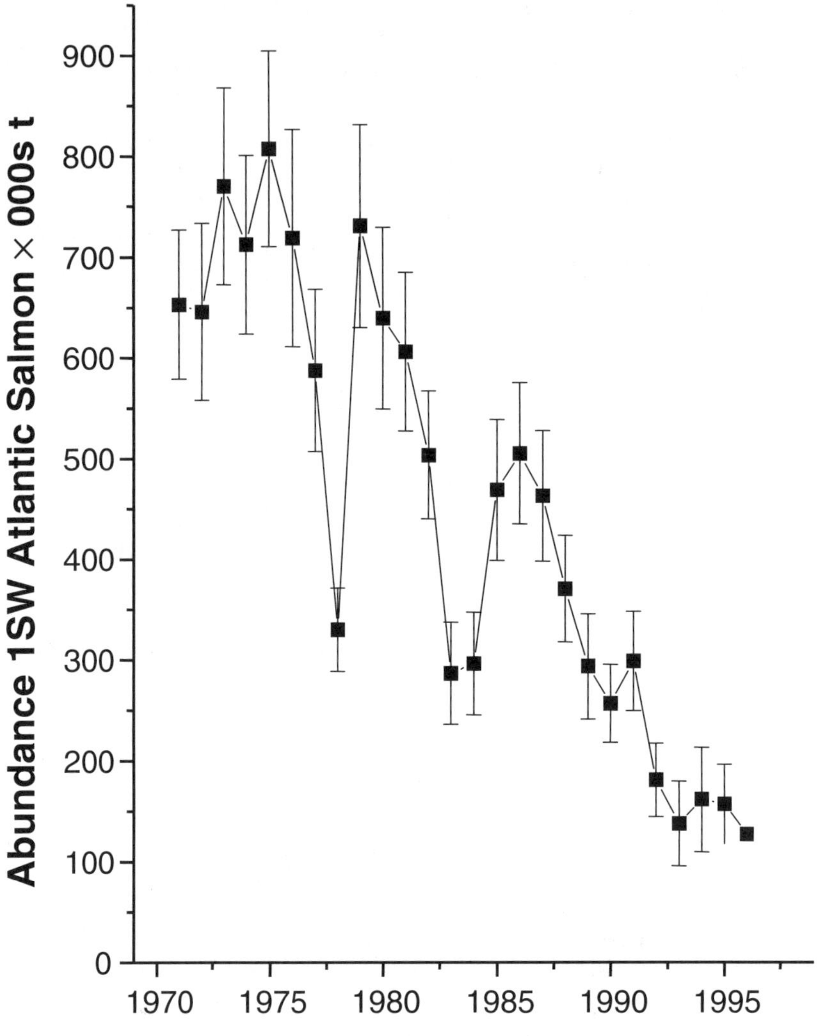

Figure 1. Estimated abundance of one sea winter (1SW) Atlantic salmon of North American origin in the North Atlantic Ocean.

levels (including those in the United States, Bay of Fundy, and Nova Scotia) were further reduced to critically low levels and threatened with extirpation.

Management of Atlantic Salmon

U.S. Hatchery Management

Hatchery management programs in the United States were initiated in the mid-1800s, several years after those in Canada. However, in 1866, shortly after the U.S. Civil War ended, a regional commission devoted to restoration of anadromous species in New England was developed among eight states (Stolte 1981; Meyers 1994; Baum 1997). Early management activities of this group included stocking of fertilized eggs from the Miramichi River, New Brunswick, into New Hampshire tributaries of the Merrimack River (Stolte 1981). By 1872, the U.S. Congress had expanded the duties of the U.S. Fish Commission to include artificial propagation of fish. Early pioneers in U.S. fisheries management, including Charles Atkins and Spencer Baird, quickly embraced the use of stocking to enhance declining Atlantic salmon stocks (Baum 1997). Prior to 1900, nearly 8 million juvenile Atlantic salmon (fry, parr, and smolts) were stocked in Maine waters. Most of these early stockings were fry (91%) or age-1 parr. The effect of these early stockings appears to have been minimal—population declines continued, and techniques frequently used during this time period (point stocking of fry) have since been found to be ineffective enhancement strategies (Wentworth and LaBar 1984; Hume and Parkinson 1987, 1988; Baum 1997). Atlantic salmon stocking in U.S. rivers continued during the first half of this century, but the focus of stocking efforts shifted to stocking of age 0+ and 1+ parr and was restricted to Maine (Baum 1997). Again, these efforts failed to halt the decline of Atlantic salmon populations under pressure from riparian development and high exploitation rates.

The modern era of salmon restoration and rehabilitation began in the late 1960s. The Maine program was revamped, and restoration programs were reinitiated in the Connecticut and Merrimack River systems. The rapid initial success of Pacific salmon programs in the Great Lakes may have provided the catalyst for renewed Atlantic salmon restoration efforts in New England. A transition from stocking 0+ and 1+ parr to releasing 2+ smolts began in the mid-1950s. It was not until new hatcheries designed specifically for smolt production started producing yearling smolts that smolt stocking increased. New England-wide smolt production approached nearly one million fish per year (USASAC 1999) in the early 1980s. Since then, smolt production has leveled off to about 700,000 age 1+ smolts annually.

In the 1990s, fry stocking (primarily unfed fry stocked prior to mid-

June) again increased throughout the New England program. The rationale for this change included the desire to make effective use of historically occupied rearing habitat, produce a more naturally reared fish, mimic natural selection within specific watersheds, accelerate downstream fish passage discussions, and balance the logistical constraints of the available production facilities (Moring et al. 1995). The final decisions that determine the river-specific structure of current stocking programs are a combination of the particular management needs in the watershed and operational funding/logistical constraints.

Canadian Hatchery Management

Hatchery management programs for Atlantic salmon in the Canadian Maritime Provinces date back to the 1860s, when Samuel Wilmot established hatcheries on the Restigouche and Miramichi rivers. Wilmot's rationale for initiating hatchery programs was his belief that the prevailing management practice of transporting naturally spawned eggs from their natural redds to hatchery environments rarely resulted in hatching rates exceeding 5% (MacCrimmon and Gots 1979). By spawning adults in the hatchery, initial culture efforts achieved 10-fold improvements in hatching rates. For the first 100 years of the hatchery program, large numbers of unfed fry, fed fry, and age 0+ summer parr were stocked into a number of river systems with virtually no assessment of the impact on wild populations (Marshall et al. 1994).

Beginning in the late 1950s and continuing through the 1960s, a growing body of scientific evidence indicated that stocking of presmolt stages of Atlantic salmon was failing to produce significant increases in adult returns. Analyses of the adult returns from hatchery unfed fry, fed fry, and age 0+ summer parr in all three Maritime Provinces from 1907 to 1947 revealed that hatchery contributions could not be adequately distinguished from natural variations in adult returns (Elson 1957). In Great Britain, hatchery releases of presmolt salmon between 1900 and 1975 also failed to produce significant increases in adult returns (Central Fisheries Board 1986). Although some stocking evaluation studies were completed, none of these studies involved a comprehensive evaluation of fry and parr stocking densities in relation to available habitat. However, experiences with fry and parr stocking were sufficient to promote a gradual shift in hatchery programs from the production of fry and parr to production and stocking of smolts during the 1960s.

Stocking of hatchery-origin fish is used to mitigate a variety of anthropogenic impacts in Atlantic salmon rivers in Canada, including hydroelectric development, silt deposition, and acid deposition. Forty percent of adult salmon returns in five rivers in New Brunswick and up to 100% of the returns in 24 rivers in Nova Scotia were of hatchery origin and were primarily stocked as smolts (Marshall et al. 1994). On the Saint John River, New Brunswick, the Mactaquac Fish Culture Station was constructed to mitigate for habitat loss due to the Mactaquac Dam. In 1998, approximately 90% of the 1SW returns and 65% of the multi-sea-winter returns were of hatchery origin (Marshall 1998). It was concluded that hatchery stocking of age 0+ fall parr in the St. John River was about four times and age-1+ smolts about nine times more effective than natural spawning in terms of the number of recruits with potential to home to areas above Mactaquac Dam (Marshall 1994). Many salmon rivers on Prince Edward Island are heavily affected by silt deposition, which resulted in reduced survival of naturally spawned eggs. To mitigate for these effects, smolt-stocking programs were initiated. By 1998, 95% of the returns to the Morell River were of hatchery origin (Marshall 1998). In 1998, along eastern and southern Nova Scotia, hatchery-origin smolts from the LaHave, Medway, and other source rivers were stocked into 15 acid-compromised rivers. Hatchery-origin fish currently make up a significant proportion of adult returns in most Nova Scotia rivers.

Fisheries issues related to indigenous peoples in Canada have resulted in increased use of hatchery-origin fish to achieve management objectives related to food fisheries. Development of indigenous peoples' subsistence fisheries has followed several divergent approaches, including augmentation of existing populations to enhance food fisheries; development of sea-ranching operations in acid-affected rivers, utilizing hatchery smolts to augment or replace historical adult returns; and provision of smolts for indigenous peoples' aquaculture projects to replace historical subsistence fishing in systems where salmon resources are now more valuable as recreational fisheries.

In 1997, the DFO terminated its responsibility for most Atlantic salmon hatcheries and ceded operational control to the provinces and private stakeholders. This resulted in transference of responsibility for the Charlo and Miramichi Salmonid Enhancement Centers and St. John Fish Culture Station in New Brunswick; the Cardigan Salmonid Enhancement Center in Prince Edward Island; and the Margaree Salmonid Enhancement Center and the Cobequid, Coldbrook, and Mersey Fish Culture Stations in Nova Scotia. The DFO continues to maintain the Mactaquac Fish Culture Station near Fredericton, New Brunswick.

In addition to large, centrally located hatcheries, in the 1990s, there was a movement to smaller, more localized rearing facilities to stock age 0+ fall parr in local watersheds. The principal goals of these satellite-stocking programs are to provide education opportunities and to produce fish for enhancement in areas where populations are depleted. More than 40 satellite-rearing stations have been established, with most producing between 3,000 and 10,000 fall parr annually.

Comparison of U.S. and Canadian Reliance on Hatchery Systems

The U.S. and Canadian strategies for the stocking of different life stages provide insights into approaches to salmon restoration efforts. From 1990 to 1998, each country spawned similar quantities of eggs: 4.6 million annually in the United States (wild returns and domestic broodstock) and 3.3 million annually in Canada. However, rearing and stocking strategies have differed greatly. U.S. strategies have evolved to greater emphasis on fry stocking, with an average of 8.8 million fry stocked annually compared with 3.1 million in Canada. Canadian stocking programs have emphasized parr production (1.9 million versus 770,000). Smolt stocking has been approximately equal (980,000 smolts annually in each country; ICES-NASWG 1999). Of course, the large discrepancy in these programs is the percentage of returning adults within each country that are of wild origin rather than hatchery origin. Although there are no estimates of the overall proportion of hatchery fish in adult returns to Canadian rivers, stocking is relatively limited compared with available juvenile habitat, and the relative contribution to overall adult returns is likely minimal. In the United States, greater than 90% of returning adults are of hatchery origin (USASAC 1999).

Homewater Fisheries in the United States

Atlantic salmon fisheries in the United States were historically located within the lower reaches of natal rivers or in closely allied coastal areas. Early historical accounts give a colorful representation of harvest methods employed, ranging from pitchforks to hook and line (Atkins 1874; Stolte 1981). The first records of commercial fisheries for Atlantic salmon in the United States are from 1628, and the first regulations were enacted by 1684 (Baum 1997). After the American Revolution, weirs became the primary gear for commercial fisheries and remained so until the closure of commercial fisheries in the United States. Though other materials, such as netting and timbers, replaced the original weirs of stakes and brush, the basic construction and use of weirs changed little over time. Baum (1997) illustrates the complexity and site-specific nature of these designs as they evolved since the 1890s. The history of the Maine fishery is similar to that of many overexploited anadromous stocks along the eastern seaboard, including the demise of Atlantic salmon in southern New England. Records of the fishery dating back to 1867 suggest a buildup of fishery effort with a peak in the late

1880s. At this time, the total catch likely exceeded 20,000 fish, primarily in the Penobscot River (Baum 1997). By the turn of the century, the harvest and populations declined rapidly through a combination of intense habitat destruction and, most likely, overharvest. The last U.S. commercial fishery was closed in 1948 (Baum 1997), marking the end of homewater commercial fisheries for Atlantic salmon within U.S. waters.

Atlantic salmon angling in Maine is rich in history and folklore and has been extensively covered in many popular works (Atkins 1887; Kendall 1935; Meister 1964). By comparison, the history of recreational angling in southern New England is poorly documented (Stolte 1981). Angling is thought to have started in the 1820s and gained popularity in the mid-1800s. The most unusual contribution of the United States to Atlantic salmon angling is surely the presidential salmon. This tradition started in 1812 with the presentation of the first fresh-run salmon caught in the Bangor Salmon Pool on the Penobscot River to President Taft (Baum 1997). With declining stocks in the 1950s and 1960s, the tradition was intermittent until 1981, when the annual event was reinstated. However, with fishery restrictions and logistical difficulties, the tradition was ended in 1993 (Baum 1997).

Salmon fishing clubs kept records on many rivers (Beland and Bielak, Chapter 3), and several of these histories date back to the 1940s. It appears that exploitation rates were quite high, ranging from 20% to 75% through the 1980s (Baum 1997). As conservation concerns grew in the 1990s, recognition and adoption of catch-and-release practices among recreational anglers increased. This behavioral switch in recreational fisheries, from harvest to catch-and-release, was voluntarily initiated by some anglers and later required by regulation for all anglers. In the early 1990s, angling regulations became increasingly stringent, with local restrictions on harvest, selective harvest of grilse, and catch-and-release-only zones. Finally, these restrictions culminated in the closure of targeted recreational fisheries (including catch-and-release) in Maine beginning with the 2000 season (USASAC 1999).

Homewater Fisheries in Canada

Canada categorizes Atlantic salmon catch (including harvested and released fish) into two size categories. The "large" salmon category is composed primarily of multi-sea-winter fish—those greater than 63 cm fork length in recreational fisheries and greater than 2.7 kg whole weight in commercial fisheries. Grilse (one-sea-winter fish) dominate the "small" salmon category, which includes sea-run salmon below the threshold sizes of the large category.

Atlantic salmon catches in Canada have shifted from historic in-river and mixed-stock ocean commercial fisheries to recreational harvest and catch-and-release fisheries as well as indigenous peoples' fisheries. Total reported catch of Atlantic salmon in Canada declined from 2,863 t in 1967 to 149 t in 1998 because of declines in abundance and management restrictions to promote stock recovery (ICES-NASWG 1999). Canadian commercial harvest declined from more than 2,400 t in 1980 to 5 t in 1998 because of a series of commercial closures and other restrictions. An Atlantic salmon stock recovery strategy initiated in 1984 resulted in a series of quota reductions extending through 1991 (O'Connell et al. 1992). In 1992, a moratorium on commercial fishing in insular Newfoundland was established that continued through 2000. The commercial fishery in southern Labrador was closed in 1997, and the closure was extended to all of Labrador in 2000. Voluntary license retirement programs have been established throughout Canada. In 1998, commercial fisheries in Canada numerically accounted for 2.3% of the small Atlantic salmon harvest and 7.8% of the large Atlantic salmon harvest.

By the late 1990s, Canadian management of recreational fisheries included restricted seasons, daily and seasonal harvest limits, areas of catch-and-release fishing only, and differential retention regulations for large and small salmon. In 1998, landing of large salmon in recreational fisheries was prohibited in all areas of Canada except in Quebec and Labrador (SFA 1 and 2). Recreational fisheries in Canada numerically accounted for 88% of the small salmon harvest and 36% of the large salmon harvest during 1998. Catch-and-release fishing has accounted for an increasing proportion of Canadian catches, and, in 1998, approximately 52% of the recreational catch was released. Harvests by indigenous peoples' food fisheries averaged 39 t between 1990 and 1998. In 1998, indigenous peoples' fisheries numerically accounted for 57% of the large salmon harvest and 9% of the small salmon harvest in Canada (ICES-NASWG 1999).

Mixed-Stock Fisheries: The West Greenland Fishery

Atlantic salmon smolts leave their natal rivers in North America in the spring and enter the marine environment. Smolts from some stocks are relatively nonmigratory, but smolts that are destined to be multi-sea-winter fish migrate north along the coast during late spring and summer, inhabiting Newfoundland and Labrador waters during their first summer, and appear to migrate back into the Labrador Sea by early in their first winter at sea.

After their first winter in the ocean, nonmaturing age 1+ Atlantic salmon have historically been targeted by marine fisheries in the Labrador Sea and West Greenland areas (Møller 1986; O'Connell et al. 1992). The historical sea-age composition of the West Greenland catch has averaged approximately 93.5% 1SW fish, 5.8% 2SW fish, and less than 1% postsmolts (Møller 1986). From 1994 to 1998, the sea-age composition of U.S. and Canadian fish was even more dominated by nonmaturing 1SW fish, averaging 96.1% 1 SW fish, 2% 2SW fish, and 1.9% previous spawners (ICES-NASWG 1999).

The fishery off West Greenland is truly a mixed-stock fishery, with tagged fish reported from the United States, Canada, Iceland, and nine European countries (Møller 1986). Continent of origin (North American or European) of salmon captured in the fishery has been historically determined by applying various discrimination techniques to samples collected from research vessel catches (1969–1982), sampling of the commercial fishery (1978–1997), and sampling of the local consumption fishery (1998–1999). Research methods employed to determine continent of origin have included analysis of scale characteristics and genetics (Reddin 1982, 1986; Reddin et al. 1987; Verspoor 1988; King et al. 2001). Estimates of Atlantic salmon captured in research vessel catches from 1969 to 1982 averaged 45% North American and 55% European origin. Estimates from commercial catches from 1978 to 1989 averaged 52% North American but have risen steadily since, averaging 73% North American during the 1990s and exceeding 80% North American in 1999 (ICES-NASWG 1999).

The West Greenland fishery for Atlantic salmon experienced rapid growth during the 1960s and peaked in 1971 with a total catch of almost 2,700 t (Møller 1986; Figure 2). Almost all of the historic catch came from the west coast of Greenland, with the fishery along the east coast accounting for less than 1% of catches from 1977 to 1998 (ICES-NASWG 1999). Since 1972, this fishery has generally been managed under a quota system. The initial quota was set at 1,100 t in 1972 and increased to 1,265 t in 1981, then gradually decreased to 840 t by 1991. The reported catches were well below the regulated quotas after 1989; the fleet was unable to harvest the quotas because of reduced catch rates resulting from depleted stocks (Møller 1986).

The allocated quotas in 1993 and 1994 (89 and 137 t, respectively) were bought out by a coalition of governmental and conservation organizations, resulting in a suspension of the fishery in those years. In 1996, after North Atlantic Salmon Conservation Organization (NASCO) members were unable to reach an agreement concerning a quota level, Greenland officials unilaterally established a quota of 174 t. However, the fishery harvested only 53% of the established quota (92 t) in 1996. From 1998 to 2000, the fishery was restricted to catches used for internal consumption in Greenland. Although NASCO did not formally establish a quota in these years, Greenland subsequently established a quota of 20 t based on its esti-

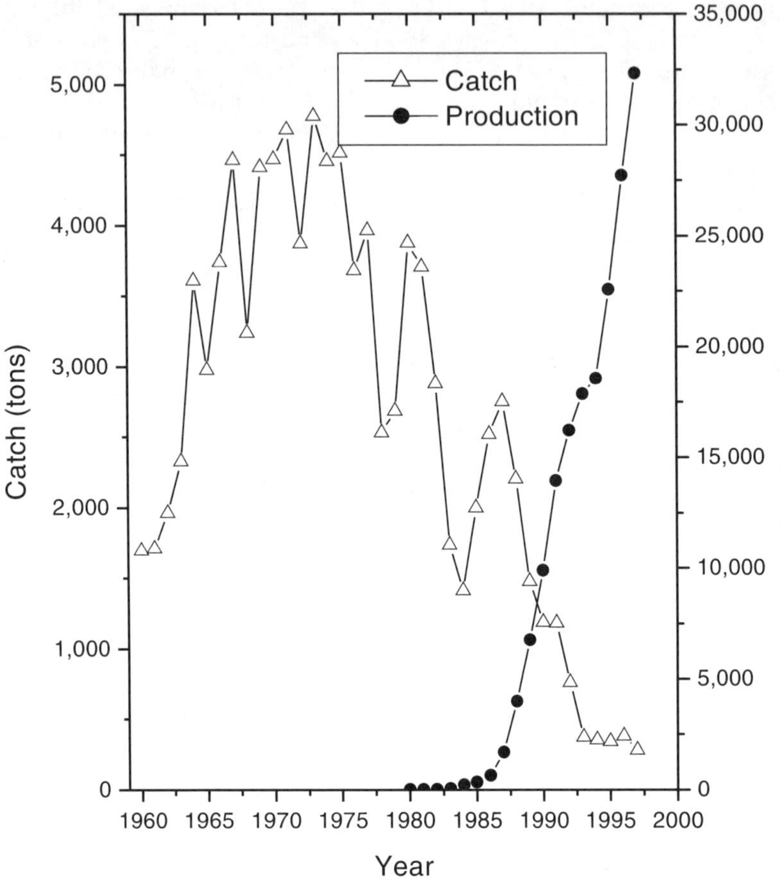

Figure 2. Total sea cage production of Atlantic salmon in the northwest Atlantic Ocean and historical catch of Atlantic salmon in West Greenland and Canada.

mate of the internal consumption of Atlantic salmon. Although catch in the Greenland fishery has been reduced substantially, concern remains about the effect of fishing mortality on stocks that remain significantly below escapement goals.

History and Status of Aquaculture Production

Private aquaculture companies have explored several rearing options for the production of Atlantic salmon and other salmonid fishes, ranging from land-based freshwater rearing facilities to sea ranching and sea cage rearing.

Sea ranching involves release of imprinted Atlantic salmon into the natural environment and recovery of returning adults at imprinting locations (Isaksson 1994; Friars et al. 1996). Most large-scale sea-ranching operations are located in Iceland (Isaksson 1994), although these have been scaled back in recent years. Mariculture in enclosures involves the release of Atlantic salmon smolts into anchored net pens or sea cages, feeding to promote accelerated growth, and harvest of fish once they reach market size. The mariculture of Atlantic salmon in floating sea cages began in Norway in the 1960s, and techniques were later developed on the west coast of the United States. Development proceeded rapidly in several countries and subsequently spread to the southern hemisphere. The total production of captive Atlantic salmon has increased from less than 4,800 t in 1980 to more than 710,000 t in 1998 (ICES-NASWG 1999). Aquaculture production in the North Atlantic area was 392,000 t in 1998, greatly exceeding the biomass of wild salmon in the North Atlantic (ICES-NASWG 1999). This explosive growth in aquaculture production has resulted in almost a complete displacement of wild-caught salmon in most commercial markets because of the relatively low cost of producing mariculture salmon.

Production in eastern North America represented only 4% of worldwide production in 1989 and increased to about 6% of the total by 1998. In North America, mariculture production was started in the late 1970s with a research facility in New Brunswick. This led to a harvest of 6 t in 1979 (Atkinson et al. 1998). By the mid-1980s, production of Atlantic salmon was expanding rapidly in the Maritime Provinces of Canada. The involvement of multinational corporations in Atlantic salmon mariculture resulted in expansion of Canadian operations across the border into eastern Maine in the late 1980s. In the United States, mariculture initially focused on the culture of Pacific salmon but ultimately switched to Atlantic salmon in the early 1980s. By 1984, the mariculture industry in Maine produced its first harvest of 20 t (Baum 1998). In the western North Atlantic, the bulk of production is located in the western Bay of Fundy and Gulf of Maine in both countries (Figure 3). Sixty-two percent of production is based in Canada, with 99.4% of Canadian production in the Maritimes and 0.6% in Newfoundland. The production trends for these facilities and the hectares occupied by marine cages have grown exponentially for the past 15 years (Figure 2). By 1998, there were at least 35 freshwater smolt-rearing facilities and 124 mariculture sites in eastern North America (Atkinson et al. 1998; Baum 1998). Despite the rapid expansion of mariculture along the Atlantic coast, expansion along the Pacific coast was even more rapid. Since the late 1990s, 68% of North American mariculture production has been raised in West Coast waters, primarily British Columbia.

Mariculture production of Atlantic salmon now exceeds production of wild Atlantic salmon by at least 10-fold in North America. From its incep-

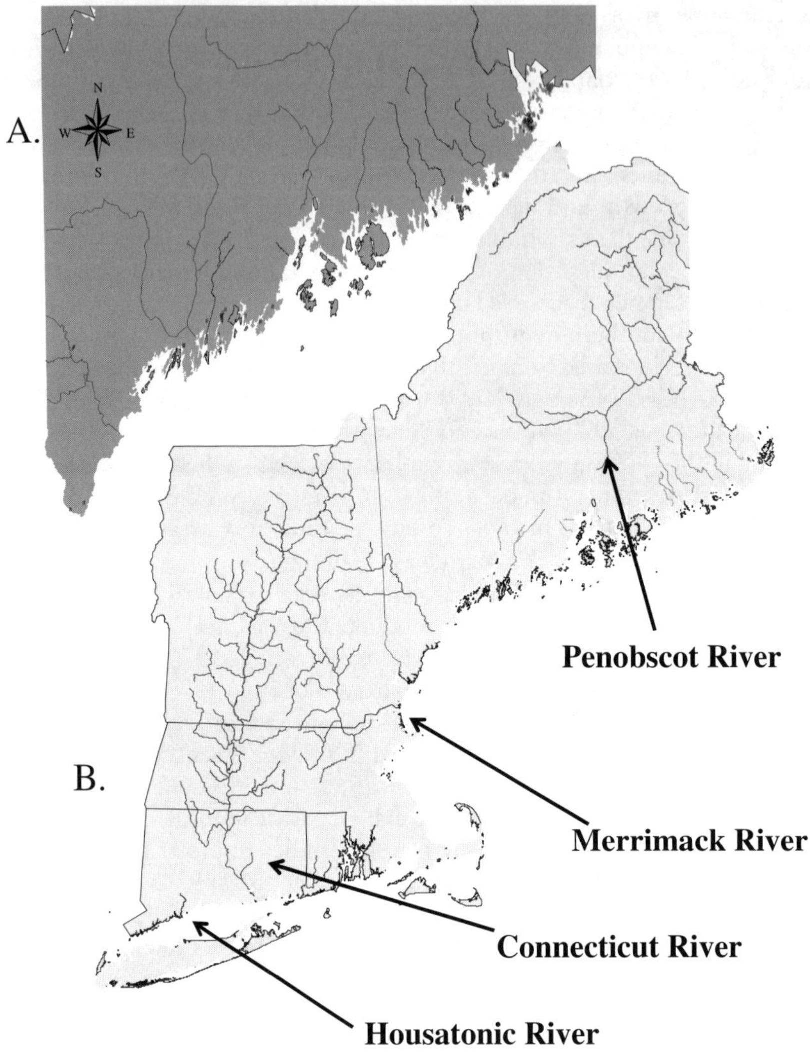

Figure 3. Map A shows locations of sea cage leases in Maine (blue) and eastern Canada (red) in relationship to restoration rivers (red) and rivers with natural runs of Atlantic salmon (blue). Map B shows the locations of Atlantic salmon populations in U.S. and transboundary restoration rivers (red) and rivers with natural runs of Atlantic salmon (green).

tion in the mid-1980s, the mariculture industry in eastern North America has grown to produce greater than 32,000 t annually since 1997 (ICES-NASWG 1999). By comparison, the total catches of Atlantic salmon in

Greenland and North America from 1960 to the present total slightly more than 99,000 t, averaging 3,100 t/year. Three years of North American harvest from mariculture production in this decade exceed the cumulative wild harvest of more than 40 years. On a worldwide scale, mariculture salmon production in the North Atlantic in 1997 totaled 602,000 t, while the total wild catch from 1960 to 1997 averaged slightly more than 17,000 t. In fact, of the Atlantic salmon harvested in the North Atlantic from 1980 to 1997, more than 99.5% were produced in artificial culture. There can be no doubt that—in aggregate as a food fish—Atlantic salmon have been transformed from an extracted resource to a domesticated animal.

If we look at the relative abundance of wild versus mariculture Atlantic salmon on the east coast of North America, there is an undeniable shift of the species from a native species locally adapted to environmental conditions to a domesticated species adapted to maximum production in mariculture environments. It may be that the evolutionary selective pressures between captive and wild populations are so divergent that this shift is inevitable unless native wild stocks are afforded adequate protection (Gross 1998). In the United States and Canada, as elsewhere, mariculture operations are utilizing selective breeding programs to produce fish suited for rearing in captivity. In New Brunswick, these efforts have focused on using family selection within the St. John strain to emphasize traits suitable for culture (Atkinson et al. 1998). The rationale for this strategy includes readily available genetic material from local stocks and superior performance relative to European and Icelandic strains (Atkinson et al. 1998). In contrast, many mariculture operators in Maine have utilized a wider variety of genetic strains of Atlantic salmon. At first, several European salmon stocks were used in Maine, but today, three stocks account for the majority of mariculture production: Penobscot River (Maine), Saint John River (New Brunswick), and Landcatch, a semidomesticated strain originally from Norway first imported via Scotland in 1989 (Baum 1998). The Maine program has hybridized these stocks, and it is estimated but difficult to document that hybridized Landcatch strains account for approximately 20–30% of fish production in Maine. Even in the absence of controlled breeding programs, it is assumed that mass selection of large numbers of fish through culling at presmolt stages and the total artificial rearing challenges of sea cage fish will result in differential selection from wild stocks (Gross 1998).

Discussion

The extirpation of local stocks and the elimination of Atlantic salmon from significant portions of its North American range have been, proportionally, as dramatic as that of Pacific salmon. The scientific and public awareness of their biological status at the end of the century, however, has not been as intensive in recent scientific literature or in the contemporary press (e.g., Gauvin 1993; Chapman 1998; Whitman 1999). This is particularly true in the United States and, to a lesser extent, in Canada. The reasons for the relatively limited attention given to the modern continental collapse and continued decline of Atlantic salmon stocks in their southern range in North America are complex. However, we believe that two factors are important for current prevailing attitudes regarding the conservation of these stocks. First, sizable populations of Atlantic salmon have not occurred within close proximity to metropolitan areas of the eastern seaboard for nearly two human generations. Second, there is an abundance of salmon in the marketplace and a lack of consumer recognition that numerous species of *Oncorhynchus* and *Salmo* are marketed as "salmon." The overwhelming production of both wild and cultured Pacific salmon and cultured Atlantic salmon creates understandable confusion among the North American public regarding the status of Atlantic salmon. How can a species that is so available at the marketplace be undergoing such a dramatic decline?

The continued reduction in range and decline in overall abundance suggest that Atlantic salmon, despite being very adaptable animals, are being pushed to the limits of their adaptive scope in many modern ecosystems. It is apparent that two distinct periods of decline in Atlantic salmon production have occurred in the past three centuries: the first coincided with the industrialization of major watersheds; the second has been the spatially broader postindustrial collapse. The first collapse is now relatively easy to understand from our modern viewpoint. The complex interaction of dams, pollution, acid deposition, habitat destruction, overfishing, and an overdependence on hatcheries as mitigation for habitat losses are common threads in the extirpation of Atlantic salmon in their southern range worldwide (Parrish et al. 1998). Many of these effects were most evident 50 or more years ago, but alterations in habitat and biotic potential of the populations continue to hinder recovery.

The mechanisms behind the recent global declines in Atlantic salmon stocks are as yet unclear because of the complicated interaction of climatic variation and anthropogenic effects. Hypotheses abound, ranging from continued impacts from habitat loss and alteration to new threats such as interactions with aquaculture fish and changing oceanic and climatic conditions (Parrish et al. 1998). At present, a continent-wide mechanism underlying climate-related effects has eluded identification, but evidence seems to point to factors occurring at global scales as causing a depression of Atlantic salmon production. At more regional scales, these declines result in the loss of unique populations and in localized extinctions. Stocks at the southern margin of the North American distribution, the United States and Bay of Fundy, appear to be at greatest risk of extirpation. Identification of the factors hindering sustainability of these populations is essential. For U.S. stocks, variations in freshwater production and marine survival have been asynchronous for many years (e.g., smolt production was low in U.S. rivers during periods of high marine survival), and synchrony is needed to facilitate recovery. This is similar to the scenario developed for coho salmon in Oregon (Lawson 1993). Because the U.S. and Bay of Fundy stocks are now at the southern end of the reproductive range in North America, they likely represent an important segment of the species with the potential to adapt to climate change. This is because stocks on the margins of the latitudinal gradient may be the first to exhibit effects from changing environmental conditions, such as warmer ocean conditions, and as such may be a source of particularly adaptive traits.

The development of the sea cage industry within the southern range of Atlantic salmon in North America has serious implications for wild populations (Gross 1998). On the positive side, increased production has reduced markets for wild fish and relieved much of the resistance to reduced commercial and recreational (food) harvest. Reduced harvest pressure on wild stocks has been an important factor in slowing the decline of Atlantic salmon populations in the past two decades. On the negative side, interactions between wild and cultured Atlantic salmon have been documented at increasing frequency, with fish that escape from aquaculture and mariculture facilities posing a threat to native populations through several mechanisms, including interbreeding, competition, transfers of disease or parasites, and behavioral interactions with migrating smolts or adults (Hindar et al. 1991; Jonsson et al. 1991; Atkinson et al. 1998; Gross 1998). Though debate over the interactions between wild and mariculture fish often becomes polarized, Gross (1998) points out that it is essential for all interests to work together. One innovative approach has been the captive rearing of wild stocks using sea cage technology and their planned release into the wild to enhance wild stocks (Land and Water Resources Council of Maine 1999). Even this program poses risks, however, because removal of selection pressures associated with large-scale marine migrations may ultimately reduce the fitness of subsequent generations. Partnerships between managers responsible for the protection of wild stocks and the mariculture industry can result in the identification of common ground between the conservation needs of wild stocks and the continued development of a sustainable aquaculture industry.

The great paradox for North American Atlantic salmon as we enter the 21st century is that their annual production is likely higher than at any time in history, yet the abundance of wild stocks is severely depressed and the future of many wild populations is precarious. The spring entry of smolts into the U.S. waters of the Gulf of Maine underscores the transition from wild game fish to captive tame fish—on an annual basis, fewer than 800,000 smolts exit rivers (>90% of hatchery origin) while 5 million to 6 million smolts are stocked into sea cages. Some believe that mariculture-reared fish are suitable substitutes for Atlantic salmon in the wild. Prominent salmonid taxonomist Robert Behnke answered this question from an esthetic viewpoint: "For a connoisseur of the arts, such comparison valuation would be simple: an original van Gogh compared with a mass produced facsimile" (Behnke 1996). Most scientific evidence suggests that selective pressures within the two rearing regimes are so diverse that Atlantic salmon in these two ecosystems must diverge and the final result may well be what Gross (1998) referred to as "a wild Atlantic salmon and a new domesticated form." Given the genetic legacy of thousands of years of evolution in the wild, the conservation of these genetic resources is in the best interest of wild fish and the aquaculture industry because it represents diverse genetic resource for natural or artificial adaptations to a changing world.

As observed historically in agriculture, wild genomes are invaluable in the development of sustainable agriculture because they are a source of genetic variation in addressing issues of productivity, disease resistance, and development of new traits valued in the marketplace. Conservation of the unique wild genome of North American origin Atlantic salmon may yet have its greatest impact on cultured salmon and, therefore, maintenance of healthy wild populations is critical to both economic and conservation interests.

References

Atkins, C. G. 1874. On the salmon of eastern North America and its artificial culture. Pages 227–335 *in* United States Commission of Fish and Fisheries. Report of the commissioner for 1872 and 1873, part II. Washington, D.C.

Atkins, C. G. 1887. The river fisheries in Maine. Pages 673–728 *in* The Fisheries and Fishery Industries of the United States. G. B. Goode and Associates, Section V(1).

Atkinson, B., J. S. Loch, and R. Chatelain. 1998. Atlantic salmon eastern Canada overview for 1997. Department of Fisheries and Oceans, Science stock status report, Ottawa, Canada.

Bailey, R. G. 1998. Ecoregions map of North America: explanatory note. Miscellaneous Publication No. 1548. Washington, D.C.

Baum, E. 1997. Maine Atlantic salmon: a national treasure. Atlantic Salmon Unlimited, Hermon, Maine.

Baum, E. T. 1998. History and description of the Atlantic salmon aquaculture industry of Maine. Canadian Stock Assessment Secretariat Research Document 98/152. Ottawa, Canada.

Baum, E. T., R. O. Bennett, A. T. Bielak, K. D. Friedland, S. Gephard, L. Marshall, and L. Stolte. 1995. Rivers Salmon New England: the need to focus on Atlantic salmon. New England Salmon Association, Amesbury, Massachusetts.

Behnke, R. 1996. New England's salmon: a fish of controversy. Trout 1996(Autumn):45–48.

Beland, K. 1984. Strategic plan for management of Atlantic salmon in the state of Maine. Atlantic Sea-Run Salmon Commission. Bangor, Maine.

Bermingham, E., S. H. Forbes, K. D. Friedland, and C. Pla. 1991. Discrimination between Atlantic salmon (*Salmo salar*) of North American and European origin using restriction analyses of mitochondrial DNA. Canadian Journal of Fisheries and Aquatic Sciences 48:884–893.

Berst, A. H., and R. Simon. 1981. Introduction to the proceedings of the 1980 Stock Concept International Symposium (STOCS). Canadian Journal of Fisheries and Aquatic Sciences 38:1457–1458.

Bley, P. W., and J. R. Moring. 1988. Freshwater and ocean survival of Atlantic salmon and steelhead: a synopsis. Biological Report 88-9. Orono, Maine.

Callicott, J. B. 1995. Conservation ethics at the crossroads. Pages 3–7 in J. L. Nielsen, editor. Evolution and the aquatic ecosystem: defining unique units in population conservation. American Fisheries Society, Symposium 17, Bethesda, Maryland

Carlson, C. 1993. The significance of Atlantic salmon in New England. The Dublin Seminar for New England Folklife. Deerfield, Massachusetts.

Central Fisheries Board. 1986. Inland fisheries strategies for management and development. Three Candles Printers Ltd., Dublin.

Chapman, D. 1998. [Review of] Pacific salmon and their ecosystems - status and future options. Fisheries 23:38–39.

Colligan, M. A., J. F. Kocik, D. C. Kimball, G. Marancik, J. F. McKeon, and P. R. Nickerson. 1999. Status review for anadromous Atlantic salmon in the United States. National Marine Fisheries Service/U.S. Fish and Wildlife Service. Gloucester, Massachusetts.

Dunfield, R. 1985. The Atlantic salmon in the history of North America. Canadian Special Publication of Fisheries and Aquatic Sciences 80:188.

Elson, P. F. 1957. The role of hatcheries in securing maritime stocks of Atlantic salmon. Canadian Fish Culturist 21.

Friars, G. W., J. K. Bailey, F. M. O'Flynn, J. W. Carr, J. M. Anderson, S. A. McGeachy, and T. F. Benfrey. 1996. Some genetic aspects of sea ranching and aquaculture. Pages 56–63 *in* S. Calabi and A. Stout, editors. A hard look at some tough issues: proceedings of the New England Atlantic Salmon Management Conference. New England Salmon Association, Newburyport, Massachusetts.

Friedland, K. D., L. P. Hansen, and D. A. Dunkley. 1998. Marine temperatures experienced by postsmolts and the survival of Atlantic salmon, *Salmo salar* L., in the North Sea area. Fisheries Oceanography 7:22–34.

Friedland, K. D., D. G. Reddin, and J. F. Kocik. 1993. Marine survival of North American and European Atlantic salmon: effects of growth and environment. ICES Journal of Marine Science 50:481–492.

Gauvin, C. 1993. Can we save America's Pacific salmon? Trout Winter 1993:12–13.

Gross, M. R. 1998. One species with two biologies: Atlantic salmon (*Salmo salar*) in the wild and in aquaculture. Canadian Journal of Fisheries and Aquatic Sciences 55:131–144.

Hindar, K., N. Ryman, and F. Utter. 1991. Genetic effects of aquaculture on natural fish populations. Aquaculture 98:259–261.

Hume, J. M. B., and E. A. Parkinson. 1987. Effect of stocking density on the survival, growth, and dispersal of steelhead trout fry (*Salmo gairdneri*). Canadian Journal of Fisheries and Aquatic Sciences 44:271–281.

Hume, J. M. B., and E. A. Parkinson. 1988. Effects of size at and time of release on the survival and growth of steelhead fry stocked in streams. North American Journal of Fisheries Management 8:50–57.

ICES-NASWG (International Council for the Exploration of the Sea-North American Salmon Study Group). 1999. Extract of the report of the advisory committee on fishery management; North Atlantic Salmon Stocks No 7. Copenhagen, Denmark.

Isaksson, A. 1994. Perspectives on Atlantic salmon ranching. Pages 64–71 *in* S. Calabi and A. Stout, editors. A hard look at some tough issues: proceedings of the New England Atlantic Salmon Management Conference. New England Salmon Association, Newburyport, Massachusetts.

Johannsson, V., J. Jonasson, S. Osakarsson, and A. Isaksson. 1998. The straying of Icelandic ranched Atlantic salmon, *Salmo salar* L: release and recapture techniques. Aquaculture Research 29(9):679–686.

Jonsson, B., N. Jonsson, and L. P. Hansen. 1991. Differences in life history and migratory behaviour between wild and hatchery-reared Atlantic salmon in nature. Aquaculture 98:69–78.

Kendall, W. C. 1894. Notes on the freshwater fishes of Washington County, Maine. Bulletin of the United States Fisheries Commission 14:43–54.

Kendall, W. C. 1935. Monographs on the natural history of New England. The fishes of New England: the salmon family. Part 2 - the salmons. Memoirs of the Boston Society of Natural History 9–1. Boston.

King, D. P. F., S. J. Hovey, D. Thompson, and A. Scott. 1993. Mitochondrial DNA variation in Atlantic salmon, *Salmo salar* L., populations. Journal of Fish Biology 42:25–33.

King, T. L., W. B. Schill, B. A. Lubinski, M. C. Smith, M. S. Eackles, and R. Coleman. 1999. Microsatellite and mitochondrial DNA diversity in Atlantic salmon with emphasis on small coastal drainages of the downeast and midcoast of Maine. U.S. Geological Survey-Biological Resources Division, Leetown Science Center. Kearneysville, West Virginia.

King, T. L., S. T. Kalinowski, W. B. Schill, A. P. Spidle, and B. A. Lubinski. 2001. Population structure of Atlantic salmon (*Salmo salar* L.): a range-wide perspective from microsatellite DNA variation. Molecular Ecology 10(4):807–821.

Land and Water Resources Council of Maine. 1999. Atlantic salmon conservation plan for seven Maine rivers: 1999 annual progress report. Land and Water Resources Council of Maine, Augusta, Maine.

Lawson, P. W. 1993. Cycles in ocean productivity, trends in habitat quality, and the restoration of salmon runs in Oregon. Fisheries 18:6–10.

MacCrimmon, H. R., and B. L. Gots. 1979. World distribution of Atlantic salmon, *Salmo salar*. Journal of the Fisheries Research Board of Canada 36:422–457.

Marshall, L. 1994. Trapping and trucking Atlantic salmon on the St. John River. Pages 158–169 *in* S. Calabi and A. Stout, editors. A hard look at some tough issues: proceedings of the New England Atlantic Salmon Management Conference. New England Salmon Association, Newburyport, Massachusetts.

Marshall, L., G. L. Farmer, and R. E. Cutting. 1994. Atlantic salmon initiatives in Scotia-Fundy Region, Nova Scotia, and New Brunswick. Pages 116–133 *in* S. Calabi and A. Stout, editors. A hard look at some tough issues: proceedings of the New England Atlantic Salmon Management Conference. Newburyport, Massachusetts.

Marshall, L. 1998. Atlantic salmon Southwest New Brunswick outer-Fundy SFA 23. Stock Status Report D3-13 (1998). Nova Scotia, Canada.

Meister, A. L. 1964. Atlantic salmon fishing—a look back. Maine Fish and Game, Spring 1964.

Meyers, T. F. 1994. The program to restore Atlantic salmon to the Connecticut River. Pages 11–21 *in* S. Calabi and A. Stout, editors. A hard look at some tough issues: proceedings of the New England Atlantic salmon management conference. New England Salmon Association, Newburyport, Massachusetts.

Møller J. 1986. Exploitation and migration of salmon on the high seas, in relation to Greenland. Pages 438–457 *in* D. Mills and D. Piggins, editors. Atlantic salmon: planning for the future. Timber Press, Portland, Oregon.

Moring, J. R., J. Marancik, and F. Griffiths. 1995. Changes in stocking strategies for Atlantic salmon restoration and rehabilitation in Maine, 1871–1993. Uses and effects of cultured fishes in aquatic ecosystems. American Fisheries Society, Symposium 15. Bethesda, Maryland.

Nielsen, J. L., and D. A. Powers, editors. 1995. Evolution and the aquatic ecosystem: defining units in population conservation. American Fisheries Society, Symposium 17, Bethesda, Maryland.

Nielsen, J. L. 1998. Population genetics and the conservation and management of Atlantic salmon (*Salmo salar*). Canadian Journal of Fisheries and Aquatic Sciences 55:145–152.

O'Connell, M. F. O., J. B. Dempson, and D. G. Reddin. 1992. Evaluation of the impacts of major management changes in the Atlantic salmon (*Salmo salar* L.) fisheries of Newfoundland and Labrador, Canada, 1984–1988. ICES Journal of Marine Science 49:69–87.

Orciari, R. D., D. Mysling, and G. Leonard. 1987. Time of migration and growth of two introduced strains of Atlantic salmon in a southern New England stream. Page 560 *in* M. J. Dadswell, R. J. Klauda, C. M. Moffitt, R. L. Saunders, R. A. Rulifson, and J. E. Cooper, editors. Common strategies of anadromous and catadromous fishes. American Fisheries Society, Symposium 1, Bethesda, Maryland.

Orciari, R. D., G. H. Leonard, D. J. Mysling, and E. C. Schluntz. 1994. Survival, growth, and smolt production of Atlantic salmon stocked as fry in a southern New England stream. North American Journal of Fisheries Management 14:588–606.

Parrish, D. L., R. J. Behnke, S. R. Gephard, S. D. McCormick, and G. H. Reeves. 1998. Why aren't there more Atlantic salmon (*Salmo salar*)? Canadian Journal of Fisheries and Aquatic Sciences 55:281–287.

Parsons, J. W. 1973. History of salmon in the Great Lakes, 1850-1970. U.S. Fish and Wildlife Service, Bureau of Sport Fisheries and Wildlife Technical Paper No. 68. Washington, D.C.

Reddin, D. G. 1982. Some general information on discriminate functions and accuracy for identifying North American and European Atlantic salmon caught at West Greenland. ICES CM 1982:/M:15.

Reddin, D. G. 1986. Discrimination between Atlantic salmon (*Salmo salar* L.) of North American and European origin. Journal du Conseil, Conseil International Pour L'Exploration de la Mer 43:50–58.

Reddin, D. G., and W. M. Shearer. 1987. Sea-surface temperature and distribution of Atlantic salmon in the Northwest Atlantic Ocean. Page 262–275 *in* M. J. Dadswell, R. J. Klauda, C. M. Moffitt, R. L. Saunders, R. A. Rulifson, and J. E. Cooper, editors. Common strategies of anadromous and catadromous fishes. American Fisheries Society, Symposium 1, Bethesda, Maryland.

Reddin, D. G., E. Verspoor, and R. R. Downton. 1987. An integrated phenotypic and genotypic approach to discriminating Atlantic salmon. ICES CM 1987/M:16.

Rostlund, E. 1953. Henry Hudson's comment on salmon in the Hudson River. Copeia 1609(September):192–193.

Ruggles, C. P., and W. D. Watt. 1975. Ecological changes due to hydroelectric development on the Saint John River. Journal of the Fisheries Research Board of Canada 32:161–170.

Saunders, R. L. 1981. Atlantic salmon (*Salmo salar*) stocks and management implications in the Canadian Atlantic Provinces and New England, USA. Canadian Journal of Fisheries and Aquatic Sciences 38:1612–1625.

Scarnecchia, D. L. 1989. The history and development of Atlantic salmon management in Iceland. Fisheries 14:14–21.

Scott, W. B., and E. J. Crossman. 1973. Freshwater fishes of Canada. Department of Fisheries and Oceans, Scientific Information and Publications Branch, Ottawa, Canada:192–197.

Ståhl, G. 1987. Genetic population structure of Atlantic salmon. Pages 121–140 *in* N. Ryman and F. Utter, editors. Population genetics and fishery management. University of Washington Press, Seattle.

Stolte, L. 1981. The forgotten salmon of the Merrimack. Department of the Interior, Northeast Region, Washington, D.C.

Stolte, L. W. 1994. Atlantic Salmon restoration in the Merrimack River Basin. 22-35.

Stolte, L. W., and J. F. McKeon. 1996. Will Atlantic salmon restoration enter the 21st century? Merrimack River, USA. ICES; Anadromous and Catadromous Fish Committee CM 1996/T:8.

Thorpe, J. E., and K. A. Mitchell. 1981. Stocks of Atlantic salmon (*Salmo salar*) in Britain and Ireland: discreteness and current management. Canadian Journal of Fisheries and Aquatic Sciences 38:1576–1590.

USASAC (U.S. Atlantic Salmon Assessment Committee). 1999. Annual Report of the U.S. Atlantic Salmon Assessment Committee. 1999/11, Gloucester, Massachusetts.

Verspoor, E. 1988. Identification of stocks in the Atlantic salmon. Pages 37–46 *in* R. H. Stroud, editor. Present and future Atlantic salmon management: measuring progress toward international cooperation. Atlantic Salmon Federation, Ipswich, Massachusetts, and the National Coalition for Marine Conservation, Inc., Savannah, Georgia.

Waples, R. S. 1991. Pacific Salmon, *Oncorhynchus* sp., and the definition of "species" under the Endangered Species Act. Marine Fisheries Review 53:11–21.

Wentworth, R. S., and G. W. LaBar. 1984. First-year survival and growth of steelhead stocked as fry in Lewis Creek, Vermont. North American Journal of Fisheries Management 4:103–110.

Whitman, D. 1999. Hook, line, and, sadly, stinker. U.S. News and World Report 22 March 1999:27–28.

Sustaining North American Salmon: Perspectives Across Regions and Disciplines, 33–60

Chapter 2

Sustaining Atlantic Salmon in the Northwest Atlantic: Considerations from a Life History Perspective

Jeffrey A. Hutchings

Introduction

Life histories reflect the ways in which individuals vary their age-specific expenditures of reproductive effort in response to physiological, developmental, genetic, and environmental influences on age-specific rates of survival and fecundity. Having direct influence on fitness, life history traits include characteristics such as age and size at maturity, number and size of offspring, and number of reproductive events per lifetime (Roff 1992). One implicit assumption of life history theory is that natural selection favours those genotypes whose age-specific schedules of survival and fecundity generate the highest per capita rate of increase relative to other genotypes in the same population (Roff 1992; Charlesworth 1994). That is, selection is assumed to maximize an individual's intrinsic rate of natural increase, r, as given by the Euler-Lotka equation

$$1 = \sum_{x=\alpha}^{x=\tau} l_x m_x e^{-rx}$$

where l_x represents survival from birth to age x, and m_x represents the age-specific fecundity of individuals spawning between ages at maturity, α, and death, τ. Thus, life history theory seeks to predict the optimal age-specific reproductive pattern (i.e., the pattern generating the highest fitness [r]) that will evolve in environments having defined effects on age-specific survival and fecundity.

In principle, life history theory can be used to predict how changes to abiotic and biotic environments influence selection on the fundamental "decisions" that genotypes face concerning reproduction, such as the age and size at which to first reproduce, the amount of energy to allocate during spawning, and the partitioning of gonadal tissue into few large eggs or many small eggs. These environmental changes can include such factors as habitat loss, surface water acidification, overexploitation, and interactions with farmed conspecifics and with introduced exotics. Each of these factors merits study in its own right, notably from the perspective of identifying specific remedial action in species and population recovery plans. But, irrespective of the proximate environmental cause, it is changes in the ability of affected populations to sustain themselves in the face of specific environmental perturbations that is of primary import to fishery managers.

It is for this reason that I will focus on life histories of Atlantic salmon *Salmo salar*, rather than on how specific environmental changes (e.g., habitat alteration) might affect the sustainability of salmon populations. The important role that life history research has to play in resource management and fish conservation lies in the fundamental link that exists between age-specific survival and fecundity and a population's maximum rate of increase, a metric that reflects the fitness (r) of all individuals in that population (see above). Among other things, rate of increase is inextricably linked to a population's ability to sustain different levels of fishing pressure, to increase after being reduced to low abundance, and to persist following the introduction of exotic species with which it might interact as competitor, predator, or prey.

Thus, the link between life history and resource management lies in the consequences that factors such as age and size at maturity, size-specific fecundity, and growth rate have on sustainable yields of recreational and commercial fisheries. Although this link has been long recognized by some fisheries scientists (e.g., Beverton and Holt 1959; Ricker 1975), attempts to incorporate life history variation directly into resource management are comparatively few (for an excellent example, see Shuter et al. 1998).

Atlantic salmon are indigenous to the basin of the North Atlantic Ocean (Berg 1948; Nielson 1961; Scott and Crossman 1973). Excluding introduced anadromous and nonanadromous populations that have persisted in locales such as New Zealand, Australia, and South America (Lever 1997) and that threaten to do so in British Columbia (Volpe et al. 2000), the range of Atlantic salmon encompasses an arc extending in the northwest from the Connecticut River, USA, and Kogaluk River, eastern Hudson Bay, Canada; north to the Kapisigdlit River, Greenland, and rivers throughout Iceland, the United Kingdom, and Ireland; northeast to the Kara River, Russia; and southward through Scandinavia to France and Spain (Figure 1). Extirpated populations of Atlantic salmon include those that once

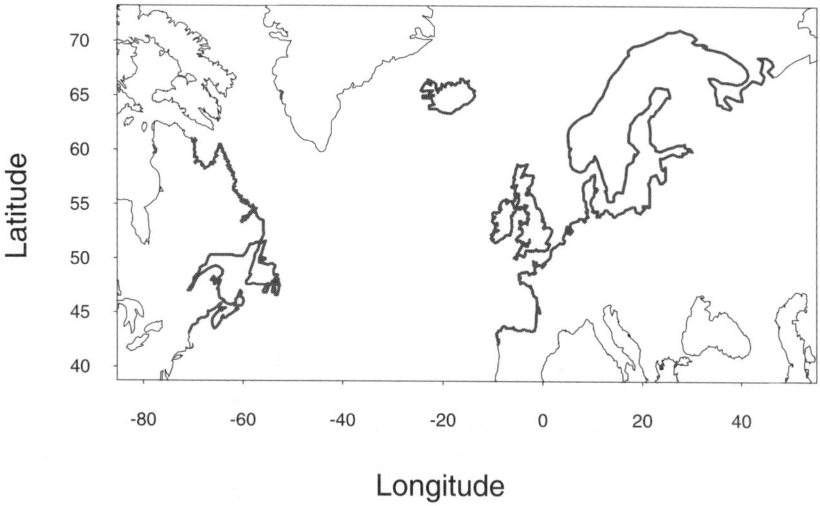

Figure 1. Distribution of endemic anadromous populations of Atlantic salmon.

spawned in the rivers flowing into lakes Ontario (Fox 1930) and Champlain (Kendall 1935).

Anadromous Atlantic salmon typically spend 20% to 50% of their life at sea; nonanadromous populations spend their entire life in freshwater. Females bury fertilized eggs in the substrate of freshwater rivers and streams in autumn. After hatching, the young obtain nourishment from a yolk sac before emerging from the substrate in spring. After periods of usually one to five years in freshwater as parr, anadromous individuals emigrate as smolts and generally spend one or two (rarely three or four) years at sea before returning to freshwater to spawn. In populations for which migration to sea is not prevented by physical barriers, females are usually anadromous, although there are numerous exceptions in Newfoundland (Scott and Crossman 1964; Leggett and Power 1969; Hutchings 1986) and elsewhere (Baglinière and Maisse 1985; also see Fleming 1996). By contrast, males often reproduce after one to four years in freshwater, after which they may or may not migrate to sea.

It has been suggested that the life history variation expressed by Atlantic salmon is matched by few fishes (Hutchings and Jones 1998). Among anadromous and nonanadromous populations, age at maturity can differ by an order of magnitude, ranging from the first year of life for male parr in France (Baglinière and Maisse 1985) and in Canada (M. W. Jones and J. A. Hutchings, unpublished data) to 10 years for females in northern Québec (Power 1969). Size at maturity varies 14-fold, from less than 7 and 9 cm for males and females, respectively, in some Newfoundland rivers (Hutchings

1986; Gibson et al. 1996), to more than 100 cm for females in Norway's River Vosso (Huitfeldt-Kaas 1946) and Canada's Margaree River (Canadian Department of Fisheries and Oceans [DFO] unpublished data). Fecundity can vary 500-fold from fewer than 40 (Gibson et al. 1996) to more than 18,000 eggs per female (e.g., Margaree River, Nova Scotia; DFO unpublished data). Offspring size, measured as egg volume, varies more than three-fold from 48 mm^3 (4.5 mm egg diameter; Gibson et al. 1996) to 171 mm^3 (6.9 mm diameter for salmon in Restigouche River, New Brunswick/Québec; DFO unpublished data). Recent reviews that address various aspects of Atlantic salmon life history include those by Gibson (1993), Fleming (1996, 1998), Hutchings and Jones (1998), and Marschall et al. (1998).

Following a brief historical treatment of Atlantic salmon research in the 19th and early 20th centuries, I describe life history variation among North American populations of Atlantic salmon and identify some management and research considerations that arise therefrom. Here I distinguish among the life histories of ouananiche, or landlocked salmon; mature male parr; and anadromous Atlantic salmon. This chapter complements the considerable research that has been directed to identifying the causal basis of population differences in the ages at which salmon migrate to and return from the sea (Schaffer and Elson 1975; Gardner 1976; Power 1981; Meerburg 1986; Myers and Hutchings 1987a; Metcalfe and Thorpe 1990) and in the incidence of early maturity among males (Jones 1959; Thorpe 1975; Thorpe et al. 1983; Myers 1984; Myers et al. 1986; Hutchings and Myers 1988, 1994; Fleming 1996). In addition, the present contribution provides a greater focus on nonanadromous Atlantic salmon, mature male parr, latitudinal changes in life history, and the implications of life history variation to the sustainability of this species in its native range.

Historical Accounts of Atlantic Salmon Life Histories

The early experimental and descriptive work on the life history of Atlantic salmon was conducted in Europe, notably in Great Britain (e.g., Brown 1862; Paton 1898; Calderwood 1906; Hutton 1924), Germany (Miescher-Rusch 1883), Finland (Nordquist 1903), and Norway (Dahl 1910). Among the earliest studies was one that invalidated the hypothesis that parr—the stage between absorption of the yolk sac and first migration to sea—constituted a separate species, *Salmo salmulus* (Scrope 1843). Shaw (1836) established the fallacy of this notion by rearing salmon ova and then following their development through the egg, parr, and smolt stages to maturity. Importantly, he also determined that the sperm of mature male parr (often identified by the misleading term "precocious") was capable of fertilizing ova from anadromous females.

Early life history accounts of North American Atlantic salmon include Adamson's (1857) treatment of Atlantic salmon in the rivers of Lake Ontario, Atkins' (1884, 1885, 1887; see Baum 1997 for more details) work on anadromous and nonanadromous populations in Maine, and Chambers' (1896) account of ouananiche in Québec. The first comprehensive description of the habits of anadromous salmon in Canadian waters (Huntsman 1931) was preceded by a series of communications from Québec (Menzies 1925; Calderwood 1927), Newfoundland (Menzies 1929), Nova Scotia (Gilpin 1866), and New Brunswick (Perley 1850) that dealt primarily with empirical and anecdotal accounts of spawning migrations.

Nonanadromous Atlantic Salmon

Atlantic salmon that mature in freshwater can be grouped into two categories: ouananiche, or landlocked salmon, and mature male parr. The former, which inhabit waters that may or may not have barriers to migration to and from the sea, include the "lake," or Sebago salmon, indigenous to four watersheds in Maine (Kendall 1935; Warner and Havey 1961); the ouananiche of Québec (Chambers 1896; Riley et al. 1989); the ouananiche, or "winnish," throughout Newfoundland (Scott and Crossman 1964; Hutchings and Myers 1985); and quite possibly the extinct populations of Atlantic salmon that inhabited the rivers of Lake Ontario (Adamson 1857). Of particular interest is the existence of watersheds in which anadromous and nonanadromous populations cohabit the same river/lake system (Leggett and Power 1969; Hutchings and Myers 1985; Couturier et al. 1986; Hutchings 1986). Maturation of parr in freshwater is limited almost exclusively to males, notwithstanding the occasional report of a mature female parr in some rivers (Baglinière and Maisse 1985; Hutchings 1986). Some mature male parr appear to spend their entire lives in freshwater, while others can migrate to sea. The postmaturation migratory behavior of mature parr within and among populations has always been difficult to quantify with certainty because of the comparatively little research that has been devoted to this question.

Ouananiche or Landlocked Salmon

The life histories of ouananiche have received comparatively little attention. Although the sizes attained by nonanadromous populations have been known for some time (e.g., Chambers 1896; Kendall 1935), there are comparatively few data on ages at maturity, age-specific fecundity (although some fecundity data were reported by Atkins 1884; Kendall 1935; Leggett and Power 1969; Barbour et al. 1979; Gibson et al. 1996), and age-specific survival (Leclerc and Power 1980).

The primary factor limiting ouananiche size at maturity would appear to be the abundance of suitable prey, rainbow smelt *Osmerus mordax*, often cited as a favored food (Scott and Crossman 1964; Leggett and Power 1969; Havey and Warner 1970; Hutchings 1986). In the presence of abundant prey, sizes of mature individuals frequently exceed 30 cm, with reproductive individuals larger than 50 cm not uncommon (Table 1). By contrast, ouananiche inhabiting small (<2 m wide) streams or lakes that do not contain high prey densities are characterized by sizes at maturity as small as 8 cm (Gibson et al. 1996) and certainly less than 15 cm; for example, Newfoundland's Ouananiche Beck (Gibson et al. 1996) and Five Mile Pond East (Sutterlin and MacLean 1984).

Management and Research Implications of Ouananiche for Sustaining Atlantic Salmon

From a management perspective, there would seem to be two issues of primary import regarding nonanadromous Atlantic salmon. The first is the need for basic age- and size-specific data on maturity, fecundity, and survival for ouananiche populations. This is particularly important for natural populations for which current exploitation rates in most regions are unknown. Unlike other freshwater salmonids, such as brook trout *Salvelinus fontinalis*, rainbow trout *Oncorhynchus mykiss*, and whitefishes *Coregonus* spp., there are remarkably few population and life history data on ouananiche, the exceptions being some populations in Maine and those adjoining Lac St. Jean in Québec.

The second management consideration pertains to systems containing both anadromous and nonanadromous forms of Atlantic salmon. In Canada, conservation targets are set only for the anadromous salmon found in a given river system. These targets are based on habitat-based metrics of juvenile salmon productivity. Of course, these targets cannot be applied separately for ouananiche and anadromous Atlantic salmon that cohabit the same system and for which barriers to seaward migration are absent. It may not be uncommon, then, for a system to have reached its maximum capacity to produce Atlantic salmon but not to have met its conservation target for anadromous salmon in the same system. Thus, when we use species-specific metrics of production to assign conservation targets for anadromous Atlantic salmon, these targets will almost certainly represent overestimates when these fish coexist with their nonanadromous counterparts.

From the research perspective, a key challenge is to identify the ecological and evolutionary mechanisms permitting the coexistence of anadromous and nonanadromous populations of Atlantic salmon in systems in which physical barriers to seaward migration are absent. The management

Table 1. Maximum reported lengths and ages of nonanadromous populations of Atlantic salmon.

River/lake	Length (cm)	Age (yr)	Reference
Ouananiche Beck, NF	12.3	4	Gibson et al. (1996)
Five Mile Pond East, NF	17.0	–	Sutterlin and MacLean (1984)
First Pond, NF	18.3	4	Scott and Crossman (1964)
Whiteheart Pond, NF	22.4	–	Scott and Crossman (1964)
Candlestick Pond, NF	25.0	7	Barbour et al. (1979)
Forest Pond, NF	26.3	3	Scott and Crossman (1964)
Bullrush Pond, NF	29.5	5	Scott and Crossman (1964)
Bluehill Pond, NF	31.2	7	Hutchings (1986)
Long Pond (Bay Bulls), NF	33.5	–	Frost (1940)
North Arm River, NF	36.5	4	Couturier et al. (1986)
Lac Aigneau, PQ	38.7	8	Power (1958)
Maranacook Lake, ME	42.7	4	Havey and Warner (1970)
Deer Pond, NF	46.0	9	Scott and Crossman (1964)
Lake St. John, NF	48.1	8	Scott and Crossman (1964)
Astray Lake, PQ	49.3	7	Power (1958)
Wings Pond, NF	52.3	7	Hutchings (1986)
Gambo Pond, NF	53.0	10	Leggett and Power (1969)
Long Lake, ME	53.3	13	Warner (1961)
West Grand Lake, ME	53.0	6	Havey and Warner (1970)
Kaegudeck Lake, NF	54.6	8	Scott and Crossman (1964)
Fish River Lakes, ME	60.2	10	Warner and Fenderson (1963); Havey and Warner (1970)
Lac St. Jean, PQ	68.5	8	Gauthier, Gilbert (pers. comm.)
Flatwater Pond, NF	70	9	Leggett and Power (1969)
Red Indian Lake, NF	70.6	11	Scott and Crossman (1964)
Sebago Lake, ME	99	6	Kendall (1935); Havey and Warner (1970)
Cold Stream Pond, ME	–	7	Havey and Warner (1970)
Rangeley Lake, ME	–	7	Havey and Warner (1970)
Mooselookmeguntic L., ME	–	8	Havey and Warner (1970)
Méo River, PQ	–	8	Leclerc and Power (1980)

implications of such research are likely to be of greatest relevance in Newfoundland, where cohabitation of the same watershed by anadromous and nonanadromous forms is widespread. To what extent, for example, are ouananiche genetically distinct from cohabiting anadromous salmon? Given the lack of evidence of temporal or behavioral isolating factors (Hutchings and Myers 1985), the probability of interbreeding between forms may be primarily attributable to spatial differences in the availability of spawning habitat. Hutchings (1986), for example, concluded that the probability of interbreeding between forms was high in a system in which spawning habitat was limited, while Couturier et al. (1986) suggested that the probability of reproductive isolation in a watershed characterized by multiple potential spawning locations may be substantial. Alternatively, if the two life history forms are a product of adaptive phenotypic plasticity (Stearns and Koella 1986; Hutchings 1996), changes to growth rate and/or survival probabilities at sea may favour an increase in the incidence of nonanadromy within such populations.

Mature Male Parr

In most populations, reproductive male Atlantic salmon can be distinguished as two maturation phenotypes (these have been examined in detail by several researchers, including Jones 1959; Thorpe and Morgan 1980; Saunders et al. 1982; Saunders 1986; Thorpe 1986; Hutchings and Myers 1988, 1994; Metcalfe 1998; Thorpe et al. 1998). Following a seaward migration as smolts, anadromous males return to their natal river and mature comparatively late in life (four to seven years) at a large size (45–90 cm). By contrast, mature male parr do not migrate to sea prior to reproduction, spawning relatively early in life (one to three years) at a comparatively small size (<7–15 cm), whereafter they may or may not migrate to sea. Prior to spawning, dominant anadromous males defend access to an anadromous female, while mature male parr establish what appears to be a size-based dominance hierarchy immediately downstream of the courting anadromous fish (Jones 1959; Myers and Hutchings 1987b). Mature male parr compete with one another, and with anadromous males, for the opportunity to fertilize eggs. Although it is clear that parr are able to fertilize eggs either on their own (Hutchings and Myers 1985) or in competition with anadromous males (Hutchings and Myers 1988; Jordan and Youngson 1992; Morán et al. 1996; Thomaz et al. 1997), recent paternity analyses, based on microsatellite DNA, indicate that parr fertilization success can vary significantly among individuals and with the level of mate competition (Jones and Hutchings, in press, a,b).

The incidence of male parr maturity ranges widely among Canadian populations (Myers et al. 1986). In a sample of 24 populations, most of which are located in Newfoundland and Québec, the incidence of maturity among male parr aged 1+ and 2+ years ranged between nil and 100% (Figure 2). However, there are significant regional differences in male parr maturity. Of the nine populations whose incidence of maturity among 1+ parr exceeds 60%, six are located in Newfoundland (Bay de l'Eau, Come-by-Chance, Little Codroy, Northwest, Southwest, Wings), two in the Maritimes (Miramichi, Margaree), and one in Québec (Matamek). Among the 12

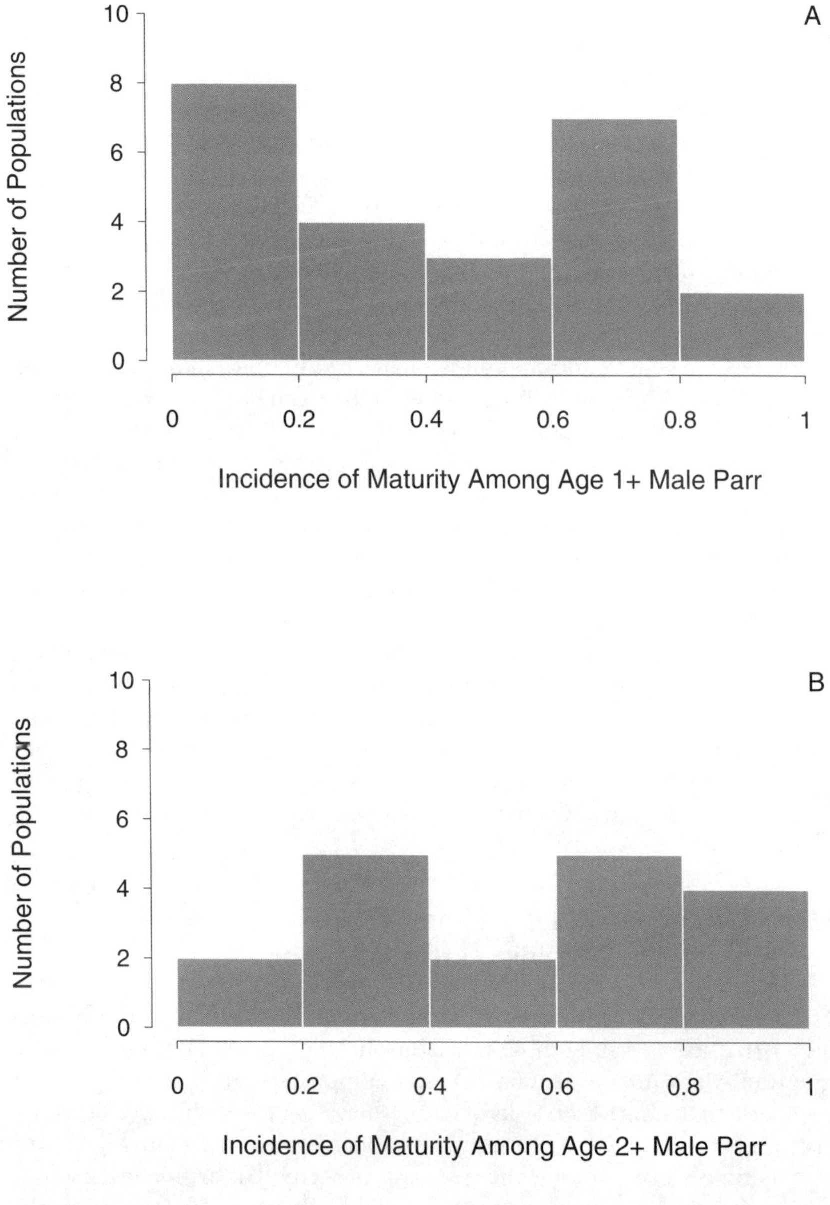

Figure 2. Incidence of maturity among male parr aged 1+ (A) and 2 + years (B) in Canadian populations of Atlantic salmon. Data are those reported by Myers et al. (1986).

populations whose incidence of maturity among 1+ parr is less than 40%, eight are located in Québec (Bouleau, Pigou, Corneille, Mistassini, Moisie, Puyjalon, St. Jean, Watshishou), three in Newfoundland (North Harbor, Rocky, Salmon), and one in the Maritimes (Restigouche). In the only American population for which comparable data appear to be available, Whalen and Parrish (1999) reported that male parr maturity ranges between 28 and 52% for parr aged 1+ years and up to 67% for parr aged 2 years.

Although population differences in the incidence of male parr maturity appear to be loosely correlated with *population* differences in average individual growth rate (Figure 3A), the frequency of the parr maturation phenotype *within* populations is also thought to be closely linked to growth rate and to age-specific survival. Assuming that parr maturation is influenced by both genetic and environmental factors (Thorpe and Morgan 1980; Thorpe 1986; Metcalfe 1998), parr maturation can be modeled as a threshold trait such that adoption of either the parr or the anadromous maturation phenotype depends on whether an individual's growth rate in early life exceeds that specified by a genetically determined growth-rate threshold (Hutchings and Myers 1994). Growth-rate thresholds for parr maturity are predicted to differ among populations (Hutchings and Myers 1994).

Theoretical representations of such growth-rate thresholds for male parr maturity are presented in Figure 3B. Sigmoid curves describe how the probability of parr maturity might increase with parr growth rate (approximated here by size-at-age) among four populations. The lengths-at-age at which 50% of the males are expected to adopt the parr strategy can be identified as parr maturation thresholds (averaged among all male parr within a population). Evidence of such sigmoid relationships exists for male parr in Newfoundland's Little Codroy River (Myers et al. 1986; Hutchings and Myers 1994).

Management and Research Implications of Parr Maturity for Sustaining Atlantic Salmon

For a number of anadromous Atlantic salmon populations in the Northwest Atlantic—certainly those in Maine and the Inner Bay of Fundy—the question is not how to sustain Atlantic salmon but rather how to rebuild populations that face high risks of extinction. A primary objective of recovery programs designed to reestablish these salmon is to maintain high levels of genetic diversity among the resulting progeny. Unfortunately, mature male parr are rarely, if ever, among the males selected to contribute sperm in these breeding programs. Genetic variation associated with salmon age at maturity, coupled with the very high frequencies of male parr maturity in many populations, underscores the necessity of including male parr in

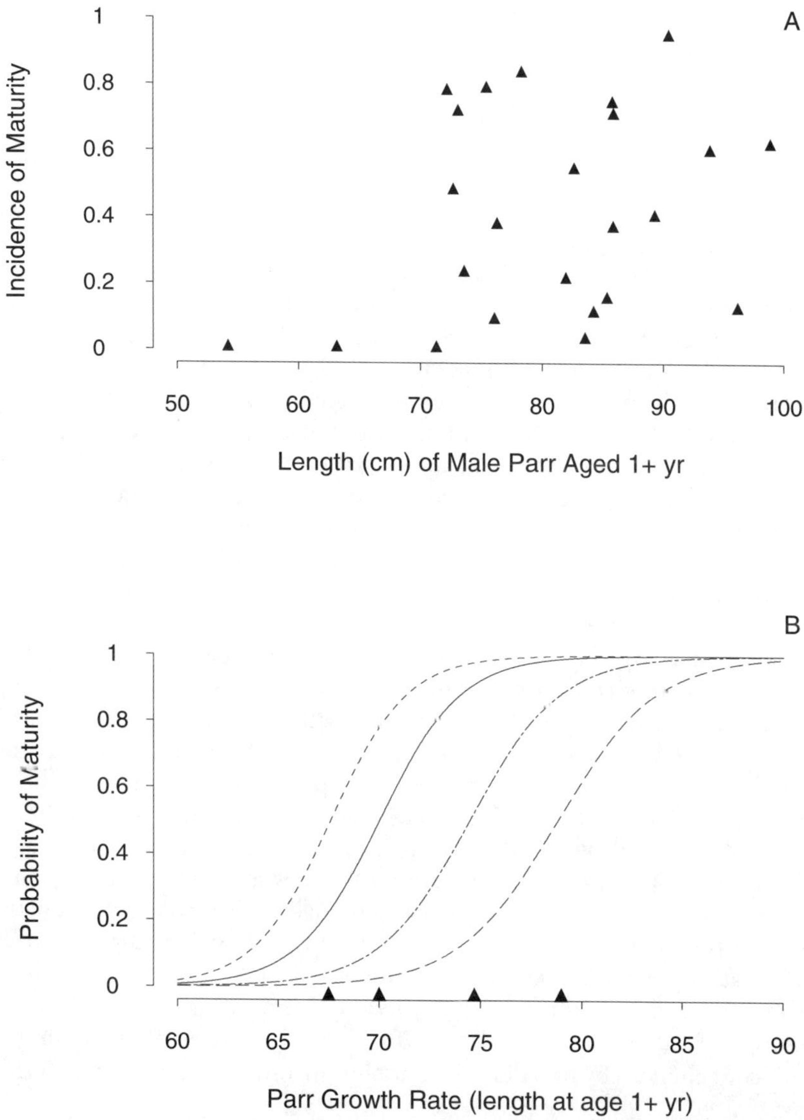

Figure 3. Incidence of male parr maturity as a function of individual growth rate, approximate by length at age 1 + yr. Empirical data (reported by Myers et al. 1986) are presented for 24 Canadian populations (A). Four theoretical growth-rate thresholds for parr maturity are depicted, which illustrate how the probability of maturation, as a function of growth rate, might differ within and among populations (B). Triangles identify the lengths-at-age at which 50% of the male parr that attain that length within a single population will mature as parr.

species recovery programs for Atlantic salmon. Failure to do so would be inconsistent with a precautionary approach to the rebuilding of depressed populations and may undermine the objective of achieving the highest levels of genetic variation possible in breeding recovery programs.

For populations not at immediate risk of extinction, the incidence of male parr maturity has at least three management considerations. First, the higher the incidence of male parr maturity, the lower the production of anadromous male salmon. This can be attributed to the higher mortality experienced by mature male parr in freshwater relative to their immature counterparts. The increase in overwinter mortality experienced by mature male parr relative to their immature counterparts has been variously estimated as 35% for an American population (Whalen and Parrish 1999) and from 19% (Myers 1984; Hutchings and Myers 1994) to 85% (Hutchings and Jones 1998) in two Newfoundland populations. This increased mortality among mature male parr, coupled with a delay in smoltification, has been hypothesized to reduce the production of anadromous males by 60% in some populations (Myers 1984). (Reduced production of anadromous males affected by male parr maturity should *not*, however, be construed as justification for any management protocol designed to eliminate mature male parr from salmon populations.)

Second, spawning contributions by mature male parr are not explicitly incorporated into stock assessments of Atlantic salmon populations. This exclusion may affect one's interpretation of the spawning escapement of males required to maintain a viable population, perhaps most notably for populations in which the incidence of male parr maturity is quite high.

Third, from a conservation perspective, it is not clear how the incidence of parr maturity influences effective population size, N_e (Jones and Hutchings, in press, a,b). Populations with low escapement of anadromous males may nonetheless have a comparably high N_e because of the genetic contributions of mature male parr. Alternatively, increased incidence of male parr maturity may be associated with increased variation in individual reproductive success, which would reduce N_e. Indeed, genetic differences among Atlantic salmon populations may be primarily attributable to genetic contributions by the relatively immobile mature male parr (Jones 2001).

Anadromous Atlantic Salmon

Life in Freshwater

The data presented here were obtained from Hutchings and Jones' (1998) compilation of published and unpublished data on life history-related variables for anadromous Atlantic salmon. Among 114 North American populations, 106 are Canadian, including 51 in Newfoundland, 36 in Québec,

10 in Nova Scotia, and nine in New Brunswick. Of American populations, many of which appear to be sustained by hatcheries (see Baum 1997), seven are in Maine and one in Connecticut. The names of each population and the sources of life history data are detailed by Hutchings and Jones (1998). In contrast to Hutchings and Jones' (1998) analysis of 275 populations throughout the species' range, the present analysis will explore latitudinal variation in life history traits along the Northwest Atlantic coast.

The ages and lengths of smolts vary significantly across the North American range of Atlantic salmon (Figure 4). Based on population means, smolt length averages 15.7 ± 0.3SE cm and differs almost twofold among populations, ranging between 11.5 cm (Rivière St. Jean, PQ) and 21.5 cm (George River, PQ). Despite the positive correlation between smolt length and latitude, this association is influenced strongly by three Ungava Bay populations, all of which are located above 55° latitude and have smolt lengths in excess of 20 cm (Figure 4A). The two American populations for which data exist (Narraguagus River, Maine, and Connecticut River, Connecticut), the most southerly of the populations, are characterized by comparatively large (17–19 cm) smolts (American populations are depicted by open triangles in Figures 4–7). As reported by Power (1981) for a smaller sample of populations, smolt age exhibits a strong positive correlation with latitude among North American populations, ranging from 1.94 years for Connecticut River salmon to 5.85 years for salmon in the George River. The average smolt age among North American populations is 3.30 ± 0.08SE years.

Population differences in smolt length suggest a minimum length (approximately 11 cm) that must be attained before salmon can migrate to sea (cf. Elson 1957). Although there is some evidence to suggest that larger smolts experience lower at-sea mortality than smaller smolts of the same population (Marschall et al. 1998), it is not clear that such reasoning can be extended to explain among-population differences in average size at seaward migration.

There is abundant empirical evidence that smolt age is negatively associated with growth rate in freshwater (Dahl 1910; Menzies 1927; Jones 1959; Power 1981; Thorpe 1986; Metcalfe and Thorpe 1990; Hutchings and Jones 1998). The adaptive significance of such an association may lie in the premise that rapid growth during prereproductive ages reduces the time required to attain a size at which reproduction is possible. By shortening this period of time, an individual would *ceteris paribus* increase its probability of surviving to reproduce, notwithstanding the possible fitness costs associated with smolting at too small a size. Thus, for anadromous salmon, the sooner an individual migrates to sea, the sooner it reaches a size at which it can reproduce. This provides one explanation for the negative association between parr growth rate and smolt age among Atlantic salmon populations.

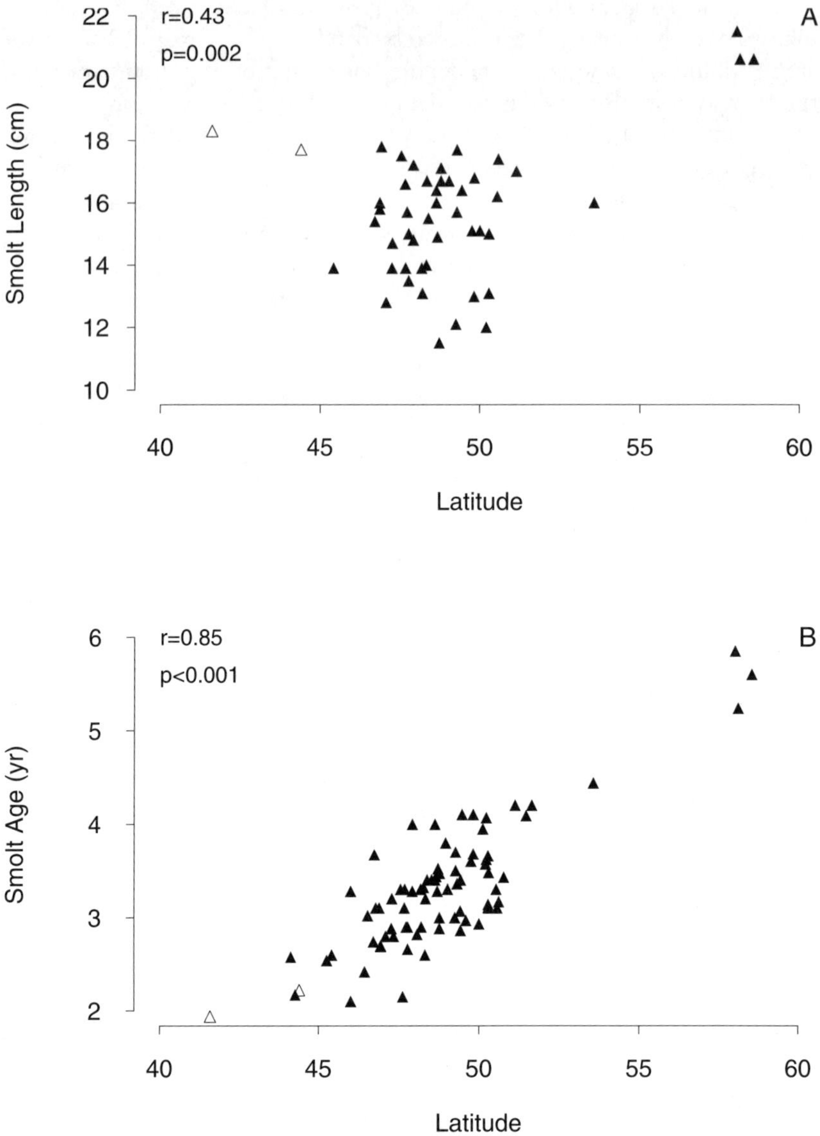

Figure 4. Latitudinal variation in smolt length (A) and smolt age (B) among North American populations of Atlantic salmon. Closed triangles represent Canadian populations; open triangles represent American populations. Data sources are presented in Hutchings and Jones (1998).

Life at Sea

Latitude appears to explain statistically significant amounts of variation in the lengths of anadromous salmon that spend one and two winters at sea—1SW and 2SW salmon, respectively—before returning to their natal rivers to spawn (Figure 5). The scatter of the data also suggests that the lengths of 1SW and 2SW salmon may be greatest at the latitudinal extremes of the species' range. However, one's interpretation of any pattern generated by these associations is highly influenced by the comparatively few data that exist for populations located north of 55°. Among all North American populations, 1SW and 2SW salmon average 54.1 ± 0.4SE cm and 74.4 ± 0.4SE cm, respectively. Among populations, mean length of 1SW salmon ranges from 48 cm (Rivière Jupiter, P.Q.) to 62 cm for salmon in northern Québec (Koksoak and Leaf rivers); population minimum and maximum lengths of 2SW salmon range from 68 cm (Conne River, Newfoundland) to 81 cm (Rivière Ste. Anne, P.Q.). Limited data suggest that American populations are characterized by comparatively large 1SW salmon (Figure 5A) but average-sized 2SW salmon (Figure 5B).

Although independent of latitude (Figure 6), growth rate at sea varies widely among North American populations. Two metrics of growth at sea are considered here: the length difference between 1SW salmon and smolts (averaging 38.4 ± 0.5SE cm; Figure 6A), and the length difference between 2SW and 1SW salmon (averaging 20.1 ± 0.4SE cm; Figure 6B). The average increase in length between the smolt and 1SW stages for the fastest growing population (Rivière St. Jean, PQ) is 25% greater than those of the populations growing slowest at sea (Campbellton River, Newfoundland; George River, P.Q.). The sole American population for which there are sufficient data appears to have a comparatively large difference in length between the smolt and 1SW stages (Narraguagus River, Maine). Population differences in the length increase between 1SW and 2SW salmon are relatively greater than those between the smolt and 1SW stages (Figure 6B). In contrast to that between the smolt and 1SW stages, the length difference between 1SW and 2SW salmon is comparatively small for American populations, the minimum length difference of 16.0 cm for Narraguagus River salmon being 62% less than the 25.9 cm difference observed for Moisie River salmon in Québec.

Time spent at sea differs more than twofold among North American populations (Figure 7A). On average, salmon spend 1.53 ± 0.05SE years at sea before returning to their natal rivers to spawn. As noted by Hutchings and Jones (1998), sea age is one life history trait with distinct regional differences. At one extreme are 23 Newfoundland populations for which sea age at maturity averages 1.15 years. By contrast, sea age at maturity in

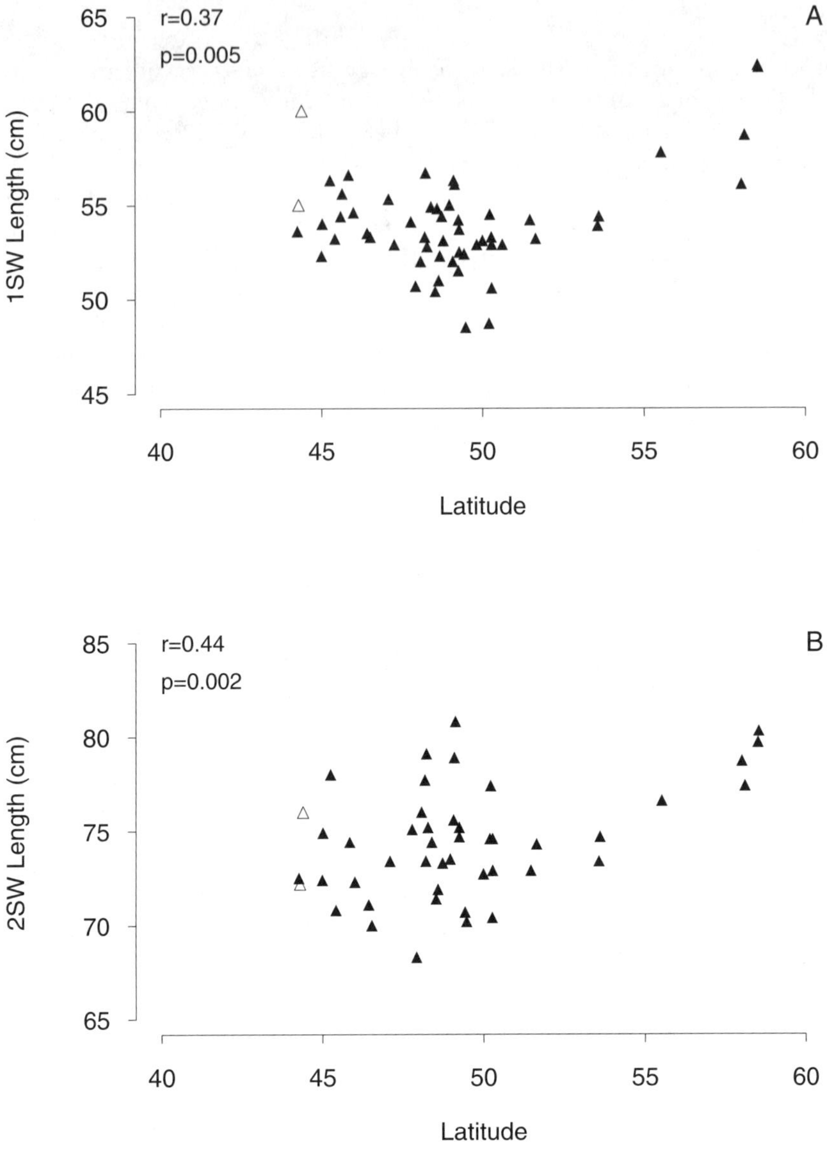

Figure 5. Latitudinal variation in the lengths of North American Atlantic salmon that spend one (A) and two (B) winters at sea prior to their return to their natal rivers to spawn. Closed triangles represent Canadian populations; open triangles represent American populations. Data sources are presented in Hutchings and Jones (1998).

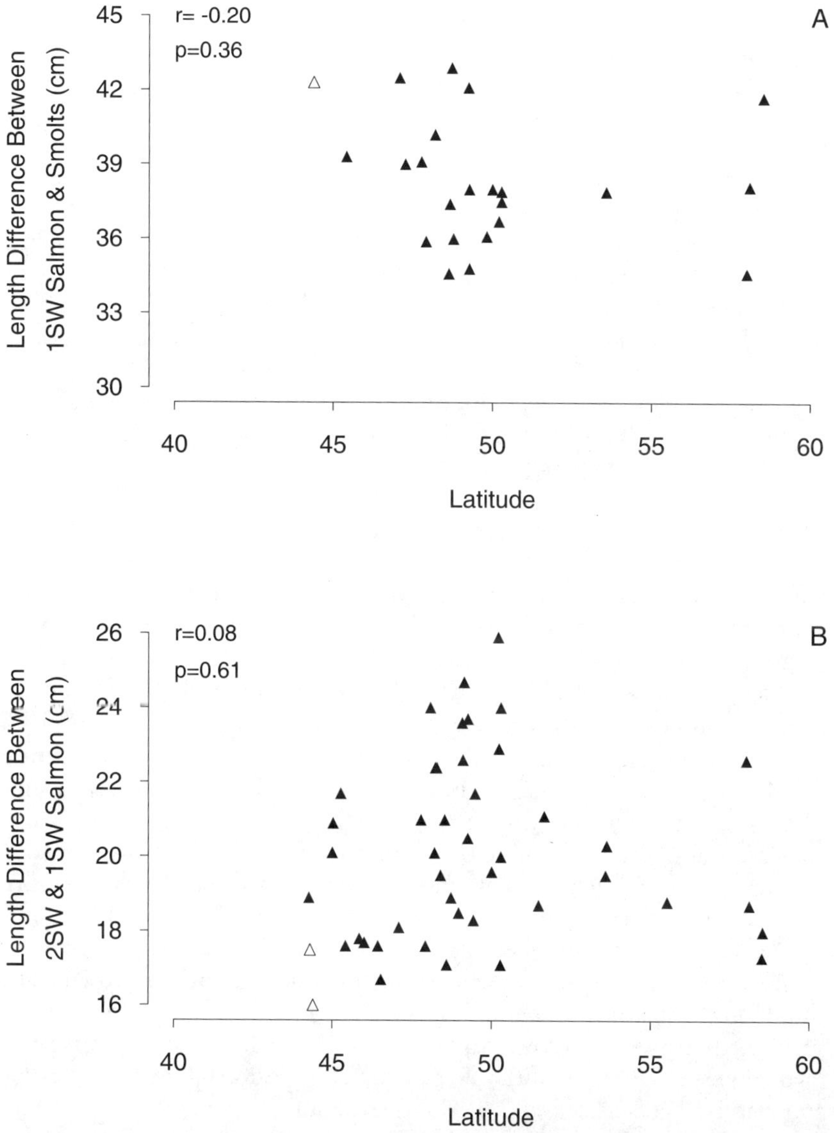

Figure 6. Latitudinal variation in two metrics of growth rate at sea among North American populations of Atlantic salmon. Growth rate is approximated by the difference in length observed between smolts and salmon that spend one winter at sea before spawning (A), and by the differences in lengths between salmon that spend one and two winters at sea before spawning (B). Closed triangles represent Canadian populations; open triangles represent American populations. Data sources are presented in Hutchings and Jones (1998).

Québec (N = 33 populations) and the United States (N = 8) averages 1.72 and 1.96 years, respectively. The population range in mean sea age at maturity is encompassed by salmon in Newfoundland's Wings Brook (1.00 year) and Western Arm Brook (1.00 year) and by those of Québec's Moisie River (2.6 years).

Sea age at maturity in Atlantic salmon has long been thought to represent an adaptive response to survival and growth rate at sea (e.g., Schaffer and Elson 1975; Gardner 1976; Power 1981; Meerburg 1986; Myers and Hutchings 1987a; Friedland and Haas 1996; Friedland et al. 1996; Marschall et al. 1998). To explore how these two variables might interact to effect optimal responses in sea age at maturity in Atlantic salmon, Hutchings and Jones (1998) estimated the fitness, r (calculated from the Euler-Lotka equation given above), associated with different combinations of sea age at maturity (1SW, 2SW), survival in freshwater and at sea, and growth rate at sea.

Given the paucity of age, size, fecundity, and survival data for the same population, Hutchings and Jones (1998) examined how changes to survival and growth rate influence optimal sea age at maturity by holding the remaining life history traits of import constant at values equal to the grand means across all populations in the northwest and northeast Atlantic for which data were available: smolt length (14.8 cm); smolt age (3.5 years); length-specific fecundity (fecundity = 0.4667 [length $^{2.2018}$]); length difference between 1SW and 2SW salmon (20 cm); and survival of 2SW salmon from the smolt stage (4%).

The empirically-based model simulations produced by this analysis provided a means of identifying optimal associations among growth rate, survival, and sea age at maturity. The primary feature of the analysis is the characterization of growth-rate thresholds for sea age at maturity (i.e., the one-year increase in length between the smolt and grilse stages that favours reproduction after one rather than two winters at sea; Figure 7B).

Survival in freshwater (i.e., egg-to-smolt survival) and at sea (i.e., smolt-to-grilse survival) appear to exert a strong influence on optimal growth-rate thresholds for maturity (Figure 7B). A decline in either freshwater or at-sea survival increases the growth-rate threshold at which early maturity is favoured over delayed maturity. For example, if egg-to-smolt survival is 3% and smolt-to-grilse survival is 1.3%, maturation after one winter at sea is favoured for individuals able to grow more than 42 cm while at sea. For the same smolt-to-grilse survival, this growth-rate threshold for maturation after one sea winter increases to about 57 cm when egg-to-smolt survival declines to 1%. And, holding egg-to-smolt survival constant at 3%, an increase in smolt-to-grilse survival to 7.4% reduces the growth-rate threshold for early maturity to approximately 27 cm. If survival is low in freshwater (e.g., less than 0.5%) and the sea (e.g., 1–2%), individuals are favoured to mature as 2SW fish, independent of growth rate. Similarly, if survival is

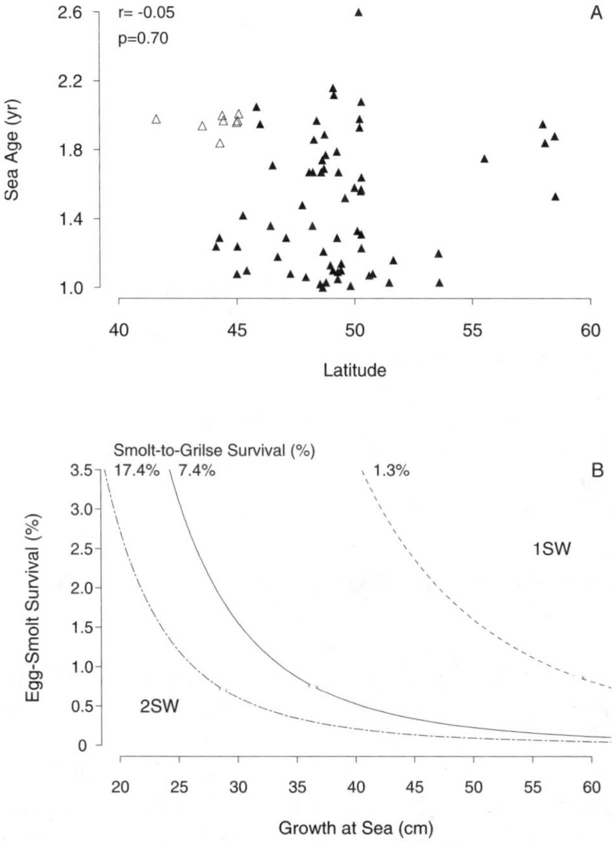

Figure 7. Sea age at maturity among North American populations of Atlantic salmon (closed triangles represent Canadian populations; open triangles represent American populations). Data sources are presented in Hutchings and Jones (1998). Panel A shows that sea age at maturity is independent of latitude. Panel B shows modeled effects of survival, in freshwater and at sea, on growth-rate thresholds for sea age at maturity (B). Growth-rate thresholds, identified by the continuous lines, specify the growth rate at sea, approximated by the difference in length between smolts and one-sea-winter salmon, that must be exceeded for maturity after one winter at sea (1SW) to be favoured (i.e., have higher fitness, *r*) over the strategy of maturing after two winters at sea (2SW). Shown here are predicted changes in thresholds affected by changes in egg-to-smolt survival (the range is that documented for Atlantic salmon throughout its range; Hutchings and Jones 1998) at rates of smolt-to-grilse survival (1.3%, 7.4%, 17.4%), which represent the minimum, mean, and maximum estimates for populations through the species' range (Hutchings and Jones 1998). Thus, combinations of egg-to-smolt survival and growth at sea that lie to the right of each line are those that favor maturity after one winter at sea; those to the left of each line favor maturity after two winters at sea.

high in both freshwater (e.g., greater than 3%) and the sea (e.g., greater than 10%), maturation after one winter at sea appears to be favored.

Thus, for a given age-specific schedule of survival and fecundity (i.e., within a single population), optimal sea age at maturity is predicted to either decline or remain constant with increases in growth rate at sea; delayed maturity appears not to be favoured by increased growth rate, all else being equal.

Management and Research Implications of Local Adaptation in Life History for Sustaining Atlantic Salmon

Perhaps the most important assumption underlying the foregoing examination of Atlantic salmon life histories is the premise that individuals are locally adapted to the environments in which they spend the freshwater portion of their lives and to the environmental conditions they experience at sea on their migration routes. This key assumption, of course, is shared by most fishery management agencies and by those charged with developing recovery plans for severely depleted populations—each river in which Atlantic salmon reside contains at least one locally adapted population.

However, considering the voluminous work that has been undertaken on this species, there has been astonishingly little research directed to the study of adaptive variation among populations of Atlantic salmon. Since the work of Schaffer and Elson (1975), Schaffer (1979), Riddell and Leggett (1981), and Myers et al. (1986), research on among-population variation in Atlantic salmon has increasingly focused on studies of genetic variability and gene flow. Important as these studies are, the dearth of ecological research on local adaptation in Atlantic salmon will almost certainly compromise efforts to identify and to restore severely depleted populations to former levels of abundance. For example, despite the importance of evaluating the consequences of genetic interactions between wild and cultured salmon, there are no data on introgression among wild populations of Atlantic salmon. Evidence of reduced fitness among hybrids produced by parents from different wild populations would be consistent with the hypothesis of local adaptation, yet such studies of outbreeding depression in Atlantic salmon have not been reported. To take another example, with the exception of work on the introduced salmon in the Connecticut River (e.g., Whalen and Parrish 1999), there is no research program in North America dedicated to the study of within- or among-population variation in the incidence of male parr maturity (i.e., male age at maturity), a variable that merits considerably more attention in fisheries management and salmon recovery programs than it has received to date.

A primary challenge faced by fishery managers is to determine the degree to which specific management actions will allow for the sustenance of fish populations. To meet this objective, however, one requires a sound knowledge base upon which to predict the demographic responses of fish populations to abiotic and biotic environmental change. In general, fish population responses to environmental changes are manifested by changes to individual growth rate and to changes in age-specific mortality. As suggested elsewhere (Hutchings and Jones 1998; Marschall et al. 1998), there are probably sufficient data on Atlantic salmon to construct an empirical framework against which environmentally-induced life history responses to changes in growth rate and mortality can be predicted. What cannot as yet be predicted with confidence, however, is the rate at which such changes might occur.

Adaptive phenotypic plasticity provides one means by which genetically-based life history responses to environmental change can be effected (Stearns and Koella 1986; Hutchings 1996; Roff 1998). Such plasticity can be represented graphically as a norm of reaction—that is, a line describing the way in which a particular character (life history, behavioral, morphological) changes along an environmental gradient (e.g., temperature, food supply, growth rate).

Herein lies one of the means by which basic research on Atlantic salmon can have direct applicability to management. Specifically, there is a general research need for studies that focus on the mechanisms for life history change in Atlantic salmon. Are population changes in characteristics such as sea age at maturity (defined here as the number of winters an individual spends at sea before returning to its natal river to spawn) and incidence of male parr maturity generally the result of changes in gene frequencies, a comparatively slow process occurring across generations, or the result of phenotypic alteration along norms of reaction, a comparatively rapid response occurring within a single generation? Future research might address additional questions: Do reaction norms, such as the predicted growth-rate thresholds for sea age at maturity, exist in Atlantic salmon? If so, do the shapes of the reaction norms differ among populations and at different spatial scales? How much additive genetic variation, and thus potential for selection, exists for the shapes of such norms of reaction? Answers to these questions would provide guidance in predicting the rate and direction of life history responses to environmental changes to growth rate and survival in Atlantic salmon.

Summary

The postulate that life history variation in Atlantic salmon is matched by few fishes seems an appropriate characterization of the data available for northwest Atlantic populations. To sustain and rebuild populations of Atlantic salmon, managers and fisheries scientists need to be cognizant of the implications of life history variability to population persistence. The coexistence of anadromous and nonanadromous salmon in the same system, for example, presents problems stemming from a lack of knowledge of the latter and the likelihood of overestimating conservation targets for the former. Regarding age at maturity, the exclusion of mature male parr from stock assessments, breeding recovery programs, and considerations of effective population size can only be considered risk-prone for the majority of populations. Appropriate research initiatives would include studies on the ability and rapidity with which salmon can respond to environmental change, on the evolutionary and ecological mechanisms underlying life history polymorphisms within populations (e.g., male parr maturity, nonanadromy), on the contribution of mature male parr to genetic variability within and among populations, and on the consequences of genetic and ecological interactions between farmed and wild salmon. The increased risks of extinction facing North American populations in the southern part of its range, coupled with comparatively low spawning population sizes elsewhere, underscore the considerable importance of directing management and research initiatives toward the ecological and genetic aspects of life history variation within and among populations of Atlantic salmon.

Acknowledgements

I thank John Kocik, Michael Jones, and Kristine Lynch for the opportunity to undertake this work. I am also grateful to the Canadian Consulate General in Detroit, the Michigan State University Canadian Studies Center, and the American Fisheries Society (AFS) for the financial assistance than enabled me to present this work at the AFS 1998 Annual Meeting in Hartford. Financial support was also provided by an NSERC research grant. Three anonymous reviewers provided helpful comments on an earlier version of this manuscript.

References

Adamson, W. A. 1857. The decrease, restoration, and preservation of salmon in Canada. Canadian Journal of Industry, Science, and Art 2:1–7.

Atkins, C. G. 1884. Notes on landlocked salmon. Transactions of the American Fisheries Society for 1884:40–56.

Atkins, C. G. 1885. The biennial spawning of Penobscot River Atlantic salmon. Transactions of the American Fisheries Society 14:89–94.

Atkins, C. G. 1887. The river fisheries of Maine. Pages 673–728 *in* G. B. Goode and Associates, editors. The fisheries and fishery industries of the United States, Section V, Volume 1. G. B. Goode, and Associates.

Baglinière, J. L., and G. Maisse. 1985. Precocious maturation and smoltification in wild Atlantic salmon in the Armorican Massif, France. Aquaculture 45:249–263.

Barbour, S. E., P. J. Rombough, and J. J. Kerekes. 1979. A life history and ecologic study of an isolated population of "dwarf" ouananiche, *Salmo salar*, from Gros Morne National Park, Newfoundland. Naturaliste Canadien 106:305–311.

Baum, E. 1997. Maine Atlantic salmon: a national treasure. Atlantic Salmon Unlimited, Hermon, Maine.

Berg, L. S. 1948. Freshwater fishes of the U.S.S.R. and adjacent countries. Akademiya Nauk SSSR Zoologicheskogo Instituta, Volume 1, 4th edition (translated by the Israel Program for Scientific Translations 1962).

Beverton, R. J. H., and S. J. Holt. 1959. A review of the lifespans and mortality rates of fish in nature, and their relation to growth and other physiological characteristics. Pages 142–174 *in* G. E. W. Wolstenholme and M. O'Connor, editors. CIBA Foundation Coloquia on Aging 5. Little and Brown, Boston.

Brown, W. 1862. The natural history of the salmon. Thomas Murray, Glasgow, Scotland.

Calderwood, W. L. 1906. Autumn migration of salmon smolts in Scotland. Reports of the Fishery Board of Scotland 1905(11):70–74.

Calderwood, W. L. 1927. The salmon of the River Grand Cascapedia, Canada. Proceedings of the Royal Society of Edinburgh 47:142–147

Chambers, E. T. D. 1896. The ouananiche, and its Canadian environment. Harper and Brothers, New York.

Charlesworth, B. 1994. Evolution in age-structured populations. Cambridge University Press, Cambridge, England.

Couturier, C. Y., L. Clarke, and A. M. Sutterlin. 1986. Identification of spawning areas of two forms of Atlantic salmon (*Salmo salar* L.) inhabiting the same watershed. Fisheries Research 4:131–144.

Dahl, K. 1910. The age and growth of salmon and trout in Norway as shown by their scales. Salmon and Trout Association, Fishmonger's Hall.

Elson, P. F. 1957. The importance of size in the change from parr to smolt in Atlantic salmon. Canadian Fish Culturist 21:1–6.

Fleming, I. A. 1996. Reproductive strategies of Atlantic salmon: ecology and evolution. Reviews in Fish Biology and Fisheries 6:379–416.

Fleming, I. A. 1998. Pattern and variability in the breeding system of Atlantic salmon (*Salmo salar*), with comparisons to other salmonids. Canadian Journal of Fisheries and Aquatic Sciences 55(Supplement 1):59–76.

Fox, W. S. 1930. The literature of *Salmo salar* in Lake Ontario and tributary streams. Transactions of the Royal Society of Canada, Section II, 1930:45–55.

Friedland, K. D., and R. E. Haas. 1996. Marine postsmolt growth and age at maturity of Atlantic salmon. Journal of Fish Biology 48:1–15.

Friedland, K. D., R. E. Haas, and T. F. Sheehan. 1996. Postsmolt growth, maturation, and survival of two stocks of Atlantic salmon. Fishery Bulletin 94:654–663.

Frost, N. 1940. A preliminary study of Newfoundland trout. Newfoundland Research Bulletin 9:1–30.

Gardner, M. J. G. 1976. A review of factors which may influence the sea-age and maturation of Atlantic salmon, *Salmo salar* L. Journal of Fish Biology 9:289–327.

Gibson, R. J. 1993. The Atlantic salmon in freshwater: spawning, rearing and production. Reviews in Fish Biology and Fisheries 3:39.73.

Gibson, R. J., D. D. Williams, C. McGowan, and W. S. Davidson. 1996. The ecology of dwarf fluvial Atlantic salmon, *Salmo salar* L., cohabiting with brook trout, *Salvelinus fontinalis* (Mitchill), in southeastern Newfoundland. Polish Archives of Hydrobiology 43:145–166.

Gilpin, J. B. 1866. On the food fishes of Nova Scotia. No. 4. The trouts and salmons. Transactions of the Nova Scotian Institute for Natural Science 1, Part 4:76–91.

Havey, K. A., and K. Warner. 1970. The landlocked salmon (*Salmo salar*): its life history and management in Maine. Sport Fishing Institute, Washington, D.C., and Maine Department of Inland Fisheries and Game, Augusta, Maine.

Huitfeldt-Kaas, H. 1946. Tribes of salmon in Norway. Nytt Magasin for Naturvidenskapene B85:115–159.

Huntsman, A. G. 1931. The maritime salmon of Canada. Biological Board of Canada Bulletin 21:1–99.

Hutchings, J. A. 1986. Lakeward migrations by juvenile Atlantic salmon, *Salmo salar*. Canadian Journal of Fisheries and Aquatic Sciences 43:732–741.

Hutchings, J. A. 1996. Adaptive phenotypic plasticity in brook trout, *Salvelinus fontinalis,* life histories. Ecoscience 3:25–32.

Hutchings, J. A., and R. A. Myers. 1985. Mating between anadromous and nonanadromous Atlantic salmon, *Salmo salar*. Canadian Journal of Zoology 63:2219–2221.

Hutchings, J. A., and R. A. Myers. 1988. Mating success of alternative maturation phenotypes in male Atlantic salmon, *Salmo salar*. Oecologia 75:169–174.

Hutchings, J. A., and R. A. Myers. 1994. The evolution of alternative mating strategies in variable environments. Evolutionary Ecology 8:256–268.

Hutchings, J. A., and M. E. B. Jones. 1998. Life history variation and growth rate thresholds for maturity in Atlantic salmon, *Salmo salar*. Canadian Journal of Fisheries and Aquatic Sciences 55(Supplement 1):22–47.

Hutton, A. J. 1924. The life history of the salmon. Aberdeen Natural History Antiquarian Society 5:1–56.

Jones, M. W. 2001. Application of genetic markers to conservation biology: salmonid fish in eastern Canada. Ph.D. thesis, Dalhousie University, Halifax, Canada.

Jones, M. W., and J. A. Hutchings. In press a. The influence of male parr body size and mate competition on fertilization success and effective population size in Atlantic salmon. Heredity.

Jones, M. W., and J. A. Hutchings. In press b. Individual variation in Atlantic salmon reproductive success: implications for effective population size. Ecological Applications.

Jones, J. W. 1959. The salmon. Collins, London.

Jordan, W. C., and A. F. Youngson. 1992. The use of genetic marking to assess the reproductive success of mature male Atlantic salmon parr (*Salmo salar* L.) under natural spawning conditions. Journal of Fish Biology 41:613–618.

Kendall, W. C. 1935. The fishes of New England. The salmon family. Part 2—The salmons. Memoirs of the Boston Society of Natural History 9:1–166.

Leclerc, J., and G. Power. 1980. Production of brook char and ouananiche in a large rapid tributary of the Caniapiscau River, northern Quebec. Environmental Biology of Fishes 5:27–32.

Leggett, W. C., and G. Power. 1969. Differences between two populations of landlocked Atlantic salmon (*Salmo salar*) in Newfoundland. Journal of the Fisheries Research Board of Canada 26:1585–1596.

Lever, C. 1997. Naturalized fishes of the world. Academic Press, London.

Marschall, E. A., T. P. Quinn, D. A. Roff, J. A. Hutchings, N. B. Metcalfe, T. A. Bakke, R. L. Saunders, and N. L. Poff. 1998. A framework for understanding Atlantic salmon life history. Canadian Journal of Fisheries and Aquatic Sciences 55(Supplement 1):48–58.

Meerburg, D., editor. 1986. Salmonid age at maturity. Canadian Special Publications in Fisheries and Aquatic Sciences 89.

Menzies, W. J. M. 1925. Salmon (*Salmo salar*) of the River Moise (Eastern Canada). Proceedings of the Royal Society of Edinburgh 45:334–345.

Menzies, W. J. M. 1927. Some aspects of the growth of salmon in river and sea as observed from scale examination of Dee (Aberdeen) and Spey salmon, 1921 to 1923 inclusive. Fisheries Scotland, Salmon Fisheries 1927, Volume I.

Menzies, W. J. M. 1929. Some Newfoundland salmon. Salmon and Trout Magasine 54:64–67.

Metcalfe, N. B. 1998. The interaction between behaviour and physiology in determining life history patterns in Atlantic salmon (*Salmo salar*). Canadian Journal of Fisheries and Aquatic Sciences 55(Supplement 1):93–103.

Metcalfe, N. B., and J. E. Thorpe. 1990. Determinants of geographical variation in the age of seaward-migrating salmon, *Salmo salar*. Journal of Animal Ecology 59:135–145.

Miescher-Rusch, F. 1883. Contributions to the biology of the Rhine salmon. Report of the U.S. Fisheries Commission 8:427–474.

Morán, P., A. M. Pendás, E. Beall, and E. Garcia-Vásquez. 1996. Genetic assessment of the reproductive success of Atlantic salmon precocious parr by means of VNTR loci. Heredity 77:655–660.

Myers, R. A. 1984. Demographic consequences of precocious maturation of Atlantic salmon (*Salmo salar*). Canadian Journal of Fisheries and Aquatic Sciences 41:1349–1353.

Myers, R. A., J. A. Hutchings, and R. J. Gibson. 1986. Variation in male parr maturation within and among populations of Atlantic salmon, *Salmo salar*. Canadian Journal of Fisheries and Aquatic Sciences 43:1242–1248.

Myers, R. A., and J. A. Hutchings. 1987a. A spurious correlation in an interpopulation comparison of Atlantic salmon life histories. Ecology 68:1839–1843.

Myers, R. A., and J. A. Hutchings. 1987b. Mating of anadromous Atlantic salmon, *Salmo salar* L., with mature male parr. Journal of Fish Biology 31:143–146.

Nielson, J. 1961. Contributions to the biology of the salmonidae in Greenland. I–IV. Meddelelser om Gronland 159:1–76.

Nordquist, O. F. 1903. Some biological reasons for the present distribution of freshwater fish in Finland. Fennia, Bulletin de la Societe de Geographie de Finlande 20:1–29.

Paton, D. N. 1898. Report on investigations on the life history of the salmon in fresh water. Annual report for the Fisheries Board of Scotland for 1898.

Perley, M. H. 1850. Report on the sea, and river fisheries of New Brunswick within the Gulf of St. Lawrence and Bay of Chaleur. J. Simpson, Fredericton, New Brunswick.

Power, G. 1958. The evolution of the freshwater races of the Atlantic salmon (*Salmo salar* L.) in eastern North America. Arctic 11:86–92.

Power, G. 1969. The salmon of Ungava Bay. Arctic Institute of North America Technical Paper 22.

Power, G. 1981. Stock characteristics and catches of Atlantic salmon (*Salmo salar*) in Quebec and Newfoundland and Labrador in relation to environmental variables. Canadian Journal of Fisheries and Aquatic Sciences 38:1601–1611.

Ricker, W. E. 1975. Computation and interpretation of biological statistics of fish populations. Fisheries Research Board of Canada Bulletin 191.

Riddell, B. E., and W. C. Leggett. 1981. Evidence of an adaptive basis for geographic variation in body morphology and time of downstream migration of juvenile Atlantic salmon (*Salmo salar*). Canadian Journal of Fisheries and Aquatic Sciences 38:308–320.

Riley, S. C., G. Power, and P. E. Ihssen. 1989. Meristic and morphometric variation in parr of ouananiche and anadromous Atlantic salmon from rivers along the north shore of the Gulf of St. Lawrence. Transactions of the American Fisheries Society 118:515–522.

Roff, D. A. 1992. The evolution of life histories. Chapman and Hall, New York.

Roff, D. A. 1998. The maintenance of phenotypic and genetic variation in threshold traits by frequency-dependent selection. Journal of Evolutionary Biology 11:513–529.

Saunders, R. L. 1986. The scientific and management implications of age and size at sexual maturity in Atlantic salmon (*Salmo salar*). Pages 3–6 *in* D. Meerburg, editor. Salmonid age at maturity. Canadian Special Publication in Fisheries and Aquatic Sciences 89.

Saunders, R. L., E. B. Henderson, and B. D. Glebe. 1982. Precocious sexual maturation and smoltification in male Atlantic salmon (*Salmo salar*). Aquaculture 28:211–229.

Schaffer, W. M. 1979. The theory of life history evolution and its application to Atlantic salmon. Symposium of the Zoological Society of London 44:307–326.

Schaffer, W. M., and P. F. Elson. 1975. The adaptive significance of variations in life history among local populations of Atlantic salmon in North America. Ecology 56:577–590.

Scott, W. B., and E. J. Crossman. 1964. Fishes occurring in the fresh waters of insular Newfoundland. Department of Fisheries, Queens Printer, Ottawa.

Scott, W. B., and E. J. Crossman. 1973. Freshwater fishes of Canada. Fisheries Research Board of Canada Bulletin 184.

Scrope, W. 1843. Days and nights of salmon fishing in the Tweed. Mercat Press, Edinburgh, Scotland.

Shaw, J. 1836. An account of some experiments and observations on the parr and on the ova of the salmon, proving the parr to be the young of salmon. Edinburgh New Philosophical Journal 21:99–110.

Shuter, B. J., M. L. Jones, R. M. Korver, and N. P. Lester. 1998. A general life history based model for regional management of fish stocks: the inland lake trout (*Salvelinus namaycush*) fisheries of Ontario. Canadian Journal of Fisheries and Aquatic Sciences 55:2161–2177.

Stearns, S. C., and J. C. Koella. 1986. The evolution of phenotypic plasticity in life history traits: predictions of reaction norms for age and size at maturity. Evolution 40:893–913.

Sutterlin, A. M., and D. MacLean. 1984. Age at first maturity and the early expression of oocyte recruitment patterns in two forms of Atlantic salmon (*Salmo salar*) and their hybrids. Canadian Journal of Fisheries and Aquatic Sciences 41:1139–1149.

Thomaz, D., E. Beall, and T. Burke. 1997. Alternative reproductive tactics in Atlantic salmon: factors affecting mature parr success. Proceedings of the Royal Society of London B:219–226.

Thorpe, J. E. 1975. Early maturity in male Atlantic salmon. Scottish Fishery Bulletin 42:15–17.

Thorpe, J. E. 1986. Age at first maturity in Atlantic salmon, *Salmo salar*: freshwater period influences and conflicts with smolting. Pages 7–14 *in* D. Meerburg, editor. Salmonid age at maturity. Canadian Special Publication in Fisheries and Aquatic Sciences 89.

Thorpe, J. E., and R. I. G. Morgan. 1980. Growth-rate and smolting-rate of progeny of male Atlantic salmon parr, *Salmo salar* L. Journal of Fish Biology 17:451–459.

Thorpe, J. E., R. I. G. Morgan, C. Talbot, and M. S. Miles. 1983. Inheritance of developmental rates in Atlantic salmon, *Salmo salar* L. Aquaculture 33:119–128.

Thorpe, J. E., M. Mangel, N. B. Metcalfe, and F. A. Huntingford. 1998. Modeling the proximate basis of salmonid life history variation, with application to Atlantic salmon, *Salmo salar* L. Evolutionary Ecology 12:581–599.

Volpe, J. P., E. B. Taylor, D. W. Rimmer, and B. W. Glickman. 2000. Evidence of natural reproduction of aquaculture-escaped Atlantic salmon in a coastal British Columbia river. Conservation Biology 14:899–903.

Warner, K. 1961. A new longevity record for the landlocked population of *Salmo salar*. Copeia 1961:483–484.

Warner, K., and K. Havey. 1961. Body-scale relationships in landlocked salmon, *Salmo salar*. Transactions of the American Fisheries Society 90:457–461.

Warner, K., and O. C. Fenderson. 1963. The salmon and trout fishery of the Fish River Lakes, Maine. Transactions of the American Fisheries Society 92:193–201.

Whalen, K. G., and D. L. Parrish. 1999. Effect of maturation on parr growth and smolt recruitment of Atlantic salmon. Canadian Journal of Fisheries and Aquatic Sciences 56:79–86.

Sustaining North American Salmon: Perspectives Across Regions and Disciplines, 61–76

Chapter 3

Atlantic Salmon Fisheries in Eastern North America: The Prince and the Pauper

Kenneth F. Beland and Alex T. Bielak

Prologue

We begin with two snapshots of Atlantic salmon angling in North America. The similarities and contrasts between these two scenes serve to illustrate the diversity found in Atlantic salmon sport fisheries in eastern North America.

SCENE 1–Penobscot River, Bangor, Maine

19 June, 3:50 a.m.

Bill's alarm goes off with an annoying electronic buzz. He rolls out of bed, takes a quick shower, and heads to the kitchen. He has finally mastered the program mode on the coffee maker, and his first cup is waiting for him. He pours the rest into a worn stainless steel thermos bottle.

4:20 a.m. (The Eddington Pool, about 5 mi away)

The morning light is just beginning to illuminate the Penobscot River as Bill pulls into the graveled parking area. His truck is the second vehicle to arrive this morning. Jerry has already donned his waders and waves to Bill as he walks down to the river. Bill follows suit, and upon reaching the bench on the stream bank, assembles his fly rod and puts it into the rack below Jerry's. They sip coffee and exchange a few words, while waiting for enough daylight to fish.

4:50 a.m.

There are now seven vehicles and a mountain bike in the parking area. Jerry and Bill are slowly fishing downstream through the pool, casting and moving in a practiced ritual that will allow them to fish all the lies in the pool in an hour's time. Fifteen-year-old Jeremy waits anxiously for the pair to move far enough downstream to allow him to begin fishing. He has been fly fishing for two years, and yesterday, he raised two salmon on a #2 Hairwing Green Butt that he tied during winter fly-tying classes at the Eddington Salmon Club. He plans to stay on the river until about 9 a.m., when he will return home to trade his fly rod for a lawn mower. He has three lawns to mow today and may return to the river this evening.

5:15 a.m.

Bill sees a salmon roll just below his fly. He lengthens his cast and has a solid strike, but the salmon comes loose on the first leap. The group of men on the bench is abuzz with commentary about Bill's near miss. There are now eight anglers waiting in rotation to fish, along with the three already in the water. This group includes four retirees, a university professor, a carpenter, a state trooper, and a dentist. Two of the anglers are from Massachusetts, on their annual salmon fishing trip to the Penobscot River. Like Bill and Jerry, two of those on the bench are planning one pass through the pool before beginning their day's work.

6:05 a.m.

Jeremy has no success with the Green Butt and changes to a Verdict, a popular local wet fly pattern. He lengthens his casts and now reaches the slot between the two submerged boulders. In a rise that seemed to take forever, the salmon takes the fly and turns downstream. Jeremy sets the hook hard and is quickly engaged in the battle that he has long dreamed about. The battered Pflueger Medalist whirrs steadily as the fish runs out line. Twelve minutes later, the salmon is in shallow water at his feet, and as the dentist takes photos, the trooper helps him release his 9-pound prize. Jeremy can hardly believe what has just happened. He pauses briefly to enjoy the attention he receives from the gallery on the bench, then hurriedly rides his bike home to share his good fortune with his family over breakfast.

8:40 a.m.

Bill and the other early birds have completed their turn through the pool and have left for the day. Those still fishing have been joined by two more regulars, and all hope for a couple of passes through the pool by noon. Another 9-pounder and a grilse will be hooked by early afternoon, and the trooper will catch and release a salmon estimated to weigh 15 lb, his fourth this season.

SCENE 2 - Moisie River, near September Isles, Québec

19 June, 3:50 a.m.

All is quiet. The big light casts a glow over Home Pool, as the camp's night guardian surveys the scene.

4:50 a.m.

The alarm goes off with an annoying electronic buzz, and the cook rolls out of bed to begin her day. She takes a quick shower and heads for the kitchen, where her morning's work awaits her. She puts wood in the old stove and fills the big pot with water and coffee. The aroma fills the kitchen as her helper enters the kitchen and hangs her sweater on the hook by the door.

6:30 a.m.

George's new electronic travel alarm wakes him to the raucous call of a jay. He lies in bed a few minutes longer, turns back the down comforter, and eventually rolls out of bed. While he showers and shaves, the cabin boy lights a fire in the stone fireplace in a corner of the cabin.

7:15 a.m.

George leaves his cabin and strolls toward the dining room in the main lodge, as he has done the third week of June for the past 23 years. He gets a cup of coffee from the urn and takes a seat on the front porch, where others are beginning to gather. The chatter is centered on the weather forecast and what flies the salmon might be taking later today. The bronze ship's bell soon rings, and people take their seats in the dining room, where a roaring fireplace takes the chill out of the morning air. The room is filled with the aromas of coffee, homemade toast, fresh fruit and eggs Benedict. The seats at George's table are filled with three regulars (a retired U.S. Army general and two Canadian bankers), as well as a couple from New York who are here for the first time. Although the lodge is in a near-wilderness setting, the food and service rival those at George's club back in Toronto.

8:40 a.m.

George's guide, François Vollant, a native Montagnais from a nearby reserve, arrives at the cabin and carries George's tackle bag and rain jacket to the 22-foot canoe anchored at the water's edge. George's rods—a 9 ½-foot

Orvis and an 11-foot double-handed beauty, custom made for his 50th birthday—have already been laid in the boat by the bowman, Maurice. As soon as George is settled in the center seat, Maurice pushes off and François motors the big canoe 2 km upstream to the Ledge Pool, which is theirs for the morning.

9:25 a.m.

Maurice raises the anchor and François eases the canoe one drop downstream. George casts with an ease attained only through years of practice. He lengthens his cast methodically and begins to cover the slick between the two submerged boulders. François sees a flash below George's fly and says "Il est venu." George gathers his line and François catches the fly, a #12 Bates killer. He gives it a good tug, checks the knots on George's leader and nods that all is in order. George waits the customary two minutes, false casts twice and lays out the coiled line. The cast is true and the fish takes the fly. George says "un grand" and the fight is on. The reel, a hand-machined Bogdan, clicks rapidly as the fish runs powerfully downstream. Maurice has already raised the anchor, and François begins to take the canoe to shore. Twenty minutes later, George hand tails the 22-lb hen salmon, and François skillfully slips the barbless hook from the corner of its mouth. The fish gleams silver and has several sea lice on her back. By late morning, two more three-sea-winter salmon will have been hooked, and one has been retained for the camp's smokehouse. At noon, François returns his sport to the lodge for a brace of cocktails, lunch, and a nap.

Introduction

This paper presents a sketch of present and historical Atlantic salmon *Salmo salar* fisheries in eastern North America. We hope to illustrate similarities and contrasts among Atlantic salmon fisheries in the United States and Canada and the fishers who participate in them. The two vignettes that we have just presented illustrate the diversity found in Atlantic salmon sport fisheries in North America. These vary from open access fisheries in Maine and Nova Scotia to the elite fisheries of rivers such as the Moisie and Restigouche in Québec and New Brunswick.

Recreational fisheries for wild Atlantic salmon have developed on rivers extending from Maine to Labrador and Ungava Bay in northern Québec, including the Canadian provinces of New Brunswick, Nova Scotia, Prince Edward Island, Québec, and Newfoundland and Labrador.

In response to declines in stock abundance throughout the region, traditional commercial Atlantic salmon fisheries have been curtailed by regulation and government buyout of commercial fishing permits. At present, with the exception of subsistence fisheries by First Nations on some Canadian rivers, there is no netting of Atlantic salmon in North American rivers or coastal waters.

Significant harvest restrictions have also been imposed on sport fisheries in the region in recent years, including a regulation change closing the catch-and-release Atlantic salmon sport fisheries in Maine, begun in 2000. Similar conservation-driven closures were recently imposed on Canadian rivers draining into the Bay of Fundy in Nova Scotia and New Brunswick.

Despite profound differences in their organization, accessibility, and participants, Atlantic salmon sport fisheries throughout the region also share many traditions and rituals.

The First Salmon Harvesters

Aboriginal North Americans (First Nations, Canada; Native Americans, United States) were undoubtedly the first human harvesters of Atlantic salmon *Salmo salar* in North America during the most recent postglacial period. Areas where salmon might concentrate, including holding pools, and obstacles to migration, such as waterfalls, were commonly selected for campsites. However, contrary to the European experience, where evidence of the prehistoric significance of the salmon can be found in the form of cave sculptures or bone carvings, there are comparatively few similar records from eastern North America. Salmon bones have been found in known encampment sites on the Miramichi, and Dunfield (1985) confirms that fishing appears to have been important to the subsistence economy for natives. In New England, little archaeological evidence of salmon use has been found in examinations of native archaeological sites (Carlson 1988).

Panusak and Proulx (1981), in an important and detailed master's thesis, "Resistance of the Montagnais to the usurping of salmon rivers by Euro-Canadians from the 17th to the 20th Century,"[1] note that rivers constituted the primary communication routes of the Montagnais. These routes were used to "access their hunting grounds and to meet their compatriots, either in the interior or along the coast." In fact, band members were identified by the names of the rivers they frequented, and an important part of their food came from the rivers and environs. The authors note that "obviously fish were harvested, but so were porcupine and water wildlife. The annual gatherings on the coast could hardly have taken place without the runs of salmon. Seals and migratory birds were also hunted near the mouths of rivers and on nearshore islands." They go on to note "till now, ethnologists have given very little attention to the role of rivers and aquatic re-

[1]All translations from the original French by the authors.

sources for the Montagnais. For example, the role of fishing locations is practically nonexistent in the …commentary on hunting territories," or in "…discussions focusing on caribou, beaver and furbearers." Carlson (1988) reviews a number of publications that also indicate that Native Americans in salmon-producing areas of New England were generalists in their use of the native fauna.

East coast First Nation traditions call for respect for the spirits of various species of animals, and parts of fish and other animals were used as auguries for hunting and fishing success. In certain cultures, such as the Nascapi in Québec, fish were accorded importance in religious observance (Panusak and Proulx 1981). Salmon may have been among the more prized species harvested, but it is clear that it was but one of many species of fish and other animals hunted and consumed during the course of the year. American shad *Alosa sapidissima*, alewife *Alosa pseudoharengus* (gaspereaux), Atlantic cod *Gadus morhua*, and striped bass *Morone saxatilis* were highly abundant and readily available in many salmon-producing areas and were probably similarly important to aboriginal peoples (Carlson 1988). Dunfield (1985) mentions that certain tribes not only harvested Atlantic salmon for immediate consumption within the family or community, but also traded salmon for other commodities, particularly after European settlers arrived on the scene. He also cites examples where smoked or dried salmon carcasses were buried with the dead to provide sustenance for the passage to the afterlife, and he notes that the First Nations along the Restigouche River reproduced images of salmon on their clothing and canoes.

Although there is evidence that Atlantic salmon were important to east coast native peoples, one does not get the same sense of all-pervasive economic, historic, cultural, artistic, and spiritual significance that appears to be the case for salmon among Pacific Northwest aboriginal cultures (Netboy 1980; Carlson 1988). These cultures developed to a much greater extent around and depended more heavily on anadromous Pacific salmon species as a staple food than did their eastern counterparts (Burgner 1991).

The Montagnais, Attikamek, and Algonquin peoples of the boreal forest bartered animal skins and fished for Atlantic salmon with nets constructed by the tribes of the St. Lawrence valley and the Gaspé Peninsula. They also used wooden hooks, principally for fishing through the ice, and fished at night by torchlight with spears from canoes. Weeks (1971) reported salmon catches as large as 900 salmon taken with spears in a single night on the Moisie River. Speck (1997) presents detailed descriptions of the spears and nets used by the Penobscot Indians to catch salmon, American eels *Anguilla rostrata*, and other fish from eastern Maine rivers. Commercially made nets were subsequently obtained through trade with European settlers and put to use in aboriginal fisheries. More recently, participation in rod and line fisheries for Atlantic salmon has become more commonplace among native peoples in Canada.

As the European settlers and their descendants established commercial salmon fisheries on North American rivers in the 18th and 19th centuries, many native peoples were precluded from traditional subsistence fisheries for salmon. As the Hudson's Bay Company and the Labrador Company expanded their resource exploitation in northeastern Canada, native peoples were often precluded from harvesting salmon in the region's rivers. Mid-18th century attempts by First Nations to assert fishing rights guaranteed in the Royal Proclamation of 1763 were generally unproductive, and exclusive fishing rights on the more productive Québec rivers were sold or leased at auction to private interests (Eidinger 1987).

Native harvests were limited to a short spearing season on rivers not leased. In 1975, the Canadian Department of Fisheries and Oceans granted subsistence fishing licenses for Atlantic salmon to First Nations in Québec and New Brunswick. The 1970s through the early 1990s represented a period of controversy over native fishing rights in parts of Canada. These expanded subsistence fisheries sparked conflicts with angler and salmon conservation groups in the region (Eidinger 1987; Hazell 1987).

In Canada, a 1990 decision by the Supreme Court (*R.* vs. *Sparrow*, 31 May 1990) reaffirmed the fishing rights guaranteed by the Royal Proclamation of 1763 and established the primacy of native fishing rights for food and ceremonial purposes over recreational and commercial fisheries, subject only to resource conservation requirements. In some cases, for fisheries on the Restigouche, Miramichi, and St. John River rivers, reassertion of fishing rights has been contentious at times and resulted in highly publicized confrontations between First Nations and law enforcement agencies. In parts of Québec, New Brunswick, and Nova Scotia (Cape Breton Island), some First Nations asserted themselves, and individuals began selling their catches of salmon and other fish in local markets. In one case involving an individual prosecuted for the illegal sale of American eels, the Canadian Supreme Court (*R.* vs. *Marshall*, 17 November 1999) affirmed a collective right of First Nations to earn a moderate livelihood from their fishing, hunting, or gathering activities. The limits to those rights have yet to be established, although the Canadian government has the right to impose regulations designed to assure resource sustainability.

Native peoples have become established as both stakeholders and active participants in Atlantic salmon management in Canada and the United States. In some instances, such as on the Grand Cascapedia River and, after a period of conflict, the Natashquan River, First Nations participate in salmon management and have established outfitting and guiding enterprises as alternate means of accruing individual and community economic benefits from the salmon resource (Bodo 1996). After periods of open disagreement and confrontation on resource allocation, whites and First Nations have become full partners in reestablishing salmon on the Escoumins River in Québec (Bruce 1995).

Although some Native Americans in Maine participated in the recently closed recreational salmon fisheries, there have been only occasional subsistence catches of Atlantic salmon in Maine by Native Americans. Despite limited current use of the salmon resource, the Penobscot Indian Nation has been an active participant in cooperative salmon research and management activities and is represented through voting membership on the interagency committee that provides technical oversight to Atlantic salmon management in Maine rivers.

Commercial Atlantic Salmon Fisheries

European settlers to New England and eastern Canada and their descendants harvested Atlantic salmon for subsistence and export. On many rivers, salmon were caught in weirs, with weir rights belonging to riparian landowners. By 1780, weirs were being used to catch salmon in the Kennebec River in Maine (Baum 1997). By the late 19th century, weir fishing was well established along most Maine salmon rivers. In 1880, a total of 233 weirs and 36 gill nets were fished on the Penobscot River alone, producing a total catch of 10,000 salmon (Baum 1997). During the late 18th century, annual salmon exports from the Miramichi River to the United States and Europe were as high as 850 metric tons (Dunfield 1985). During the first half of the 19th century, agents of the Hudson's Bay Company enjoyed a monopoly on salmon harvesting on many rivers on the North Shore of the St. Lawrence in Québec (Eidinger 1987).

Commercial salmon fisheries in Newfoundland developed by the mid-19th century and were important (although less than Atlantic cod) to local economies through the 1980s. Although cumulatively significant in terms of resource exploitation, most Atlantic salmon commercial fisheries in New England and eastern Canada were part-time seasonal enterprises, practiced by individuals rather than corporate entities. Commercial fisheries on the east coast of Canada were practiced with gill nets or trap nets set from small and medium-sized boats (<50 ft in length). These fisheries harvested locally produced fish and intercepted migrating salmon from other regions as they passed through coastal waters of Newfoundland and Labrador.

The availability of greater numbers of salmon over a longer period off the east coast of Newfoundland enabled these fishers to derive a greater portion of their annual income from salmon. The eastern North American commercial salmon catches peaked around 1930 at about 6,000 metric tons and were much smaller than counterpart Pacific Northwest and Alaskan salmon catches.

Dam construction and overfishing from the late 18th through the early 20th centuries diminished salmon abundance in New England, such that the final vestiges of commercial salmon fisheries in Maine had been eliminated by 1948 (Baum 1997). In addition to local fisheries, Atlantic salmon from North America were also subjected to interception fisheries in the coastal waters and high seas off Greenland. Catches of Atlantic salmon in the Greenland fisheries rapidly escalated through the 1960s, peaking at nearly 2,600 metric tons in 1971 (Baum 1997). By this time, the combined Canadian and Greenland catches far exceeded the sport fishing catches, and salmon anglers and conservation groups pressured governments for curtailment of commercial salmon fisheries.

Wealthy salmon anglers on both sides of the Atlantic exerted their considerable political influence to affect public opinion and pressure governments into complex international negotiations (Buck 1993). As salmon stocks in the region declined through the 1970s into the 1990s, commercial salmon fisheries in eastern Canada and Greenland were incrementally reduced and ultimately curtailed, culminating in the closure of the Québec North Shore and Labrador salmon fisheries in 1997–1998.

Many of the Canadian commercial salmon fishery closures were accompanied by government buyouts of licenses from several thousand displaced fishers, at a cost of many millions of dollars (Bruce 1992). The Greenland Atlantic salmon fishery has been reduced to a minor fishery for local use only.

Atlantic Salmon Recreational Fisheries

Atlantic salmon recreational fisheries in North America date back to the early 19th century. British military officers and officials of the Hudson's Bay Company fished the St. Marguerite River as early as 1825 (Haight 1977). In the United States, fly fishing for Atlantic salmon was first documented in 1832 on the Dennys River in eastern Maine (Bartlett and Robinson 1988). By the 1870s, sport fisheries for salmon were established throughout the region. For example, the St. Lawrence Salmon Fishing Company promoted steamer excursions along the North Shore of the St. Lawrence in Québec to fish for salmon and trout in the region's rivers (Hallock 1873). By this time, Atlantic salmon sport fisheries had also become established along other Maine rivers.

The large fish, the high catch rates, the relative inaccessibility, and the undeveloped character of many Canadian rivers attracted the wealthiest anglers. They secured angling rights along some of the best waters, either by outright purchase or lease of riparian lands from the provincial governments. Permanent fishing camps and lodges were built near the mouths of many rivers, including such well known rivers as the Moisie, Restigouche, and Grande Cascapedia.

The clubs, leaseholders, and outfitters were often instrumental in eliminating local subsistence and/or commercial netting in Canadian rivers by outright purchase of netting rights. Weeks (1971) reports more than 27 km of nets were fished in the Moisie River in the early 20th century, with average annual catches of approximately 8,000 salmon. The Moisie Salmon Club purchased the netting rights from many fishers in 1913 and hired local residents as river keepers to curtail poaching.

Losses in local fishing income were also offset by other employment opportunities associated with salmon fishing camps and lodges, such as camp staff and guides. Many well established British and Scottish salmon angling traditions carried over to the new camps and lodges on Canadian rivers. The fishing day was often rigidly structured, each angler fished his assigned beat, and the camps often kept meticulous records of effort, catches, and fish weights.

Fly fishing was generally the favored method in these elite fisheries and eventually became established by legislation or regulation throughout eastern Canada and the United States as the only legal means to angle for Atlantic salmon.

Access to these elite fisheries requires more than the ability to pay the monetary costs, which can be considerable. For example, a week during the peak season on the Moisie River costs many thousands of dollars per angler for fishing and accommodations, not including travel costs from all over the world. Access to these fisheries also requires the proper family, social, or business connections to obtain an invitation to fish the waters of a particular club, and often the patience to spend years on the waiting list for any available week of fishing, let alone prime time.

Status among anglers in the elite fisheries accrues to those who have the means to fish regularly and who provide fishing access for their friends and business associates. The angling exploitation rates on these elite waters have traditionally been relatively low, often 10 percent or less, with only a handful of anglers fishing over an entire run of fish.

Individual catches during periods of high salmon production were at times spectacular, surpassing even those reported in recent years from newly developed elite fisheries, such as on the Kola Peninsula in Russia. On the Godbout River in Québec, Comeau (1923) reported that one angler landed 46 salmon averaging 9.25 lb in one day's fishing. That catch translated into an average of one fish every 10.5 min, based upon the traditional fishing times at that particular club.

Among the traditions established on the elite rivers was the institution of the guide. To this day, salmon fishing at the elite camps and clubs is a partnership between angler and guide. The angler's success is influenced as much by his/her ability to develop a trusting and respectful long-term relationship with the guide as by the ability to cast a fly or play a hooked fish (Anderson 1989). When fishing from an elite club or lodge, the host or senior member of an angling party works with the head guide to match each angler with the guide best suited to the angler's skill level and temperament. In the eyes of the guide, the angler with the greatest stature is the one who not only casts well and plays hooked fish skillfully, but also understands and acknowledges that the guide's skill is as important as his/her own to their mutual success. Exchanges of letters and gifts to the guide at Christmas are common ways of sustaining these relationships over the winter months.

Because of current legal requirements for guides on many Canadian rivers, including some public waters, the use of guides now extends beyond the elite camps and lodges. This has led to the evolution of a new class of entrepreneurial guides, who are independent operators hired directly by visiting anglers to guide them as they fish unfamiliar waters (Anderson 1989).

In sharp contrast to Atlantic salmon fisheries on the elite Canadian rivers are the low-cost open fisheries of Maine, Nova Scotia, and Prince Edward Island. Angling rights on these salmon rivers are in the public domain, and the participants include anglers from a wide variety of socioeconomic backgrounds. On a given day, the anglers at a local salmon pool might include local residents of very limited means, working-class individuals and local professionals, as well as visitors with the means to take more expensive trips on a regular basis.

Despite the accessibility of these fisheries to all, the fisheries show parallel developments in organization and structure to the more elite Canadian fisheries. Many anglers on these rivers learned about salmon fishing while on a trip to Canada, from angling acquaintances, or by reading about salmon angling in books or magazines. The equipment used is similar in function but differs widely in cost and attention to detail. The flies and fishing tactics used are generally similar, whether on the Penobscot River in Maine, the Margaree River in Nova Scotia, or the Moisie River in Québec.

High angler demand for the most productive sites is one consequence of a fishery with open access to a relatively scarce angling opportunity. When angler demand exceeds the capacity at prime angling sites, a common response is the establishment of a rotation system to share the waters. Anglers fish downstream through the pool at a regular pace, with the order in which anglers place their rods on the poolside rack determining the order in which they fish the pool.

The rotation areas and protocols at each salmon pool are established by consensus among the regular anglers, and peer pressure is applied as needed to enforce compliance. An angler who is fishing at an unacceptably slow pace can expect to hear "rotate!" shouted out by someone waiting his turn to fish. Guides are not required and are rarely used by visiting anglers.

Newcomers to the river typically learn where salmon lie and the local customs through observation and conversations with anglers waiting in rotation for their turn through the pool (Lee 1992). An angler's status in these fisheries accrues not only on the basis of fishing skills, but also from social skills exhibited during time spent waiting in rotation with others. Some anglers with highly developed fishing skills are disdained by the group for their inability to abide by local angling customs or their inability to manage interactions with others while waiting to fish (Lee 1992).

Atlantic salmon fisheries in Newfoundland and many Québec rivers are intermediate in access, cost, and social development. In parts of Québec, in addition to the private clubs or outfitters mentioned previously, access to salmon fisheries may be controlled by local watershed stewardship groups, administering a system called the Zone d'Exploitation Contrôlée—Controlled Exploitation Zone (ZEC).

This system provides a moderate cost, fee-based system of allocating limited access to natural resources, including salmon fishing (Davignon 1996; Zajak 1999). Use permits for specific waters are purchased at local checkpoints. Access to most river sections is allocated by lottery, but access is not limited on some rivers or sections. This system is founded on the principles of equality of access and local control. Entrepreneurial guides familiar with local waters are often available and can greatly increase angler success, but their use is not generally mandatory.

The Québec and New Brunswick system of private ownership of fishing waters and clubs did not develop in Newfoundland. Access to salmon fisheries in Newfoundland is open to any licensed angler, with nonresidents required to have a guide in most areas. The guide requirement and the relative inaccessibility of Newfoundland make fishing there more expensive than in Maine, Nova Scotia, or Prince Edward Island for nonresidents, but still less costly and more accessible than the elite waters.

Although catch rates on some Newfoundland rivers can approach those for the elite waters, the salmon tend to be smaller. Visiting anglers may be sharing pools with regulars from local communities. Traditions such as pool rotation are not as well established in Newfoundland as in Maine or Nova Scotia. The angling protocols and social interactions among anglers are highly variable among Newfoundland rivers, as well as along individual river reaches. At some locations, anglers are friendly to visitors, and they may offer advice on fly selection or share prime angling sites. At other locations, competition among anglers at productive pools that receive heavy use can be intense, and verbal or physical fisticuffs occasionally break out among anglers fishing in close quarters. Such events are not restricted to

Newfoundland fisheries—physical confrontations are not unknown at some salmon pools in Nova Scotia.

Salmon fisheries in the province of New Brunswick are a juxtaposition of Crown Open (public) waters open to all licensed anglers (with a nonresident guide requirement); private waters controlled by individuals, clubs, or private lodges; and the Crown Reserve system.

Crown Reserve waters can provide angling quality comparable to the best private waters at modest cost, although these fisheries lack the amenities of the exclusive clubs or the more expensive commercial lodges. Access to these publicly owned waters is by lottery limited to residents of New Brunswick, who pay a modest daily fee for exclusive fishing access.

Competition for the best sites at prime times can be keen. Because of the juxtaposition of public and private access at a variety of prices, the salmon fishery on a large river such as the Miramichi includes participants from a broad socioeconomic spectrum. As a result, the rituals and behavior of salmon anglers on a river can be quite variable, depending upon where one is fishing.

At clubs and lodges on private waters, the rituals are similar to those observed in the elite fisheries, as is the symbiosis between angler and guide. Crown Reserve fisheries are usually solitary or small group experiences, with social interactions limited to those within the angling party. Angling in Crown Open sections can be similar to fisheries in Newfoundland or Nova Scotia, with pool-specific rituals for sharing waters. One popular pool may be fished in rotation, whereas another pool might be fished on a first-come, first-served basis, with those fishing on either side of an angler seemingly as immovable as the rocks upon which they are perched.

The remote salmon rivers of Labrador and northern Québec provide high quality wilderness angling at a relatively high cost. Access to these fisheries is generally obtained through commercial outfitters, who contract with the provincial governments to obtain exclusive concessions for specific territories. Access is typically by float plane through undeveloped wildlands.

In some instances, First Nations may hold the fishing rights or the outfitter concessions. Because the salmon season overlaps with the hunting season for caribou, some outfitters offer combination hunting and fishing excursions.

These fisheries have generally developed since World War II, and the traditions that have evolved are less rigid than those at private clubs or long-established commercial lodges, although the amenities may be first-class. Within an outfitter's territory, the angler density is low, and the need for well-established rituals to manage interactions among competing anglers is limited. The clientele on these trips may range from a middle-income angler who has saved for the trip of a lifetime to professionals and business people with the means to travel to such destinations on a regular basis.

Conclusion

In this chapter, we tried to convey an overview of the wide diversity of Atlantic salmon fisheries that have existed in eastern North America. First Nations have reasserted their treaty rights and net Atlantic salmon on some Canadian rivers.

Although commercial Atlantic salmon fisheries were widespread in the region at one time, there have been progressive closures in response to regional declines in salmon abundance. Recreational fisheries still occur on many salmon rivers throughout eastern Canada. Although these fisheries are extremely diverse in their participants, the accessibility of fishable waters, the costs of participation, their traditions, and angler etiquette, the equipment and methods used share many similarities. Across all fisheries, contemporary anglers are more likely to release a captured salmon than was the case in years past, even where this is not mandated by law.

The rivers upon which these fisheries occur and the sizes and numbers of fish present are also quite diverse. Those who participate in these fisheries represent a broad spectrum of American and Canadian society, as well as international visitors to the region. Atlantic salmon anglers include an economic royalty with the means to fish anyplace in the world, as well as individuals of far more modest means who are fishing public waters on the river closest to their home.

References

Anderson, G. 1989. Les bons guides (The good guides). Atlantic Salmon Journal 38(2):25–27.

Bartlett, E., and R. Robinson. 1988. Salmon on the Dennys: 1786–1988. The Ellsworth American, Ellsworth, Maine.

Baum, E. 1997. Maine Atlantic salmon: a national treasure. Atlantic Salmon Unlimited, Hermon, Maine.

Bodo, P. 1996. An interracial partnership that really works. Atlantic Salmon Journal 45(1):46–49.

Bruce, H. 1992. Bless the ban. Boost the buyout. Atlantic Salmon Journal 41(2): 14–17.

Bruce, H. 1995. Salmo salar: keeper of the peace. Atlantic Salmon Journal 44(2): 16–19.

Buck, R. 1993. Silver swimmer: the struggle for the survival of the Atlantic salmon. Lyons and Burford, New York.

Burgner, R. L. 1991. Life history of sockeye salmon. Pages 1–117 *in* C. Groot and L. Margolis, editors. Pacific salmon life histories. University of British Columbia Press, Vancouver.

Carlson, C. C. 1988. Where's the salmon? A reevaluation of the role of anadromous fisheries in aboriginal New England. Pages 47–79 *in* G. P. Nicholas, editor. Holocene human ecology in Northeastern North America. Plenum, New York, New York.

Comeau, N. A. 1923. Life and sport on the North Shore. 2nd Edition. Québec Telegraph Printing Company.

Davignon, J-F. 1996. The ABCs of ZECs in Québec. Pages 103–108 *in* Symposium proceedings: toward a better sportfishing future in New Brunswick. Fredericton, New Brunswick. Atlantic Salmon Federation, St. Andrews, New Brunswick.

Dunfield, R. W. 1985. The Atlantic salmon in the history of North America. Canadian Special Publication of Fisheries and Aquatic Sciences 80.

Eidinger, J. 1987. Rites and wrongs. Atlantic Salmon Journal 36(3):13–17.

Haight, F. 1977. Notes on the founding of the Ste. Marguerite Salmon Club. The Anglers Club Bulletin 56(3).

Hallock, C. 1873. The fishing tourist. Harper and Brothers, New York.

Hazell, S. 1987. Letter vs. spirit: the Indian fishery. Atlantic Salmon Journal 36(3):8–9.

Lee, A. 1992. How to fish public waters without being bellowed at. Atlantic Salmon Journal 41(3):33–35.

Netboy, A. 1980. Salmon: the world's most harassed fish. Andre Deutsch, London.

Panusak, A. M., and J. R. Proulx. 1981. La résistance des Montagnais à l'usurpation des rivières à saumon par les Euro Canadiens du 17e au 20e siècle. Master's thesis. Université de Montréal, Montréal.

Speck, F. G. 1997. Penobscot man. The University of Maine Press, Orono, Maine.

Weeks, E. 1971. The Moisie salmon club. Barre Publishers, Barre, Massachusetts.

Zajak, B. 1999. Québec's magic middle ground. Atlantic Salmon Journal 48(2): 44–48.

Sustaining North American Salmon: Perspectives Across Regions and Disciplines, 77–92

Chapter 4

The Atlantic Salmon Recreational Angling Industry: Economic Benefits

F. G. Whoriskey and J. Glebe

Introduction

Humans and Atlantic salmon *Salmo salar* have been interacting in North America for thousands of years. Human populations have grown; by contrast, both the productive capacity and the range of wild Atlantic salmon have been drastically reduced since European colonization (Watt 1988, 1989). Many factors have contributed to these declines, including overfishing, habitat destruction, blockage of migration routes with dams, and poor forestry and agriculture practices. Pressures continue on the remaining salmon populations, and returns of two-sea-winter (2SW) maiden Atlantic salmon to North America, which provide most of the eggs needed to seed the rivers, are now believed to be at the lowest levels ever recorded (ICES [International Council for the Exploration of the Sea] 2000; Figure 1). In 2000, in Maine, the U.S. Department of the Interior announced the listing of the Downeast River populations of salmon as endangered under the Endangered Species Act. In Canada, an endangered designation by the Committee on the Status of Endangered Wildlife in Canada is being sought for populations in rivers draining into the Inner Bay of Fundy (Kenchington 1999).

Economics can be a powerful force for conservation. People conserve the things from which they draw benefit. There is nothing new in this idea; it has been a central tenet of wildlife management since the early years of the science (Leopold 1933). The chances of successfully conserving and restoring species are greatly enhanced when people derive direct economic benefits from the species. The ideal situation occurs when market and conservation forces support each other.

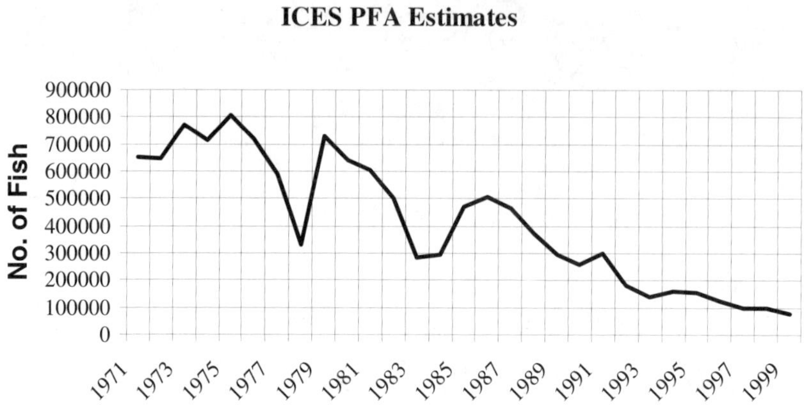

Figure 1. Estimates of the prefishing abundance (PFA) of North American origin salmon present in the ocean about seven months before they would return to spawn in their home rivers as multi-sea-winter salmon. Data from ICES (2000).

Given the importance of economic considerations as part of a comprehensive approach to salmon fisheries conservation, we were tasked with reviewing the literature and trying to assess the current value of the recreational fishery for Atlantic salmon. We are not economists, and this may be why we were asked to undertake the study. What was requested was an outsider's perspectives of the degree to which the Atlantic salmon functions as an economic engine, descriptions of what resources are being devoted to the engine's upkeep, and thoughts about the role of economics in salmon conservation. Here we attempt to answer three questions:

1) What is the recreational fishery worth as calculated by economists?
2) Has this economic value peaked, or is there room for expansion?
3) What other conservation investments have been made that promote recreational fishing economic benefits?

Estimating Economic Values and Benefits

We reviewed literature on the economic values and benefits of salmon angling. Our analysis focused on Canada—in particular, the Canadian provinces of Quebec and New Brunswick. In these areas, salmon populations are well monitored, large recreational angling fisheries occur, and more economic information is available. The recreational fishery for Atlantic salmon *Salmo salar* is pursued in fresh water and river estuaries; we are unaware of any sport fishing for the species at sea. Salmon angling is limited to fly fishing, and voluntary live-release angling is growing in popularity. In some jurisdictions, for conservation reasons, anglers are required to release alive any salmon larger than 63 cm fork length.

New Brunswick has kept a significant part of its salmon fishing resources in the private or quasi-private sector. Briefly, in an 1882 court judgment (*The Queen vs. Robertson 1882*, confirmed by the UK's Privy Council in 1898), authority over freshwater fishing rights was accorded to the Canadian provinces. Following this decision, each province then had the option of converting the resources into public property or treating them as private property and letting individuals or organizations own, buy, sell, or lease the exclusive right to fish in freshwater, either together with or separate from the ownership of the underlying land. The economic importance of resource ownership decisions by provinces is that public property resources are often disassociated from market demands. Prices are set depending on government objectives and policy (e.g., provide cheap public recreational opportunities). By contrast, private resources are subject to market pressures and will better reflect the true economic value (Tuomi 1980, 1987). Private resources therefore provide a much better indicator of the value accorded by the public to the Atlantic salmon sport fishery when the fishery has to be compared with competing water uses.

Recreational Atlantic salmon fisheries, as public resources, are also more difficult to study because of a lack of economic information. In Maine, the only remaining state with self-sustaining populations of wild Atlantic salmon (a public resource in the United States), stocks are severely depressed and all recreational angling fisheries for these salmon are closed (see Kocik and Brown, Chapter 1; Rideout and Ritter, Chapter 5). In the absence of a recreational fishery, few, if any, economic benefits are being realized. Instead, huge costs are being borne by taxpayers to provide restoration programs. A casualty of the fishing closure is a loss of volunteers who, in the past, would have worked to help conserve wild Atlantic salmon. Interest in the species is dwindling, and membership in local salmon associations is declining.

To address the question of the value of the recreational angling fishery in Canada, we present information about the capital value of salmon fishing properties (camps) in New Brunswick and on the money that anglers spend to go fishing. To determine whether the value of salmon fishing has peaked, we reviewed studies that examined the potential for making more money from salmon fishing. Some of these studies were regional in nature; others were plans for specific rivers. To document what we are doing to conserve the economic benefits of the recreational salmon fishing industry, we describe federal, provincial, and private-sector investments in promoting Atlantic salmon recreational fishing. Some of these are annual investments; others are one-time programs.

Studies of the economics of Atlantic salmon resources have been done in different places at different times. Unless otherwise noted, all values in this chapter are reported in Canadian dollars. Where considerable time has elapsed since a study was done, we have adjusted reported dollar values for rates of inflation. Inflation rates were calculated as

$$\text{Inflation (\%)} = (\text{CPI}_{i+1} - \text{CPI}_i)/\text{CPI}_i \times 100,$$

where CPI is the Bank of Canada Global Consumer Price Index for year *i*. Rates are posted on Statistics Canada's web site (www.statcan.ca/english/Pgdb/Economy/Economic/econ46.htm).

Results

What is the Recreational Angling Industry Worth?

In New Brunswick, Tuomi (1980) attacked the problem of determining the value of New Brunswick's recreational salmon fishery by aggregating the capital value of salmon angling waters as established by the market, rather than the annual revenues and expenditures from salmon fishing (i.e., the calculation of consumer and producer surpluses). Tuomi took this approach because information was available on the capital value of angling waters (lease and sale prices); information on revenues and expenditures was much poorer. A second reason was to focus on the fact that the value of the fishery is really future-oriented. That is, this information reflects what people are willing to pay in the expectation of future fishing; therefore, property value trends become more reliable predictors of future economic activities.

Using this approach, Tuomi (1980) assessed the market value for private water in 1978 as $74.8 million. Correcting for annual rates of inflation (Bank of Canada review of Annual Consumer Price Index), the 2000 capital value of the fishery would be about $194.4 million.

Additional economic value accrues to New Brunswick from the annual spending by anglers to go fishing. At present, there is no direct estimate available of what this value is for Atlantic salmon anglers. A crude value can be estimated using information from the latest Survey of Recreational Fishing in Canada (Economic and Policy Analysis Directorate 1997). This survey is updated at five-year intervals. In Canada, resident and nonresident anglers in 1995 spent on average $523 and $666 per angler per year, respectively. Multiplying these values by the number of resident and nonresident salmon licenses sold in New Brunswick in 1995 (21,949 and 6,030, respectively) gives values of about $11 million and $4 million, respectively, for a total of $15 million per year. This is probably a minimum figure because costs of salmon fishing are above average and because it does not include monies spent on capital items such as boats and motors that are purchased for angling purposes. Direct expenditures by anglers on all fish species in New Brunswick in 1995 were estimated at about $27 million (Economic and Policy Analysis Directorate 1997).

In Quebec, Hydro-Québec commissioned a study of the value of the Atlantic salmon sport fishery for the province of Quebec (Michaud 1990). Here, information was available on the number of anglers and their annual spending on Atlantic salmon. Michaud (1990) based his analysis on a "travel cost" approach, in which the amounts individuals must spend to equip themselves for and undertake an annual salmon fishing trip are tallied. His goal was to model the dollar values that average anglers annually assigned to the practice of salmon fishing and sum these to arrive at a global estimate for the province. For 1988, this value was calculated as $97 million. Correcting for inflation up to 2000 gives a value of $131 million.

The present and future economic potential of these recreational fisheries seems to be eroding as salmon populations have declined. The numbers of salmon angling licenses in both New Brunswick and Quebec have been trending downward (Figure 2), although far more slowly than the salmon populations have declined (Figure 1).

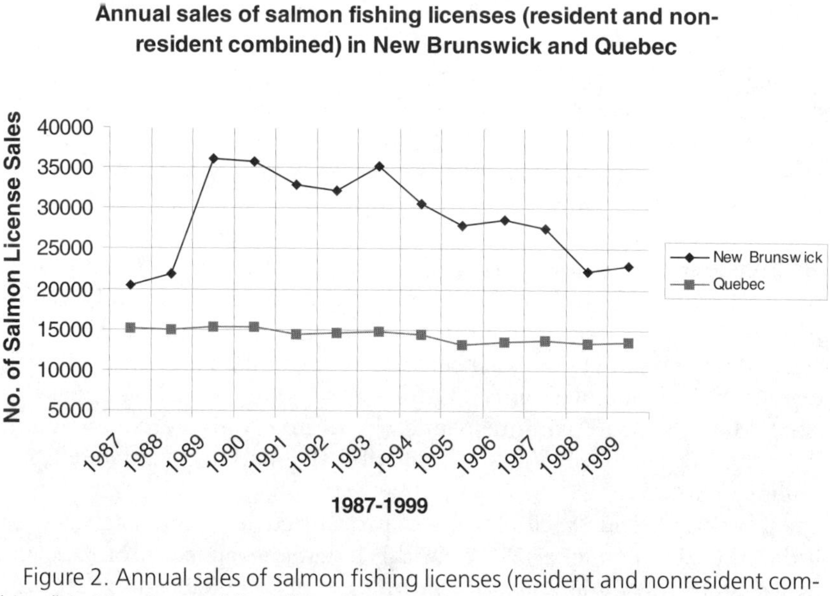

Figure 2. Annual sales of salmon fishing licenses (resident and nonresident combined) in New Brunswick and Quebec. Data provided by New Brunswick's Department of Natural Resources and Energy and Quebec's Ministère de l'Environnement et de la Faune.

Can Economic Benefits from the Recreational Fishery be Expanded?

Many of the sources we consulted believe there is potential to expand recreational fisheries and generate more money from them. A number of targeted investments by public and private-sector partners have been undertaken in recent years, with the specific goal of expanding recreational fishing in general and salmon fishing in particular. Geographic scales of the efforts varied, ranging from those at the regional level to local endeavors in specific river systems.

Regional approaches

Quebec's salmon plan. The economic plan for the development of salmon (Programme de dévelopment économique du saumon, or PDES) was an offshoot of a larger agreement between the Quebec provincial government and the Canadian federal government to foster the economic development of the more isolated regions of Quebec (École nationale d'administration publique 1997a, 1997b). Of significance here is the fact that the Atlantic salmon fishery was identified as an economic engine with significant regional growth potential.

Over a five-year period (1991–1996/1997), the PDES made matching funding available for projects that would improve or restore habitat and runs in salmon rivers, develop infrastructure (making rivers more accessible to anglers), and improve sustainable management of the species. Worthy projects would receive grants of up to 85% of their total costs. At the project's end, about $31.4 million had been spent or committed, of which $3.3 million (10%) was a match from beneficiary organizations.

The bulk of the government funding for the program came from government agencies concerned with economic development rather than with environment or fisheries. Canada's Regional Development Office for Quebec contributed $12 million (48%), and the province's Regional Development Office (Secrétariat au devélopment des régions) contributed an additional $3.5 million (11%). A smaller fraction of the money was provided by more fishery-oriented sources: Quebec's Environment and Wildlife Ministry (Ministère de l'Environnement et de la Faune, or MEF; $7.6 million, 24%) and the Quebec Wildlife Federation (Fondation de la faune; $2.1 million, 7%).

The PDES had specific, well-defined objectives (École nationale de l'administration publique 1997a,1997b). Its projects aimed to increase annual capture rates of salmon in the province from 16,000 captures (five-year average of captures in the 1983–1988 period) to a new high of 24,000, and to increase the annual number of person-days of fishing by 80,000. This was expected to translate into additional annual revenues of $12 million and create 200 to 250 new jobs in the fishery (e.g., guides, camp workers, etc.).

To date, the anticipated successes have not been fully realized. Recent North Atlantic-wide downturns in salmon populations have affected the rivers targeted under the PDES (e.g., Michaud 1990; Tremblay et al. 2000). The anticipated increase in salmon numbers following PDES efforts that could have generated additional fish captures did not materialize. However, the number of angler days has remained at the pre-PDES level, and about 40% of the targeted new jobs have been created (École nationale de l'administration publique 1997a). The partners in the program believed that it would take about two salmon generations for the full benefits of the PDES to be realized. Thus a full accounting for the benefits of the program will not be available until about 15 years after all the work is finished.

The Atlantic Sportfishery Enhancement Program. In 1992, the five-year Atlantic Sportfishery Enhancement Program for the Canadian Atlantic Provinces (Newfoundland and Labrador, Prince Edward Island, Nova Scotia, and New Brunswick) was initiated. The Atlantic Canada Opportunities Agency, an economic development arm of the Canadian federal government, contributed $57.3 million towards this initiative. The four Atlantic provincial governments provided an additional $21.6 million. The money was used to fund recreational fisheries cooperation agreements (Gardner Pinfold Consulting Economists Limited 1996).

The Atlantic salmon was one of the species targeted by this program. We were unable to determine how much money was spent specifically on Atlantic salmon in New Brunswick, Nova Scotia, and Prince Edward Island. However, figures are available for Newfoundland, where the sport fishery is generally limited to salmonids, primarily the Atlantic salmon and the brook trout *Salvelinus fontinalis*.

In Newfoundland, the federal and provincial governments developed a five-year (1992-1996) Cooperative Agreement on Salmonid Enhancement/Conservation (CASEC). In this period, $18.4 million was spent in the province to foster the health and economic benefits drawn from salmonid fisheries. The money was used for assessments of populations, enhancement, habitat enhancement, enforcement, and strategic planning and industry development. Atlantic salmon projects received the lion's share of the money (80% of assessment expenditures, 87.5% of enhancement, 84% of habitat, about 90% of enforcement, and 90% of economic development monies, respectively; D. Ivany, Atlantic Salmon Federation Regional Director for Newfoundland, personal communication).

Individual river approaches

Newfoundland's Humber River. Gardner Pinfold Consulting Economists Limited (1997) examined the present economic value of the sport fishery for Atlantic salmon on the Humber River and studied various scenarios to draw additional value from it. The scenarios involved enhancing the value of the fishery's potential economic benefits by extending the annual fishing season later into the autumn, controlling access to the fishery (limit the daily numbers of anglers and duration of angling), increasing stock size, offering trophy fish (killing of large fish is currently forbidden), and creating river-specific angling licenses.

The Humber River is the second largest salmon river in Newfoundland. Its watershed encompasses about 7,679 km^2, and its salmon run includes both grilse or one-sea-winter (1SW) fish and large multi-sea-winter (MSW) fish. The Gardner Pinfold (1997) report points out, "The Humber River is classified as a Zone 1 system by the North Atlantic Salmon Conservation Organization (NASCO) protocols. Rivers in this category are defined as 'pristine rivers, with no significant, man-made habitat alterations, no history of transfers of fish into the watershed, and no fish rearing operations in the watershed.'" The Humber's NASCO designation is curious, given that the main branch of the Humber River watershed (about 78%) was developed for hydroelectric generation by Deer Lake Power in 1924, eliminating it as productive fish habitat (Humber River Allies fact sheet, 15 May 1997).

The consultants estimated that in 1996, anglers spent about $1.5 million dollars to fish this river, with $0.7 million coming from Newfoundland residents and $0.8 million from nonresidents, creating about 78 jobs.

Gardner Pinfold staff queried anglers to ask them what things could be done to get them to fish more (and spend more) on the Humber River. With this information, they evaluated a series of scenarios involving combinations of the actions described above to see if economic benefits could be increased.

In their best scenario, they estimated that nonresident expenditures could be improved by $2.7 million to $3 million per year (a maximum of +375%) and an additional 96–113 (maximum of +144%) jobs created. The keys to success lay in increasing the size of the river's salmon population, so that more people could and would want to fish, and in prolonging the annual season. Potentially, this could also permit anglers to take home large (trophy) fish. By contrast, scenarios using a river-specific licensing system or adding access limitations were estimated to reduce total angler expenditures on the river by about $20,000 per year. Finally, an aggressive marketing campaign would be required to let potential nonresident clients know of the fishing opportunity.

The consultants actually felt that their approach was conservative. By "packaging" salmon fishing with other activities (e.g., hiking in national parks, etc.), they felt that much more money could be made.

Nova Scotia's Margaree River. GTA Consultants (1993) evaluated the economic benefits of Atlantic salmon angling in the Margaree River, Nova Scotia. The Margaree's salmon run principally arrives over a short period during the autumn. GTA (1993) estimated that, in 1992, 2,100 license holders (both resident and nonresident) fished the Margaree River and spent about $986,258 to do so (95% confidence limits $700,000–$1,300,000). This created 55 jobs.

The consultants cautiously suggested that enhancing present stocks of the river could increase revenues by about $250,000 per year. Their caution stemmed from the fact that these benefits would be difficult to realize, because the short time period in which the run enters the river occurs when many of the fishing spots are already full. However, GTA felt that an enhancement that altered run timing, so that more fish arrived in the summer could also increase spending by about $250,000. The biological consequences of tinkering with the genetics of the fish were not discussed.

How do Conservation Investments Aid Recreational Fisheries?

Buyouts of commercial licenses

Commercial interceptory fisheries in the oceans off Greenland and along the coasts of Canada had enormous impacts on North American Atlantic salmon populations through the 1960s and 1970s. Recognizing this and the fact that—on a per fish basis—much more value was realized from an angled fish than a gill-netted fish, the Canadian government began to buy out existing commercial licenses in the 1970s (Tuomi 1987). At a cost of more than $40 million, more than 5,200 commercial salmon licenses were retired throughout Atlantic Canada between 1972 and 1996 (Pat Chamut, assistant deputy minister, Department of Fisheries and Oceans, speech to a November 1996 Atlantic Salmon Federation conference).

The private North Atlantic Salmon Fund, based in Iceland, also intervened to buy out the Greenland fishery. In 1993 and 1994, this organization purchased the 201-metric ton annual commercial quota (about 122,000 maturing salmon) allocated to Greenland by the North Atlantic Salmon Conservation Organization at an annual cost of US$400,000 (contract between Kalallit Nunaanni Aalisartut Piniartullu Kattuffiat [the Association of Hunters and Fishermen of Greenland] and the North Atlantic Salmon Fund). The agreement permitted a subsistence fishery of 12 metric tons in each of these years. Money for the purchase came from private sources, corporations, and various governmental agencies. The contract called for a three-year renewal, starting in the 1995 season. For various reasons, however, this was never achieved.

A voluntary two-year program to permanently buy out the limited commercial salmon fishery (84 licenses; 1,285 salmon caught in 1999) that was still being pursued in the Lower North Shore region of Quebec (Gulf of St. Lawrence) was initiated in 1997. The $1.4 million cost of the buyout was split between the Canadian federal government and the province of Quebec. Commercial salmon catches declined about 39% between 1998 and 1999 as a result of this effort, and the 1999 catches were 88% lower than the average of the 1994–1998 period (Tremblay et al. 2000).

Education Efforts

Investments have been made in education programs that help conserve fish populations and promote recreational fisheries. In Atlantic Canada, a $5.7 million, five-year (1992–1996) education and public awareness program (EPAP) was undertaken by the Atlantic Salmon Federation to heighten public awareness of recreational fishery resources and public involvement in the fishery itself (Gardner Pinfold Consulting Economists Limited 1996; Anderson and Dominy 2000). This program had four components: public awareness, education, public involvement, and native involvement. Measuring the effectiveness of such programs is difficult. An independent evaluation of EPAP's success concluded that the program positively influenced people's attitudes and, to some extent, their behavior toward recreational fisheries (Gardner Pinfold Consulting Economists Limited 1996; Anderson and Dominy 2000). It was not possible, however, to determine if this translated into an increase in expenditures by the public on recreational fishing.

More sustained educational efforts come from salmon interpretation centers that have been established in New Brunswick (the Miramichi Salmon Museum in Doaktown and the Wilfred M. Carter Atlantic Salmon Interpretive Centre in St. Andrews) and Quebec (the Centre d'interpretation du saumon atlantique in Ste. Flavie). Built as tourist attractions, these three centers collectively generate annual revenues of at least $150,000 per year, provide seasonal employment, and are an important tool for promoting the conservation of the species and the sport fishery. Other similar centers exist or are under development in other regions.

Discussion

We have estimated current direct expenditures on Atlantic salmon fishing in Quebec and New Brunswick to be about $146 million per year. Recreational salmon fishing is also pursued in Newfoundland and Labrador, Nova Scotia, and Prince Edward Island. We do not have figures for salmon angling for these areas, but in 1995, anglers in Newfoundland and Labrador, Nova Scotia, and Prince Edward Island spent $48 million, $17.6 million, and $1.8 million, respectively, to go fishing for all species (Economic and Policy Analysis Directorate 1997). If only 30% of this was for salmon fishing, the Maritimes and Quebec Atlantic salmon fishery would inject $166 million into the Canadian economy. New Brunswick's capital value for salmon waters could now be worth $194 million.

There is great interest in increasing the economic value obtained from all recreational fisheries, including salmon. A flurry of strategic plans have been developed, including those for New Brunswick (Loftus et al. 1993; Economic Planning Group of Canada 1993), Prince Edward Island (Ace Consultants, Inc., and Island International Consultants, Inc. 1992), Newfoundland and Labrador (Buchanan et al. 1994), Quebec (Michaud 1992), Nova Scotia (Canadian Fishery Consultants Limited 1994), and for the Maritimes in general (Hickling Corporation 1994). The common thread of these plans is that the sport fishery has significant regional economic benefits and that, with enhancements to present populations and targeted marketing, a lot more money could be generated.

The studies for Atlantic salmon reviewed in this chapter suggest that there is significant potential to generate more money from salmon angling in North America. The key lies in improving salmon returns to rivers to levels well above those occurring today (Figure 1). Quebec's regional economic development plan for salmon has not yet achieved its goals, at least in part because of poor returns to the rivers.

It is fair to ask what ecological impacts have resulted from the various economic initiatives. The buyouts of commercial sea nets have stopped mixed-population fisheries that may have resulted in salmon overharvesting, especially of populations in smaller rivers. The closure of the commercial fisheries in North America may have slowed but failed to reverse the declines in returns to river. The buyouts have not, therefore, achieved their hoped-for effect. However, education and public awareness programs (Anderson and Dominy 2000) have successfully promoted catch-and-release fishing for salmon. Research (Tufts et al. 2000; Whoriskey et al. 2000) has shown that released fish survive and go on to spawn. Catch-and-release fishing provides many economic benefits with minimal biological impacts upon salmon populations. Despite the implementation of catch-and-release fishing and the other economic initiatives, however, salmon runs to North American rivers have continued to decline.

Although it is far from clear what has caused the downturn in salmon returns, poor survival of salmon at sea seems to be a major determinant (Department of Fisheries and Oceans 1998; ICES 2000). In addition, threats to freshwater salmon production capacity continue. These seem to have started when Samuel de Champlain installed the first dam on a salmon stream, a tributary of the Annapolis River, Nova Scotia, around 1606 (Morison 1972). At present, we are aware of registered proposals for hydrodevelopment in at least seven Atlantic salmon watersheds: the Moisie, St. Jean, and Romaine rivers in Quebec; and the Northwest, Southwest, and Humber rivers and Garia Bay in Newfoundland. Pollution, habitat destruction, and poor forestry and agriculture practices are contributors to the decline in salmon returns (Watt 1988, 1989). These activities also have significant economic potential, and conflicts on rivers are increasingly focusing on which activities bring the greatest economic and social gains with the least environmental damage.

A number of social, nonmonetary, and/or indirect benefits also flow to humans from wild salmon populations. For example, in *Ronald Edward Sparrow vs. Her Majesty the Queen* (1990), Canada's Supreme Court upheld an aboriginal right to fish under the Canadian Constitution Act of 1982. In this decision, the court ruled that allocations to the fisheries were to be made according to the following priorities: 1) conservation of the fishery, 2) aboriginal fisheries for food, social, and ceremonial purposes, and 3) other fisheries. Thus, fishing was recognized as having an enormous social value in native cultures, including those that fish Atlantic salmon in Atlantic Canada. But what is the dollar value of a constitutional right?

Aesthetic values are also difficult to quantify. For many anglers, the actual catching of salmon is a small portion of the fishing experience. Enormous pleasure and value can be drawn from being out of cities on wild rivers. This value exists and, to a degree, compensates for poor angling capture rates. If it did not, given present capture rates of as low as 0.5–0.1 Atlantic salmon/angler/day (e.g., Tremblay et al. 1998), the sales of salmon licenses would be far poorer than they actually are (Figure 2). However, present and future economic benefits from salmon angling are eroding, and they will continue to erode if the downward trends in license sales continue.

A third example of hard-to-quantify benefits that flow from wild salmon populations is in the salmon farming industry. Canadian east coast Atlantic salmon farming is variously estimated as having annual revenues of $115 million to $140 million. The industry started with broodstock from wild fish. The many river-specific populations of Atlantic salmon are potential sources of problem-solving genetic characteristics (e.g., disease resistance) that could improve aquaculture production in the future. It is difficult to estimate what these economic benefits might be.

In summary, the Atlantic salmon is important as a current and potential economic engine in the North Atlantic region, and it has other significant social or nonmonetary values. Despite the concurrence of economic, ecological, and social values of salmon, populations of the species remain depressed and, in some places, in danger of biological extinction. By examining the economic (and noneconomic) values and benefits of this fishery, however, we can better understand the importance of conserving this species, and we can develop more comprehensive and effective conservation approaches.

References

Ace Consultants, Inc., and Island International Consultants, Inc. 1992. A recreational fisheries development plan for Prince Edward Island. Prepared for the Morell River Management Co-op, Inc., Morell, Prince Edward Island, Canada.

Anderson, J. M., and C. L. Dominy. 2000. Socio-economic issues. Pages 225–234 *in* F. G. Whoriskey and K. E. Whelan, editors. Managing wild Atlantic salmon: new challenges–new techniques. Atlantic Salmon Federation, St. Andrews, New Brunswick.

Buchanan, R. A., T. A. Pinfold, L. S. Carter, J. Baird, and J. Christian. 1994. Strategy for the development of recreational fisheries in Newfoundland and Labrador. Prepared for the Canada-Newfoundland Cooperative Agreement for Salmonid Enhancement/Conservation. October 1994. Available from the Canadian Department of Fisheries and Oceans, St. John's, Newfoundland.

Canadian Fishery Consultants Limited. 1994. Recreational fisheries strategic plan for the freshwater and marine fish species of Nova Scotia. Project No. B514. Prepared for the Nova Scotia Department of Fisheries and the Department of Fisheries and Oceans, Halifax, Nova Scotia.

Department of Fisheries and Oceans. 1998. Atlantic salmon abundance overview for 1997. DFO Science Stock Status Report DO-02 1998. Dartmouth, Nova Scotia.

École nationale d'administration publique. 1997a. Évaluation du programme de dévelopment économique du saumon. Rapport final. Groupe d'évaluation des politiques de programmes publics, École nationale d'administration publique. Août 1997. Ste-Foy, Québec, Canada.

École nationale d'administration publique. 1997b. Évaluation du programme de dévelopment économique du saumon. Annexes techniques. Groupe d'évaluation des politiques de programmes publics, École nationale d'administration publique. Août 1997. Ste-Foy, Québec, Canada.

Economic and Policy Analysis Directorate. 1997. 1995 survey of recreational fishing in Canada. Economic and Commercial Analysis Report No. 154. Canadian Department of Fisheries and Oceans, Ottawa, Ontario, Canada.

Economic Planning Group of Canada (Halifax Office). 1993. Strategy for New Brunswick's Outfitting Sector. Report prepared for the New Brunswick Outfitters Strategy Steering Committee. Halifax, Nova Scotia, Canada.

Gardner Pinfold Consulting Economists Limited. 1996. A final assessment of the Education and Public Awareness Program. November 1996. Available from the Atlantic Salmon Federation, St. Andrews, New Brunswick, Canada.

Gardner Pinfold Consulting Economists Limited. 1997. The current and potential economic value of the Atlantic salmon fishery on the Humber River. February

1997 report prepared for the Salmonid Council of Newfoundland and Labrador, St. John's, Newfoundland.

GTA Consultants, Inc. 1993. Economic impact of the recreational fishery on the Margaree River. March 31, 1993. Prepared for the Department of Fisheries and Oceans, Dartmouth, Nova Scotia.

Hickling Corporation. 1994. Recreational fishing and economic development in Canada: policy foundations and options for the Maritime provinces. Hickling Report No. 4660. Hickling Corporation, Ottawa, Ontario.

ICES (International Council for the Exploration of the SEA). 2000. Report of the Working Group on North Atlantic Salmon. ICES Headquarters, Copenhagen, Denmark, 3–13 April 2000. ICES CM 2000/ACFM:13.

Kenchington, E. 1999. Proceedings of the Inner Bay of Fundy Salmon Working Group Regional Advisory Process Maritimes Region. Canadian Stock Assessment Proceedings Series 99/29, Dartmouth, Nova Scotia, Canada.

Leopold. A. 1933. Game management. The evolution of orderly mechanisms for the marketing of game crops and thus activating the financial incentive to produce them. C. Scribner and Sons, New York.

Loftus, K. K., L. G. Kreig, T. A. Pinfold, M. Kilfoil, and J. D. Meisner. 1993. Comprehensive development strategy for the anadromous and inland recreational fisheries of New Brunswick. Canadian Manuscript Report of Fisheries and Aquatic Sciences No. 2216, St. Andrews, New Brunswick, Canada.

Michaud, L. 1990. Évaluation de la valeur economique du saumon Phase II. Estimation des valeurs reliées aux utilisations directes. Rapport préparé pour la vice-presidence Environnement d'Hydro-Québec. Décembre 1990. 2 volumes. Hydro Québec No. A-49, HQ ENVI 90 414.

Michaud, Luc. 1992. Valeur économique de la pêche récréative au saumon. Résumé. Hydro-Québec, Vice-Presidence Environnement. Quebec City, Quebec, Canada.

Morison, S. E. 1972. Samuel de Champlain. Father of New France. Atlantic Monthly Press, New York.

Tremblay S., F. Caron, C. Groleau, and D. Deschamps. 2000. Bilan de l'exploitation du saumon au Québec en 1999. Société de la faune et des parcs du Québec, Direction du développement de la faune. 177 pp. Quebec City, Quebec, Canada.

Tufts, B. K. Davidson, and A. T. Bielak. 2000. Biological implications of "catch and release" angling of Atlantic salmon. Pages 195–224 *in* F. G. Whoriskey and K. E. Whelan, editors. Managing wild Atlantic salmon: new challenges–new techniques. Atlantic Salmon Federation, St. Andrews, New Brunswick, Canada.

Tuomi, A. L. W. 1980. Canada's Atlantic salmon: an economic evaluation of the New Brunswick salmon sportfishery. The International Atlantic Salmon Foundation (now the Atlantic Salmon Federation), Special Publication Series No. 8, St. Andrews, New Brunswick, Canada.

Tuomi, A. L. W. 1987. Canada's Atlantic salmon recreational fisheries and their future: an economic overview. Atlantic Salmon Federation, Special Publication Series No. 14, St. Andrews, New Brunswick, Canada.

Watt, W. D. 1988. Major causes and implications of Atlantic salmon habitat losses. Pages 101–111 *in* R. H. Stroud, editor. Present and future Atlantic salmon management. Atlantic Salmon Federation, Ipswich, Massachusetts.

Watt, W. 1989. The impact of habitat damage on Atlantic salmon (*Salmo salar*) catches. Pages 154–163 *in* C. D. Levings, L. B. Holtby, and M. A. Henderson, editors. Proceeding of the national workshop on effect of habitat alterations on salmonid stocks. Canadian Special Publication, Fisheries and Aquatic Science 105.

Wheaton, F. E. 1995. Co-operation agreement on recreational fisheries development. Resource development workshop. Rapporteur's report. Fredericton, New Brunswick. February 15, 1995.

Whoriskey, F. G., S. Prusov, and S. Crabbe. 2000. Evaluation of the effects of catch-and-release angling on the Atlantic salmon (*Salmo salar*) of the Ponoi River, Kola Peninsula, Russian Federation. Ecology of Freshwater Fish 9:118–125.

Sustaining North American Salmon: Perspectives Across Regions and Disciplines, 93–116

Chapter 5

Canadian and U.S. Atlantic Salmon Institutions and Politics: Where are the Fish and Who Cares?

Stephen G. Rideout and John A. Ritter

Introduction

The Atlantic salmon *Salmo salar* is frequently touted as the "king of fishes" or the "fish of kings." In North America, its exploitation in commercial fisheries has virtually given way to the higher valued recreational fishery and to aboriginal communities exercising their constitutional right to fish for food, social, and ceremonial purposes. A 1988 economic survey of Canada's Atlantic salmon anglers estimated expenditures attributable to Atlantic salmon angling in Canada at $74.1 million (DFO 1988). *Salmo salar* is held in the highest esteem by anglers throughout North America and Europe. Within its historical range, the general public generally regards the salmon as a barometer of the health of the natural environment.

Management of the Atlantic salmon in eastern North America falls under federal, provincial, and state jurisdictions, with each having its own roles and legislative authorities (Figure 1; Table 1). In Canada, the federal government works with all five eastern provinces—Quebec, Newfoundland, New Brunswick, Nova Scotia, and Prince Edward Island—in managing Canada's Atlantic salmon resource. The U.S. federal government works with all six New England states, primarily through state and interstate management entities: the Maine Atlantic Salmon Commission, the Connecticut River Atlantic Salmon Commission, a state-federal management consortium for the Merrimack River, and, to a limited degree, the Rhode Island

Figure 1. Canadian provinces and U.S. states encompassing current and historical range of North American stocks of Atlantic salmon.

Division of Fish and Wildlife for the Pawcatuck River. International management is through the North Atlantic Salmon Conservation Organization (NASCO), with scientific support provided by the International Council

Table 1. Fisheries jurisdictions for Atlantic salmon in eastern North America.

Canada

Jurisdiction	Geographic range	Primary authority
Department of Fisheries and Oceans Canada (except in Province of Quebec: the province administers management program)	Fresh and marine waters	Conservation of stocks
Provinces of Quebec, New Brunswick, Nova Scotia, Prince Edward Island, Newfoundland (including Labrador)	Fresh water	Recreational fishing access

United States

Jurisdiction	Geographic range	Primary authority
National Marine Fisheries Service and U.S. Fish and Wildlife Service	Fresh and marine waters	Preservation of stocks listed under the Endangered Species Act
New England Fishery Management Council acting through the Secretary of Commerce	Marine waters from 3 to 12 nautical miles	Conservation of stocks
Maine Atlantic Salmon Commission	State of Maine fresh and coastal (<3 nm) waters	Conservation of stocks
States of Connecticut, Massachusetts, New Hampshire, Rhode Island, and Vermont	Fresh and coastal waters except main stem of Connecticut River	Conservation of stocks
Connecticut River Atlantic Salmon Commission	Main stem Connecticut River	Conservation of stocks

for the Exploration of the Seas (ICES). Locally, stakeholders such as the Atlantic Salmon Federation, provincial organizations of the Canadian Wildlife Federation, local chapters of Trout Unlimited, and the Connecticut River Salmon Association invest considerable time and money toward salmon conservation.

Institutional Arrangements

Canada

By virtue of Canada's constitution, legislative responsibility for conservation of Canada's Atlantic salmon resource rests with the federal government. Such responsibility applies to all fisheries resources in both inland (fresh) and marine (including tidal) waters. Only the federal government has authority to enact laws to regulate the fisheries for Atlantic salmon conservation.

The provincial governments legislate on matters pertaining to property and civil rights, including those pertaining to fisheries. They regulate the terms and conditions of access to the recreational freshwater fisheries. For example, licensing in the recreational fishery is exercised under provincial legislation. Similarly, the granting of special privileges to limit access to certain waters falls under provincial jurisdiction. Ownership of the riverbed and control of water rights are areas of provincial responsibility, while the fish within the waters and the conditions under which they may be removed are under federal jurisdiction. Exceptions to provincial jurisdiction over access to fish in fresh waters are on federally-owned lands such as national parks, where the federal government retains authority, and on riparian privately-owned waters, where the landowner owns the river bed adjoining his or her property and may control access.

The federal government has primary responsibility for administering the management program for Atlantic salmon in Canada except in the Province of Quebec, where administration was delegated to provincial authorities by virtue of a 1922 agreement between the federal government and the Province of Quebec. All provinces have legislation that extends well beyond regulating access to the recreational fisheries, including areas such as environmental protection and aquaculture development.

Canadian federal and provincial governments also have responsibilities to uphold constitutional and treaty rights of aboriginal peoples pertaining to fishing for the Atlantic salmon and other fisheries resources. The Supreme Court of Canada, in its 1990 ruling in R. vs. Sparrow, found that an aboriginal right exists to fish for food, social, and ceremonial purposes, and that aboriginal people exercising that right have priority over other users of the resource, after conservation needs have been met. In September 1999, the Supreme Court, ruling in R. vs. Marshall, affirmed the right of Mi'kmaq, Maliseet, and Passamaquoddy peoples to a "moderate livelihood" from wildlife, fish, and resources traditionally "gathered" in an aboriginal economy. This latter decision conferred the right to fish commercially. The court later confirmed that the federal government has the authority to regulate the treaty right, that the right is communal in nature, and that it does not have to be satisfied ahead of other users.

The lead government agency for managing the Atlantic salmon resource in Canada is the federal Department of Fisheries and Oceans (DFO). DFO works with other federal (e.g., Environment Canada) and several provincial agencies that have supporting legislated and administrative responsibilities related to the management of fisheries resources in general.

Canada's Fisheries Act is the principal regulatory instrument used to manage the fisheries, fish stocks, and fish habitat. The act pertains to fish only or factors potentially affecting them. Its authority pertains to fish stock conservation. The term "fish" under the act has a broad definition and includes all finfish, aquatic invertebrates, marine mammals, and marine plants. The act and the regulations under it cover such things as the alteration or destruction of fish habitat; the requirement that effective fish passage be provided at man-made obstructions to migration; the protection of the fish, which generally includes activities pertaining to their destruction, removal, or introduction; and the release or discharge of deleterious substances into the water.

United States

Unlike Canada, responsibility for conservation and management of Atlantic salmon in U.S. coastal (3 nautical miles) and freshwater habitats rests primarily with the individual states. Federal responsibility supercedes state authority for population segments listed for protection under the federal Endangered Species Act. Federal authority also prevails in marine waters beyond 3 nautical miles for all anadromous salmonids throughout their migratory ranges, except while they are within a foreign nation's territorial sea or fishery conservation zone. This federal authority is exercised by the Secretary of Commerce pursuant to the New England Fishery Management Council's 1987 Fishery Management Plan (FMP) (New England Fishery Management Council 1987). The council's primary purpose for the FMP was to provide for the marine area between the coastal 3 nautical-mile limit of the states and the U.S. territorial sea beyond 12 nautical miles that is recognized by NASCO.

Prior to the passage of the Anadromous Fish Conservation Act (AFCA) of 1965, anadromous Atlantic salmon management in the United States was primarily under the jurisdiction of the Maine Atlantic Sea Run Salmon Commission, established by the Maine state legislature in 1947 (Baum 1997). Atlantic salmon had been extirpated from most of New England by the middle of the 19th century, including the Connecticut River by the end of the late 18th century (Foster 1991) and the Merrimack River by the

1830s (Stolte 1981). Consequently, the only remaining wild runs of Atlantic salmon in the United States were in Maine when the state, through the creation of the Atlantic Sea Run Salmon Commission in 1947 (currently named the Maine Atlantic Salmon Commission), began regulating the taking, possession, and management of salmon on a statewide basis. The federal government, through the U.S. Fish and Wildlife Service, also played an active role in salmon conservation and restoration and continues to do so today, primarily through the stocking of salmon produced at its Green Lake and Craig Brook national fish hatcheries and participation in the Maine Atlantic Salmon Technical Advisory Committee, formerly the Salmon Research Committee (Baum 1997).

Atlantic salmon restoration programs in Maine, stimulated by the Anadromous Fish Conservation Act, resulted in the remaining New England states becoming active in Atlantic salmon management by the late 1960s and early 1970s. States outside Maine prohibited the taking of salmon in an effort to increase the probability of successful restoration. These new restoration efforts were focused on the Connecticut, Merrimack, and Pawcatuck rivers. Management authority remained vested with the individual states until congressional consent (P.L. 98–138, October 1983) of the Connecticut River Basin Atlantic Salmon Compact established the Connecticut River Atlantic Salmon Commission (CRASC) with management authority for the main stem of the Connecticut River.

The 1980 Maine Indian Claims Settlement Act established a treaty between the United States and both the Passamaquoddy Tribe and the Penobscot Indian Nation. The settlement defined a special relationship between the state of Maine and the Passamaquoddy Tribe and Penobscot Nation. It also created a nine-member intergovernmental Maine Indian Tribal-State Commission (MITSC) with authority to promulgate fishing rules concerning certain ponds, rivers, and streams adjacent to or within Indian territory (MITSC 1995). Some of these areas encompass current or historical anadromous Atlantic salmon habitat.

International Approaches

The principal organization for international cooperation in management of north Atlantic salmon is the North Atlantic Salmon Conservation Organization (NASCO). This organization was established through a 1983 convention to further the conservation, restoration, enhancement, and rational management of salmon stocks in the North Atlantic. A main function of NASCO has been to provide a forum for consultation and cooperation in cases where activities undertaken or proposed by one of its members affect salmon originating in the rivers of another of its members.

Prior to the formation of NASCO, there was a widely held opinion within the salmon-producing nations that the interception fisheries—the fishery at West Greenland, in particular—were a major threat to Atlantic salmon conservation in both North America and Europe. The perceived threat of these fisheries and the concern that control over them was not being adequately addressed were the main reasons underlying the formation of NASCO.

This organization consists of a council and three regional commissions: the North American Commission (NAC), the West Greenland Commission (WGC), and the North-East Atlantic Commission (NEAC). Canada and the United States constitute the NAC, and they serve with Denmark (in respect of the Faroe Islands and Greenland) and the European Union to make up the WGC. Each contracting party to NASCO has a single vote on council matters, and similarly each member of a commission has one vote in any proceedings of that commission. Unanimity is required for decisions within the commissions. Although the council requires only a three-fourths majority of the members present and casting an affirmative or negative vote, the council has generally operated by unanimity as well. In addition, Canada and the United States may submit and vote on proposals for regulatory measures concerning salmon stocks originating in their respective rivers and occurring off East Greenland. The waters of East Greenland lie within the NEAC area, and neither Canada nor the United States holds membership on the NEAC. A reciprocal arrangement exists for the European Union on the NAC concerning their stocks that may be found in the NAC area.

NASCO has regulatory powers over fishing activities in international waters and the territorial waters of Greenland and the Faroe Islands. Such measures are binding on affected members 60 days after the date of notification, unless a member to whom the measures would apply lodges an objection within those 60 days. Should an objection be lodged, the measures would not be binding on any member. Within the NAC area, salmon fisheries are not permitted to be altered in a manner that would result in an increase in catches of salmon originating in the rivers of another party, except with the consent of that party.

Contracting parties to NASCO have agreed to prohibit salmon fishing in marine waters beyond the jurisdictions of coastal states (countries). They also have agreed to prohibit salmon fishing within coastal states' jurisdictions beyond 12 nautical miles, with exceptions for West Greenland and the Faroe Islands. NASCO has been successful, particularly since the 1980s, in reducing the harvest of salmon in interception fisheries to low levels. The principal examples of this are the fisheries at West Greenland (nominal catch down from about 900 t in the late 1980s to 19 t in 1999) and in the area of the Faroe Islands (down from about 600 t in the mid-1980s to no catch in 1999). These results have been achieved through negotiations within the NASCO framework and, in some instances, with the payment of financial incentives to the fishermen that were made independent of NASCO initiatives.

The focus of NASCO continues to broaden beyond the marine interception fisheries for salmon in the north Atlantic. For instance, in 1994, it adopted what is commonly referred to as the Oslo Resolution, a resolution "to minimize impacts from salmon aquaculture on wild salmon stocks" (NASCO 1994a). At its 1998 annual meeting, NASCO and its contracting parties agreed to adopt and apply a "precautionary approach" to the conservation, management, and exploitation of salmon to protect the resource and preserve the environment in which it lives (NASCO 1998; see also Friedland, Chapter 6). Consistent with the precautionary approach, the parties agreed that the absence of adequate scientific information would not be a reason for postponing or failing to implement conservation and management measures. Locally, the North American Commission adopted in 1992 and amended in 1994 Protocols for the Introduction and Transfer of Salmonids for its commission area (northeastern United States and Canada) (NASCO 1992, 1994b). The protocols were developed to minimize adverse genetic, fish health, and ecological effects from introductions and transfers of aquatic life. They are currently the subject of considerable controversy and discussion among wild salmon advocates and aquaculture industry proponents.

The International Council for the Exploration of the Sea (ICES) is another international forum that provides advice and guidance in management of North Atlantic salmon. The ICES, through its Working Group on North Atlantic Salmon, provides scientific advice to NASCO on wild, enhanced, and aquaculture salmon management issues that transcend individual national boundaries. The interception fisheries for salmon in the North Atlantic is such an issue. ICES has also developed a code of practice that recommends procedures and practices to diminish the risks of detrimental effects from intentional introductions and transfers of aquatic life, including Atlantic salmon and other organisms that could potentially harm salmon (ICES 1988).

The United Nations has conventions, agreements, and codes of conduct that provide guidance to it members, including Canada and the United States, to manage wild fish stocks. These include the conventions on the Law of the Sea in 1982 and Biodiversity in 1992; the Agreement on Straddling Fish Stocks and Highly Migratory Fish Stocks (yet to be ratified), which advances the precautionary approach (FAO 1995a, 1996); and the Food and Agriculture Organization's (FAO) Code of Conduct for Responsible Fisheries (FAO 1995b).

The Key Issues

Changes in Abundance and the Fisheries

The single greatest problem confronting North American Atlantic salmon is the dramatic decline in the overall abundance of adult salmon. Since the 1970s and early 1980s, the numbers of one-sea-winter (1SW) and two-sea-winter (2SW) salmon of North American origin have declined from well over 1 million fish to approximately 500,000 (DFO 1998a; ICES 2000). Most dramatic in their decline are the 2SW salmon (including those harvested as nonmaturing 1SW salmon), which in recent years have numbered fewer than 100,000 fish, down from well over 500,000 fish during the 1970s and early 1980s. It is the 2SW salmon upon which North American salmon are most dependent for egg deposition and stock conservation. Primary exceptions to this general tendency are the Newfoundland and inner Bay of Fundy rivers, which are largely dependent upon 1SW salmon, or grilse, for stock conservation. The decline in returns to rivers has been most severe at the southern end of the species range (DFO 1998a). Returns to rivers along the Atlantic coast of Nova Scotia, rimming the Bay of Fundy, and of New England are currently at critically low levels, with several stocks already lost and many others threatened with extirpation (Atlantic Salmon Biological Review Team 1999; DFO 2000a; USASAC 2000).

River returns of salmon of all sea ages are lower than expected across the entire range of North American salmon. Such low returns persist, in spite of relatively high juvenile levels in many rivers and the closing of almost all marine interception fisheries (DFO 1998a; ICES 2000). These fisheries, off the west coast of Greenland and along the Canadian coast from Labrador southward, intercepted large numbers of migrant salmon originating from almost all North American production areas, including New England (Ruggles and Ritter 1980; Pippy 1982; Ritter 1989; Baum 1997). Collectively, these fisheries harvested well over 500,000 salmon annually during the 1970s (Ritter 1993; ICES 2000). By the late 1990s, most commercial fishing effort directed toward North American Atlantic salmon had been curtailed. Nominal catches of North American salmon in interception fisheries (i.e., both commercial and subsistence) in 1999 were as low as 5,700 salmon at Greenland, about 4,000 in Canadian waters and 2.3 t or roughly 1,000 fish in the French waters of the Islands of Saint-Pierre and Miquelon (ICES 2000).

Low returns of salmon to North American rivers in recent years have resulted in the complete closure of all directed salmon fisheries on many rivers. Such closures encompassed southern New Brunswick and mainland Nova Scotia, except for rivers along the Northumberland Strait shore. Even Canadian aboriginal communities, with first right of access to fish for food, social, and ceremonial purposes, have been denied access to fish rivers in these areas. Everywhere in eastern Canada, low salmon returns are precluding aboriginal communities from exercising their treaty rights to harvest salmon for a moderate livelihood. Similarly, in the United States, the Maine Atlantic Salmon Commission closed its sport fishing for anadromous Atlantic salmon in 2000 because of low returns. Recreational fisheries for salmon have yet to be allowed in the other New England states.

The low returns have also precluded hook-and-release angling for salmon on many rivers. A major drop-off in stakeholder participation and growing apathy for salmon conservation have been associated with the closing of recreational fishing opportunities. This drop-off is particularly apparent in the membership rolls of the river-specific conservation organizations for those rivers on which recreational fishing for salmon has been prohibited for a number of years. Growing apathy for salmon conservation is of great concern because stakeholder participation and public interest and involvement in salmon conservation are critical to the success of stock recovery programs.

Endangered Species Protection

Salmon of the Inner Bay of Fundy rivers were declared endangered by the Committee on the Status of Endangered Wildlife in Canada (COSEWIC) on 3 May 2001. COSEWIC, an organization operating at arm's length from government, has the responsibility in Canada for determining the national status of Canadian wildlife species, subspecies, and populations suspected of being at risk. Though the designations of COSEWIC have no legal standing, they are usually accorded special consideration by the responsible governments. This particular designation applies to a special grouping of salmon populations that may have occupied at least 32 rivers rimming the upper Bay of Fundy (22 in Nova Scotia and 10 in New Brunswick) and with annual historic catches averaging 1,061 fish in the commercial fishery (1970–1984) and 2,059 fish in the recreational fishery (1970–1990) (DFO 2001). Juvenile surveys conducted in 2000 found no salmon in many of these rivers, and where salmon were detected, populations were at trace levels only (DFO 2001).

The Canadian government is introducing new legislation to prevent wildlife species from being extirpated or becoming extinct; to provide for the recovery of wildlife species that are extirpated, endangered, or threatened as a result of human activity; and to manage species of special concern to prevent them from becoming endangered or threatened. The proposed Species at Risk Act (SARA) would prohibit the killing, harming, harassing, capturing, or taking of wildlife officially listed as threatened or endangered. It would also prohibit the destruction of the habitat of threatened, endangered, and extirpated populations. SARA would apply to all aquatic species but not necessarily all terrestrial species. Under SARA, there would be a mandatory requirement for developing recovery strategies and action plans for endangered and threatened species as well as management plans for species of special concern. According to the current wording of the proposed legislation, COSEWIC would be given legal status to assess and classify the risk status of wildlife in Canada, and its designations would become the SARA designations, subject to the concurrence of the federal government. Potential reasons for rejection of a particular COSEWIC designation would be unique socioeconomic considerations or disagreement on the interpretation of the science (e.g., disagreement on the status or the genetic uniqueness of the population).

Four of the five Canadian provinces in which Atlantic salmon are present already have endangered species legislation, and the fifth, Newfoundland, is enacting such legislation. Actual protection for Atlantic salmon under such legislation of the provinces is yet unclear; at least some provinces are considering deferral of responsibility for all fish species, including those inhabiting inland waters, to the federal government. Within those provinces where the provincial government may choose to include aquatic species under their legislation, such as Nova Scotia, the potential exists for species such as Atlantic salmon to be afforded additional protection.

In the United States, the U.S. Fish and Wildlife Service and the National Marine Fisheries Service announced on 13 November 2000 that they would protect distinct population segments (DPS) of wild Atlantic salmon in Maine by listing them as endangered under the federal Endangered Species Act. With this decision came the listing of eight river populations (i.e., Dennys, East Machias, Machias, Pleasant, Narraguagus, Ducktrap and Sheepscot rivers, and Cove Brook). This action was preceded by petitions to the U.S. Fish and Wildlife Service and the National Marine Fisheries Service to list the Atlantic salmon in 1993, followed by two stock status reviews and development of a major conservation plan for seven of the eight rivers now listed.

With the exception of prohibiting the taking or possession of Atlantic salmon in federal waters by the New England Fishery Management Council in 1987 through the establishment of an Atlantic Salmon Fishery Management Plan, this is the first federal action in U.S. Atlantic coastal and fresh waters to prohibit the taking of salmon. Among several reasons given for the listing action, the two federal agencies cited "the low number of adult salmon returning to spawn in the eight rivers" and "low survival rates for young salmon in the rivers."

The listing does not pertain to a number of salmon rivers in Maine and the rest of New England, whose stocks were extirpated and subsequently have been the target of restoration efforts since the 1960s and 1970s. In spite of extensive efforts to restore runs to these other rivers (the St. Croix in Maine and New Brunswick; the Union, Penobscot, Androscoggin, and Saco in Maine; the Merrimack in Massachusetts and New Hampshire; the Pawcatuck in Rhode Island; and the Connecticut in Connecticut, Massachusetts, New Hampshire and Vermont), they, too, are showing declines in adult returns similar to and over the same time period as recorded for Canadian rivers. The predominance of 2SW salmon in the returns to U.S. rivers (>80%) shows a striking difference from the runs to neighboring Canadian rivers, which generally have a much higher component of 1SW salmon (Atlantic Salmon Biological Review Team 1999).

The American Fisheries Society (AFS) endorses the distinct population segment (DPS) concept as defined by the U.S. Fish and Wildlife Service and National Marine Fisheries Service, and it has incorporated the concept in its new system of categorization for Marine Stocks at Risk of Extinction (MSRE) (Musick et al. 2000). Under the AFS system, the Atlantic salmon of the Gulf of Maine (i.e., the stock of the United States), the interior Bay of Fundy (i.e., Canada's inner Bay of Fundy stocks), and the eastern and southern Nova Scotia rivers (i.e., the area of Nova Scotia affected by acid rain, the Southern Upland) are listed as endangered. AFS has also designated the southwestern New Brunswick stocks (i.e., the Saint John River and west of it to the St. Croix) as threatened and salmon of the Miramichi River, New Brunswick, as vulnerable.

Freshwater Production

Freshwater productive capacity for Atlantic salmon continues to decline. Seemingly most obvious of the factors responsible for past losses has been dam construction often associated with hydroelectric development. Dam construction has generally reduced salmon production, more so at the southern end of the species' range—in Nova Scotia, southern New Brunswick, and New England (Watt 1988; Ritter and Marshall 1992; Baum 1997; USASAC 2000). Principal impacts have been loss of freshwater habitat due to impoundment formation, reduced efficiency in upstream and downstream fish passage, and delays in migration, which, in the case of smolts, can be expected to reduce their subsequent survival (Ruggles 1980). Man-made dams have been cited as historical obstacles to the full utilization of spawning and rearing habitat in the DPS rivers in Maine, but they are not considered to be a major factor in the recent decline of these stocks (Atlantic Salmon Biological Review Team 1999). Inefficiencies in both upstream and downstream passage at dams continue to lower Atlantic salmon production in both Canada and the United States (Ruggles 1980; Watt 1988; Ritter and Marshall 1992; Baum 1997; USASAC 2000).

Perhaps most subtle of the factors working to reduce freshwater productive capacity are those activities that have served to alter runoff patterns and contribute to sediment deposition in streams and rivers. Agricultural and forestry practices, road construction, and urban development are prominent contributing factors. Such activities have, to varying degrees, heightened runoff peaks, which have frequently caused nonreversible widening and straightening of streams and rivers. Associated with such events and physical change have been more severe low flow conditions and higher river temperatures during summer. Water extraction for agricultural purposes (primarily irrigation of blueberry lands) was cited by Baum (1997) and the Atlantic Salmon Biological Review Team (1999) as a significant threat to salmon habitat in some of Maine's listed rivers. Clogging of the substrate from sediments in the runoff also has affected productive capacity by reducing reproduction success. Freshwater habitats damaged by sediments can be expected to recover if sedimentation is halted or significantly reduced, although recovery is unlikely to occur quickly or be complete.

Documented lower than expected overwinter survival of presmolting parr was cited by federal reviewers in their 1999 status review of the Maine stocks for listing under the endangered species legislation (Atlantic Salmon Biological Review Team 1999). Kocik et al. (1999) reported a 99% probability that overwinter survival rates were below 30% (actual estimates ranged from 11% to 24%) and, as such, lower than previously accepted survival rates for Maine rivers. Whalen et al. (2000) also reported mean rates of 18% for a Vermont tributary of the Connecticut River. These recent estimates of poor overwinter survival and subsequent low smolt survival have led researchers and managers to question whether there is a long-term and possibly irreversible downward trend in freshwater production throughout New England rivers.

Acid rain is having a major effect on the salmon stocks of Nova Scotia's Southern Upland (i.e., the area extending the full length of the Atlantic coast of mainland Nova Scotia) (DFO 2000b). The grouping of stocks in this area is designated by the American Fisheries Society as endangered (Musick et al. 2000). Naturally-reproducing salmon are no longer present in many of the 65 rivers in the area and are at reduced levels in all other rivers (DFO 2000b). Because of acid rain, salmon reproduction is no longer possible in at least 14 of the rivers and impeded to varying degrees in most others. The effects of acid rain on Southern Upland salmon are worsened by low marine survival, common to all North American salmon stocks (DFO 1998a, 2000b; ICES 2000). Current low marine survival is hastening the decline and further loss of Southern Upland stocks.

The principal acidifying agent in acid rain is wet sulfate (DFO 2000b). Approximately 60% of the sulfate in the rain falling on the Southern Upland originates from human activities, and roughly three-quarters of it is from eastern U.S. sources. Although a significant decrease in the deposition of acidifying substances to the Nova Scotia environment has occurred since the mid-1980s, as a result of reductions in emissions from sources in both the eastern United States and Canada, acid concentrations in the Southern Upland rivers are not decreasing, and significant reductions are not anticipated in the near future (DFO 2000b). Further major reductions in sulfate emissions are required before conditions can be expected to be suitable for salmon restoration in many of the acidified rivers of Nova Scotia's Southern Upland.

Salmon Farming Industry

The Atlantic salmon farming industry in eastern Canada and Maine produces more than 30,000 t annually, or, in actual numbers of market size fish, about 8 million (DFO 1999; ICES 2000). Approximately 95% of the production comes from marine cages concentrated in the western part of the Bay of Fundy; the remainder is from Nova Scotia and Newfoundland. Maine accounts for about 40% of the production. Evidence exists that farmed salmon escape from marine cages and some of these escapees ascend rivers (Atlantic Salmon Biological Review Team 1999; DFO 1999). Also, juvenile salmon are escaping from local commercial hatcheries and migrating to sea (DFO 1999).

The salmon farming industry was initially viewed by many wild salmon advocates as the saviour of the wild Atlantic salmon resource. These advocates saw the aquaculture industry providing salmon for the consumer and thereby reducing the viability of and justification for the rather extensive local and high seas commercial salmon fisheries existing at that time. The ready availability of salmon from the farming industry was also expected to reduce the market demand for illegally taken salmon. The intended suppression of both the interception fisheries and the illegal take of salmon as well as the growth of the salmon farming industry were, to these advocates, the intuitive answers to "having your cake [or salmon] and eating it." Today, we must acknowledge that the main expectations of these earlier advocates have been met—the salmon farming industry has provided an abundance of salmon flesh for the consumer and the interception fisheries have been virtually eliminated, the latter in part seemingly attributed to the increased market availability of farmed salmon flesh. What the advocates had not foreseen was the current view by many that the salmon farming industry poses major threats to wild salmon stocks.

The potential threats to wild salmon stocks from salmon farming are well documented (NASCO 1993; Atlantic Salmon Biological Review Team 1999; DFO 1999). Among those concerns frequently expressed are competition for food and habitat, the disruption of wild salmon redds, enhanced predator attraction, the spread of diseases and parasites, and interbreeding with wild salmon. Of these, the transfer of diseases and parasites to wild salmon must be viewed as highly threatening. For example, infectious salmon anemia (ISA) has recently become a threat to salmon farms in the western Bay of Fundy. Recently, this viral disease agent has been diagnosed in both escaped farm salmon and wild salmon. Similarly, infection with sea lice *Lepeophtheirus salmonis* is unavoidable to the salmon farming industry. Infected farmed fish are recognized internationally as contributing to elevated sea lice infections within wild salmonid stocks (DFO 1999).

Perhaps the greatest threat to wild salmon stocks from salmon farming is genetic introgression. Wild stocks low in abundance and situated proximate to the farming industry are particularly vulnerable to genetic swamping by farmed escapees. The dangers of genetic introgression by escapees are heightened by the presence of salmon of European origin in the eastern North American industry. Pure and hybridized strains of European genetic influences are reported to account for approximately 30–50% of the current production of fish in Maine (Baum 1998). Reproductively viable salmon of European strains and their hybrids are presumed not to be present on the neighboring farms in Canada because they have been legally prohibited from entering eastern Canada and its aquaculture industry throughout the life of the industry. The presence of European genetic influences in Maine's industry are largely the result of fertilized eggs being brought into the country before the federal government closed its borders to such imports. One known exception was a recent importation of salmon milt of European origin, which gained access because of a loophole in the federal regulations.

It seems reasonable to conclude that the salmon aquaculture industry, as currently practiced, is not a threat to the overall survival of wild salmon in North America. It could, however, contribute to the loss of weakened wild stocks in some rivers, particularly those proximate to on-going aquaculture operations. It is a definite threat to the recovery of weakened wild stocks, particularly those in Maine and the Canadian rivers that empty into the Bay of Fundy.

Restoration Challenges

Despite several decades of restoration activities in a number of New England rivers, much is unknown about the freshwater factors affecting restoration success. Baum (1997) notes that, despite use of some of the latest technological advances in fish passage and fish culture techniques, no sustainable naturally reproducing runs of salmon have been established in the former New England Atlantic salmon range. Nevertheless, in part because of differences in restoration approaches—particularly for the larger New

England rivers such as the Connecticut, Merrimack, and Penobscot—continuing research is producing information that is expected to lead to improved freshwater survival. Many question whether stocks that have been lost for so many generations, such as those formerly present in central and southern New England rivers, can ever be restored. Successful restoration depends in large part on the ability of the reintroduced stock to evolve and adapt to local conditions. Recent research on sockeye salmon *Onchorynchus nerka* suggests that rapid evolution (within 13 generations) may occur under conditions of reproductive isolation from introduced stocks (Hendry et al. 2000). Unfortunately, the sockeye salmon displaying this evolution timetable did not face the same complex array of anthropogenic factors that are currently reducing freshwater survival in New England Atlantic salmon rivers (Baum 1997; Atlantic Salmon Biological Review Team 1999; Kocik et al. 1999; USASAC 2000).

Government leadership and support have been inadequate to address the current state of the Atlantic salmon resource. Other fisheries resources currently having higher commercial or recreational value have drawn the primary time and attention of senior resource managers in both the United States and Canada. Concerned Atlantic salmon interests generally are of the opinion that federal expenditures on Atlantic salmon programs are inequitably low in comparison with their expenditures on other fisheries resources for which they have management responsibilities.

Leadership and increased financial support from governments are required to research the cause of the two-decade decline in North American Atlantic salmon abundance and to develop and implement recovery programs for the unique stocks of salmon threatened with extinction. Opportunity exists to secure significant financial and in-kind assistance from stakeholders and private industry, but such is available only if governments are willing to increase their commitment. Recreational fishing interests, such as the Atlantic Salmon Federation and its benefactors, are the main source of both current and increased support from the private sector.

Required Action

As a first priority, stronger leadership and increased financial support are required from the government agencies responsible for managing the Atlantic salmon resource in North America. Both leadership and support are essential to attract the participation and investment by nongovernment organizations (NGOs) and private industry that are necessary to research the cause for the current low marine survival and to carry out the recovery programs for the threatened and endangered stocks. European support for the research effort can be expected because their stocks are also experiencing higher than normal losses while at sea. DFO and NASCO have each hosted an international workshop to initiate the development of research plans to address the low sea survival problem.

The harvest of North American Atlantic salmon in marine interception fisheries, though currently low, will continue to be an issue that requires international cooperation and agreement. Threats of increased harvests still remain at Greenland and in the narrow corridor of water belonging to Saint-Pierre and Miquelon, although the latter is probably not a major threat because its nominal catch has not exceeded 3.4 t during the past 17 years for which catches have been reported (ICES 2000). The Greenland fishery continues to be managed by quota that is annually negotiated through NASCO. This fishery, though managed as a subsistence fishery in both 1999 and 2000, could revert to a commercial fishery in the future. The coastal fishery off Canada's Labrador coast is not expected to be a threat in the future because the commercial fishery has been closed and only a small subsistence fishery remains. Although the threats of interception fisheries in the future seem rather insignificant, the efforts of NASCO will continue to be needed to ensure the protection of the resource from the expansion or reestablishment of such fisheries.

Virtual loss of access to the resource by aboriginal peoples and recreational users on many rivers is currently challenging fisheries managers to be more innovative in their approach to managing the resource. This situation applies to many rivers and, perhaps of greatest significance, to extensive geographic areas, such as most of mainland Nova Scotia, all of southern New Brunswick (including the Saint John River), and all of New England. Fisheries managers must recognize that conservation of the resource implies protection for sustainable use rather than simply protection of a species or a particular population. The need for stakeholder involvement in the management program to ensure stock conservation and implement recovery programs is further motivation to explore ways that will permit some use of the salmon resource while it remains at low levels. Hook-and-release angling is one possible form of use under such circumstances, but not before accommodating the rights of aboriginal peoples to fish for food, social, and ceremonial purposes. In Canada, a hook-and-release fishery is considered an allocation, in that it provides access to fish and will result in some mortality, albeit generally minor (DFO 1998b). Hence, the communal rights of aboriginals to fish must take precedence unless they concur otherwise. Governments in Canada have a fiduciary responsibility to aboriginals to ensure that their rights are upheld.

An important aspect of any management regime for salmon is respect for the constitutional and treaty rights of aboriginal peoples. Such respect is required from all participating parties, not just governments. The success of future management and recovery programs for Atlantic salmon will depend upon mutual respect, cooperation, and participation of all interested parties. An enhanced effort and united approach is required by aboriginals, recreational fishing interests, and all levels of government to ensure Atlantic salmon stock conservation in the future.

More and different types of effort will be required to ensure the recovery of stocks at risk of becoming extinct. Unfortunately, many stocks appear to fit this category, and many others are on the verge of joining them unless marine survival improves. Special measures will be required to protect what remains of these stocks, to determine the cause of their weakened status, and to implement the corrective measures necessary to ensure stock recovery. The implementation of recovery programs will require changes in thinking and approaches from those utilized to manage the stocks when salmon abundance was greater and managing the harvestable surplus was generally the main focus.

With the commitment to initiate recovery programs comes the requirement to develop criteria by which to identify stocks at risk of extirpation and those that are genetically unique and thereby worthy of special preservation measures. Protocols are required for the operation and use of live gene banks to preserve population genomes and to restore them to natural habitats. Customization of the conservation limit for salmon, currently defined as 2.4 eggs m^{-2} of fluvial habitat (DFO 1991), is required to accommodate differences in habitat productive capacity. Progress is being made with respect to refining the conservation limit (e.g., in Quebec), but further effort is urgently required in view of the current low stock levels in many areas.

A more proactive approach to salmon habitat protection and, in particular, habitat restoration is required because the productive capacity of freshwater salmon habitat continues to be eroded in spite of the extensive and combined efforts of governments and stakeholders to protect it. Restoration activities, such as removing man-made obstructions, improving upstream and downstream fish passage efficiency at dams, reducing acidified emissions, cleaning pollution-bearing industrial effluents, and reconstructing stream habitat, are required to compensate for the continuing loss in freshwater production capacity.

Further precautionary measures are required to minimize the potentially harmful effects of aquaculture on wild salmon stocks. Some of the principal measures required are developing and implementing of an effective containment code of practice, improving fish health management, using local stocks only unless rendered sterile, eliminating European strains and their hybrids from the industry, and marking all salmon released into cages. To effectively implement such measures, a concerted and joint effort is required by the United States and Canada. A uniform approach to managing New England and Atlantic Canadian farming industries is necessary because of the close proximity of American and Canadian salmon farms in the western Bay of Fundy and the fact that these farms combined account for about 95% of the farmed salmon production in eastern North America (DFO 1999; ICES 2000).

Introductions and transfer policies and complementing regulations require upgrading, in both New England and eastern Canada, to close loopholes and improve enforceability. Compliance monitoring and increased enforcement are required to prevent unauthorized movement and release of fish. Currently, compliance monitoring of aquaculture operations is generally lacking. The dramatic spread of harmful freshwater competitor species, such as smallmouth bass *Micropterus dolomieu* and chain pickerel *Esox niger*, between major watersheds in Nova Scotia and New Brunswick also demonstrates the inadequacy of current regulations in Canada.

A bilateral forum is required to coordinate the implementation of Atlantic salmon recovery programs in Maine and Canadian rivers emptying into the Bay of Fundy. The Maine program involves 17 rivers and streams, including the eight populations that the U.S. federal agencies concluded to be in danger of extinction (Atlantic Salmon Biological Review Team 1999). Canada's current main focus is on the salmon recovery program for the 32 or more inner Bay of Fundy rivers, but recovery programs are also likely to be required for the other Bay of Fundy rivers, particularly the Saint John River and rivers situated west of it (DFO 2000a). The salmon farming operations in the western Bay of Fundy are a common concern to salmon recovery programs in both Maine and Canadian Bay of Fundy rivers (Atlantic Salmon Biological Review Team 1999; DFO 1999; Doubleday 2001).

Canada's proposed Species at Risk Act would provide additional protection to Atlantic salmon in danger of becoming extinct or extirpated, such as the inner Bay of Fundy stocks. The application of such legislation would heighten the visibility of the endangered populations and potentially ensure some funding support toward their recovery. Accordingly, every effort possible should be made to ensure the expedient passage of this new Canadian legislation.

In the United States, the legislation already exists (i.e., the Federal Endangered Species Act) and the eight stocks of salmon in Maine that are in danger of being lost are listed under the act. The onus now rests with the management agencies to ensure the efficient delivery an effective recovery plan for the listed stocks (collectively defined as a distinct population segment by the act). Canada faces a similar challenge now that the inner Bay of Fundy stocks are listed as endangered by COSEWIC.

Where have all the salmon gone? The short answer is "no one knows," or at least there is little agreement among scientists on the principal cause of the dramatic decline in salmon abundance over the past two decades (ICES 2000). What is known is that the decline, though accentuated by local anthropogenic factors, is largely attributable to exceptionally high mortality of salmon while in the sea (DFO 1998a; ICES 2000). The decline appears to be most severe at the southern end of the species' range (DFO 2000; USASAC 2000), and fishing mortality does not appear to be the cause (DFO 1998a).

Who cares about the Atlantic salmon? The recreational fishing interests, such as the Atlantic Salmon Federation, are passionate about the salmon and its conservation. These interests constitute an exceptionally large and diversified stakeholder constituency and a relatively strong political lobby. In spite of this, the Atlantic salmon is generally viewed by the lead management agencies to be of lower importance than the marine fisheries resources upon which the northeastern North American commercial fishing industry is dependent. The fact that the principal fishery for Atlantic salmon is recreational, which is viewed in Canada to be more of a provincial government responsibility, is a further hindrance to increased support from the Canadian federal government. Salmon interests, including many government bureaucrats working on the species, are adamant that government commitment and support for the Atlantic salmon resource are inadequate in both the United States and Canada.

Future efforts to restore Atlantic salmon hinge upon a consolidated effort by all interested parties and a strong political lobby to secure the much needed new resources. Unfortunately, the current situation is somewhat of a conundrum. Strong government leadership and financial support are required to secure significant private-sector involvement and investment, and such support from governments will be secured only if there are both strong stakeholder commitment and a powerful political lobby. In Canada, one could question the insistence by government that the private sector contribute to fundamental research and the implementation of the much needed recovery programs when such activities could be deemed a fundamental duty of government according to the parliamentary act that created the Department of Fisheries and Oceans, the responsible government agency. In the United States, the current inability of state and federal salmon management agencies to agree on the uniqueness of wild stocks in Maine has caused that state, supported by a coalition of fish farmers, blueberry growers, and the forest products industry, to sue the federal government over the Endangered Species Act listing. This lack of agreement among those directly responsible for the conservation and management of Maine's salmon stocks is inhibiting effective mobilization of all interested parties to implement a common suite of conservation measures. Regardless of the differing perspectives on who should pay and what should be done, we urge all interested parties to pay heed that failure to invest in recovery programs and supporting research *now* yields the high risk of disaster for many North American Atlantic salmon populations and the resource as we have known it.

Acknowledgements

We wish to acknowledge and thank Ed Baum (senior fisheries scientist with the Maine Atlantic Salmon Commission before his retirement in 2000 and author of *Maine Atlantic salmon: a national treasure*) and Robert Jones (retired chief of fisheries, Connecticut Department of Environmental Protection and current U.S. commissioner to NASCO) for their very helpful comments on an early draft of the manuscript. We also appreciate the excellent comments received from the three anonymous referees selected by the editors to review the chapter and a fourth reviewer who also was very helpful. Special thanks are also owed to John Loch (manager, Maritimes Aquatic Species at Risk Office, Department of Fisheries and Oceans Canada) for his assistance with the sections pertaining to Canada's Species-at-Risk matters; David Meerburg (senior policy advisor, Diadromous Fish, Fisheries Research Branch, Department of Fisheries and Oceans Canada) for reviewing and advising on the text regarding NASCO's involvement in salmon management; and Dr. Fred Whoriskey (vice president, Research and Environment, Atlantic Salmon Federation) for assisting with the interpretation of the American Fisheries Society's listing of North American Atlantic salmon stocks at risk of extinction.

References

Atlantic Salmon Biological Review Team. 1999. Review of the status of anadromous Atlantic salmon (*Salmo salar*) under the U.S. Endangered Species Act. Prepared by the Anadromous Atlantic Salmon Biological Review Team. U.S. Department of the Interior, U.S. Fish and Wildlife Service, U.S. Department of Commerce, National Oceanic and Atmospheric Administration, Woods Hole, Massachusetts.

Baum, E. T. 1998. History and description of the Atlantic salmon aquaculture industry in Maine. DFO Canadian Stock Assessment Secretariat Research Document 98/165.

Baum, E. T. 1997. Maine Atlantic salmon: a national treasure. Atlantic Salmon Unlimited, Hermon, Maine.

DFO. 1991. Definition of conservation for Atlantic salmon. DFO Canadian Atlantic Fisheries Scientific Advisory Committee Advisory Document 91/15.

DFO. 1988. 1988 survey of Atlantic salmon anglers in Canada. DFO Economic and Commercial Analysis Report No. 63.

DFO. 1998a. Atlantic salmon abundance overview for 1997. DFO Science Stock Status Report D0-02(1998).

DFO. 1998b. Effects of hook-and-release angling practices. DFO Science Stock Status Report D0-03(1998).

DFO. 1999. Interaction between wild and farmed Atlantic salmon in the Maritime Provinces. DFO Maritimes Regional Habitat Status Report 99/1E.

DFO. 2000a. Atlantic salmon Maritime Provinces overview for 1999. DFO Science Stock Status Report D3-14(2000).

DFO. 2000b. The effects of acid rain on the Atlantic salmon of the Southern Upland of Nova Scotia. DFO Maritimes Regional Habitat Status Report 2000/2E.

DFO. 2001. Atlantic salmon Maritime Provinces overview for 2000. DFO Science Stock Status Report D3-14(2001).

Doubleday, W. G. 2001. Is Atlantic salmon aquaculture a threat to wild stocks in Atlantic Canada? ISUMA: Canadian Journal of Policy Research 2(1):114–120.

FAO. 1995a. Precautionary approach to fisheries. Part 1: guidelines on the precautionary approach to capture fisheries and species introductions. FAO Fisheries Technical Paper No. 350, Part 1. FAO, Rome.

FAO. 1995b. Code of conduct for responsible fisheries. FAO, Rome.

FAO. 1996. Precautionary approach to fisheries. Part 2: scientific papers. FAO Fisheries Technical Paper No. 350, Part 2. FAO, Rome.

Foster, C. H. W. 1991. Yankee salmon: the Atlantic salmon of the Connecticut River. CIS, Needham, Massachusetts.

Hendry, A. P., J. W. Wenburg, P. Bentzen, E. C. Volk, and T. P. Quinn. 2000. Rapid evolution of reproductive isolation in the wild: evidence from introduced salmon. Science 290:516–518.

ICES. 1988. Code of practice and manual of procedures for consideration of introductions and transfers of marine and freshwater organisms. ICES Cooperative Research Report No. 159.

ICES. 2000. Extract of the report of the Advisory Committee on Fisheries Management (pertaining to North Atlantic salmon stocks) to the North Atlantic Salmon Conservation Organization. ICES, Copenhagen, Denmark.

Kocik, J. F., K. F. Beland, and T. F. Sheehan. 1999. Atlantic salmon overwinter survival and smolt production in the Narraguagus River. O-99-NEC-1. Woods Hole, Massachusetts.

MITSC. 1995. Fish and wildlife provisions under the 1980 Maine Indian Claims Settlement. Maine Indian Tribal-State Commission. Augusta, Maine.

Musick, J. A., M. M. Harbin, S. A. Berkeley, G. H. Burgess, A. M. Eklund, L. Findley, R. G. Gilmore, J. T. Golden, D. S. Ha, G. R. Huntsman, J. C. McGovern, S. J. Parker, S. G. Poss, E. Sala, T. W. Schmidt, G. R. Sedberry, H. Weeks, and S. G. Wright. 2000. Marine, estuarine, and diadromous fish stocks at risk of extinction in North America (exclusive of Pacific salmonids). Fisheries 25(11):6–30.

NASCO. 1992. Protocols for the introduction and transfer of salmonids. NASCO/NAC (92)24.

NASCO. 1993. Impact of salmon aquaculture. NASCO/CNL (93)29.

NASCO. 1994a. Resolution by the parties to the Convention for the Conservation of Salmon in the North Atlantic Ocean to minimize impacts from salmon aquaculture on the wild salmon stocks. NASCO/CLN (94)53.

NASCO. 1994b. Amendments to the protocols for the introduction and transfer of salmonids. NASCO/NAC (94)14.

NASCO. 1998. Agreement on adoption of a precautionary approach (incorporating minor amendments). NASCO/CLN (98)46. Annex 15, Report of the Fifteenth Annual Meeting of the Council:167–171.

NASCO. 2000. Report on the activities of the North Atlantic Salmon Conservation Organization. 1998–1999. NASCO, Edinburgh, Scotland.

New England Fishery Management Council. 1987. Fishery management plan for Atlantic salmon. New England Fishery Management Council. Massachusetts.

Pippy, J. 1982. Report of the Working Group on the Interception of Mainland Salmon in Newfoundland. Canadian Manuscript Report of Fisheries and Aquatic Sciences No. 1654.

Ritter, J. A. 1989. Marine migration and natural mortality of North American Atlantic salmon (*Salmo salar* L.). Canadian Manuscript Report of Fisheries and Aquatic Sciences. No. 2041.

Ritter, J. A. 1993. Changes in Atlantic salmon (*Salmo salar*) harvests and stock status in the North Atlantic. Pages 3–25 *in* D. Mills, editor. Salmon in the sea and new enhancement strategies. Fisheries News Book, Oxford.

Ritter, J. A., and T. L. Marshall. 1992. Management of the Atlantic salmon (*Salmo salar*) resource of the Saint John River, New Brunswick. Pages 205–218 in R. H. Stroud, editor. Fisheries management and watershed development. American Fisheries Society, Symposium 13, Bethesda, Maryland.

Ruggles, C. P. 1980. A review of downstream migration of Atlantic salmon. Canadian Technical Report of Fisheries and Aquatic Sciences No. 952.

Ruggles, C. P., and J. A. Ritter. 1980. Review of North American smolt tagging to assess the Atlantic salmon fishery of West Greenland. Rapports et Proces-Verbaux des Réunions Conseil International pour l'Exploration de la Mer 176:82–92.

Stolte, L. W. 1981. The forgotten salmon of the Merrimack River. U.S. Fish and Wildlife Service, Washington, D.C.

USASAC (U.S. Atlantic Salmon Assessment Committee). 2000. Annual report of the U.S. Atlantic Salmon Assessment Committee. USASAC, Gloucester, Massachusetts.

Whalen, K. G., D. L. Parrish, M. E. Mather, and J. R. McMenemy. 2000. Cross-tributary analysis of parr to smolt recruitment of Atlantic salmon (*Salmo salar*). Canadian Journal of Fisheries and Aquatic Science 57:1607–1616.

Watt, W. D. 1988. Major causes and implications of Atlantic salmon habitat losses. Pages 101–111 *in* R. H. Stroud, editor. Present and future Atlantic salmon management: measuring progress toward international cooperation. Atlantic Salmon Federation and National Coalition for Marine Conservation, Inc., Ipswich, Massachusetts, and Savannah, Georgia.

Sustaining North American Salmon: Perspectives Across Regions and Disciplines, 117–136

Chapter 6

Emerging Science and Management Issues with Atlantic Salmon

Kevin D. Friedland

Introduction

Atlantic salmon *Salmo salar* science and management will have to adapt to times most recently colored by the unprecedented decline in salmon stocks across the entire North Atlantic. Whereas, in the not too distant past, salmon scientists were totally preoccupied with understanding the migration of salmon to distant water fishing grounds, they are now consumed with the factors that affect the natural mortality and maturation of salmon. Whereas managers and user groups at one time hoped that the development of salmon aquaculture would reduce the harvest pressure on wild stocks, they must now resolve issues of sea cage aquaculture site selection and fish farm escapees that threaten wild stocks with extinction instead of overexploitation. Whereas managing salmon stocks usually meant dealing with issues of allocation between user groups, we now increasingly see user groups calling for the protection of the resource in fear of its further demise and extirpation. This fundamental change in the priority of issues that the salmon community faces will reorder the direction of salmon science and management for decades to come. It is also plain to see from the historical record of salmon stock abundance, the pace with which reviewed scientific evidence becomes available to the management community, and the time frame in which management intent is translated into regulatory action that future change in salmon populations will likely occur on the same decadal time scales as past change. And for that matter, we cannot be confident we will find a practicable solution to the problem of contemporary salmon stock decline, simply because its root cause is unlike anything we, or the fish, have previously experienced. Instead, we may be cast in the role of observers to some of the most dramatic changes in the health and utilization of this natural resource.

Have all the Issues in Freshwater Habitats Already Emerged?

The battle for salmon survival began in riverine and freshwater habitats literally centuries ago, when alterations such as dams clearly curtailed or ended runs of salmon. Since the fish are in close contact with human activity while in freshwater, many interactions are of potential or realized importance. Issues such as pollutants, water use, and forestry practices have been shown to have some linkage to salmon populations. Most of the issues in freshwater have emerged and remain important areas of research and management. However, one area of concern that I believe is taking on a larger role is the issue of climate change impacts on salmon life history and survival in freshwater, particularly in relation to the overwintering success of parr (Cunjak et al. 1998). The changes we are experiencing in climate are not likely to produce "Miami in Maine" conditions, but they are going to affect weather patterns in perhaps less noticeable ways (e.g., the frequency of storms, the amplitude of winter thermal variations, or the amount of regional rainfall. The added source of mortality variability associated with freshwater, in concert with changes in the marine environment, may prove to be the undoing of many salmon stocks, particularly at the extreme ends of the range for the species.

Continental Abundance Trends and Climate

The overall abundance of salmon in the North Atlantic has declined dramatically over the past three decades. Midpoint estimates of the abundance of prefishery recruits, or the number of one-sea-winter (1SW) salmon before any fisheries, totaled about 9 million fish during the early 1970s (Figure 1). This estimate of abundance accounts for removals by fisheries and includes spawning escapements for both grilse and two-sea-winter salmon (Anonymous 1999). Regionally, the added abundance of repeat spawning fish and multi-sea-winter spawning may be important; however, the abundance of these groups has not been sufficient to compensate for the dramatic decline in the predominant age groups ascending rivers. In recent years, this estimate has been nearly halved, to less than 5 million fish.

This declining trend is a common feature for both North American and European continental stock complexes, despite the different abundance levels observed for the respective groups. The scatter relationship between

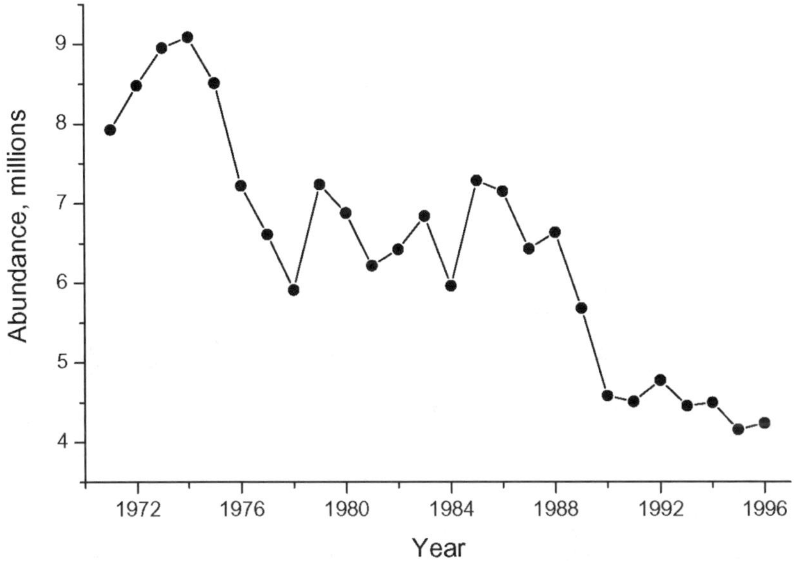

Figure 1. Total prefishery abundance of North American and European stocks of Atlantic salmon.

prefishery recruitment for North American versus European stock complexes shows a linear trend (Figure 2). The European complex tends to be more abundant than the North American complex, reflecting the greater number of salmon-producing rivers in Europe and the fundamental differences in ocean and freshwater environments between the eastern and western sides of the ocean basin. In addition, the slope of the plot for the two stocks is not 1, reflecting a faster rate of decline for the North American stock complex. In fact, the decline of the North American complex has been on the order of a fourfold decrease, whereas the European stock complex has declined by only twofold.

The dominant climate indicator in the North Atlantic, the North Atlantic Oscillation (NAO), has shown multidecadal patterns that undoubtedly affect salmon in many ways. The NAO is often represented by the atmospheric pressure difference between observing stations in Iceland and Portugal (Figure 3). These stations have been making weather observations for more than a century, thus allowing calculation of the NAO index back to the 1800s. This index represents the oscillation between two dominant states of weather and storm patterns over the North Atlantic. Though the index is computed for all months, the index that appears to be of greatest interest for salmon is the winter index (December through March), which shows highly systematic decadal trends. In recent decades, the index has

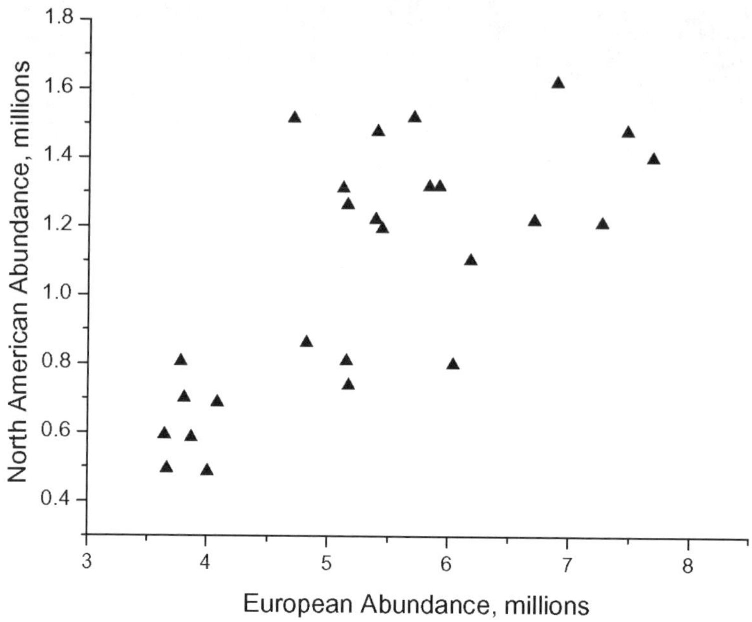

Figure 2. Prefishery abundance of North American versus European stock complexes.

been increasing. This increase generally matches the pattern of declining salmon abundance and has been applied to salmon life history with respect to the expected fate of salmon postsmolts in spring (Dickson and Turrell 2000).

The NAO is not internally coherent in that seasonal change is not synchronous; thus, the change in climate conditions in summer may be different from those observed in winter or spring. The ramifications of these differences are obvious for fish species such as Atlantic salmon *Salmo salar* that have varying climate requirements at different life cycle stages. In the case of Atlantic salmon, postsmolts are most vulnerable in the spring and early summer, when they enter the marine environment (Friedland et al. 2000). Maturing fish would likely be affected by overwintering conditions; thus, changing climate conditions in spring may be uncorrelated (Martin and Mitchell 1985; Friedland et al. 1998b). Before events in the ocean, presmolt numbers may be affected by winter conditions in the year prior to the migration to the sea (Whalen et al. 1999). The impact of climate may influence salmon survival during a number of critical life history stages, so researchers will have to be mindful of the cummulative effects of climate that sequentially affect a single cohort.

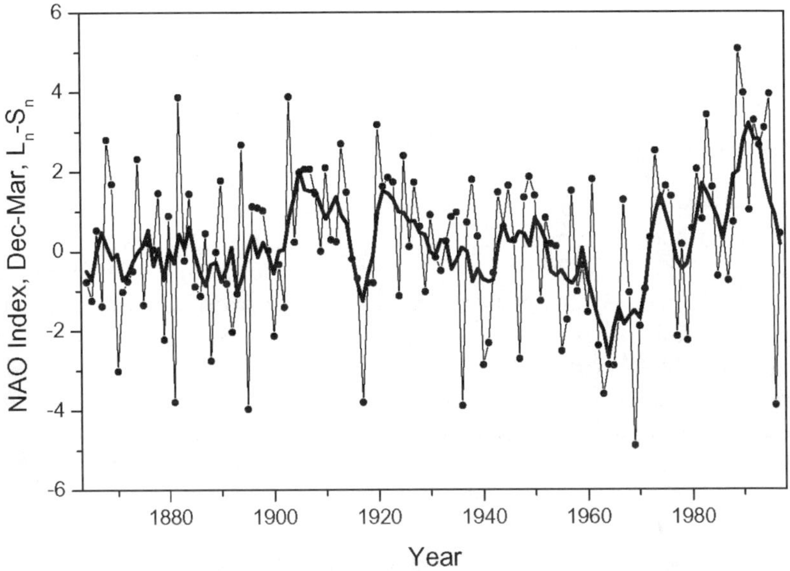

Figure 3. North Atlantic Oscillation (NAO) pressure differential index for December through March. Index is computed on the basis of sea level atmospheric pressure between Lisbon, Portugal, and Stykkisholmur, Iceland (from Hurrell 1995). Smooth curve is 5-point moving average.

Local-Scale Climate Interactions

In studies focusing on two salmon stocks in the North Sea area, the Figgjo in Norway and the North Esk in Scotland, researchers have found evidence to link climate conditions with growth and survival of salmon. The Figgjo is located along the southwestern coast of Norway and is a small, unimpacted salmon river that supports a run of mostly one-sea-winter salmon. The North Esk, located on the east coast of Scotland, is of similar size to the Figgjo and has similar run characteristics. Tagging studies conducted over a period of nearly three decades show that the return rates of salmon to these rivers are synchronous—following similar patterns of survival (Friedland et al. 1998a).

Analysis of thermal habitat in the northeast Atlantic reveals some interesting trends that may reflect the vulnerability of salmon postsmolts to ocean conditions when they enter the marine environment. The amount of coldwater habitat in the postsmolt area, as defined by the thermal limit of 5–7°C, was negatively correlated with survival of Figgjo and North Esk salmon (Friedland et al. 1998a). This correlation was strongest in May, the first month the fish are at sea, and showed the greatest spatial contrast in the North Sea area, which would have the greatest impact on these stocks . A reciprocal relationship was also found in the amount of warm thermal habitat, as defined by the distribution of 8–10°C water, which was positively correlated with survival. The inference that can be drawn from these analyses is that when salmon smolts enter warm water conditions in the ocean, survival is high; whereas, when they enter cold ocean conditions, their survival is lower.

A number of survival factors would be expected to covary with sea surface temperature conditions. One of the most important temperature-dependent factors for juvenile fish survival is growth. Growth-mediated predation and survival effects respond directly to variation in water temperature. Mortality rates of pelagic fishes are inversely related to body size (Peterson and Wroblewski 1984); thus, growth variation in early life history stages is broadly identified as critical to the recruitment process (Anderson 1988; Pepin 1991). In salmonids, growth-related mechanisms have also been associated with survival patterns (Furnell and Brett 1986; Mathews and Ishida 1989; Holtby et al. 1990; L'Abee-Lund et al. 1993; Friedland et al. 1996). Postsmolts from the Figgjo and North Esk advected into the Norwegian coastal current would be subjected to prevailing temperature conditions in a given year. In years when water temperatures were high enough to support rapid growth, the postsmolts would be expected to attain large body sizes and thus avoid predation more effectively than in colder years. Conversely, colder temperature conditions would depress growth and likely keep juvenile salmon under predation pressure for a longer period. This represents the most direct effect that sea surface temperature distributions could have on postsmolts. Other factors associated with sea surface temperature trends may be affecting growth and survival through indirect mechanisms.

The variation in water temperature in the northeast Atlantic may be indicative of changes in ecosystem productivity due to weather and oceanographic conditions. This habitat variability may affect the supply of prey species and thus influence feeding opportunities for postsmolts. Recent studies with Pacific salmonids point to the importance of variation in diet in the recruitment process (Healey 1991; Brodeur et al. 1992). Prey availability is often related to oceanographic processes and structural features in the water column (Levings 1994). Juvenile salmonids have been shown to modify their diet in response to restrictions imposed by thermal structure (Reddin 1985; Pearcy et al. 1988) or opportunities created by the concentration of food along fronts (Brodeur 1989; St. John et al. 1992). Variation in ocean productivity has also been hypothesized to act on postsmolts by shifting predation pressure from other species onto the postsmolts (Fisher and Pearcy 1988).

Salmon undergo extensive marine migrations and must use a number of cues to successfully orient or navigate during their migrations. The conditions that postsmolts encounter when they enter the sea may impose additional swimming on the fish as they respond to migration cues important to their overall survival (Salminen et al. 1994). Such additional swimming may create energy deficits for the fish as they swim harder to avoid undesirable thermal conditions and thus affect their growth and survival.

Until we understand the mechanisms by which climate affects survival and maturation of regional stocks, we have little hope of developing the predictive capability to calibrate fisheries management or other protective actions. Fisheries management is all too often quite content to react to stock condition. However, what is perhaps particular to salmon is that climate change will likely go beyond defining changes in the fishing grounds and instead determine patterns of regional stock collapse and extinction.

Growth-Mediated Effects

The connection between survival and oceanographic conditions is intriguing, but it is essential to tie all climate mechanisms to the biology of the fish. Retrospective growth analyses from scales offer a way of describing that link. Salmon scales are remarkably descriptive in the way they record growth patterns and life history events characteristic of a stock. The measurements from scales are sensitive enough to capture annual variability in growth. Salmon biologists have worked extensively with two types of scale growth data. First, one type—which can be characterized as low-resolution data—is collected with conventional aging equipment and measures the distances between annual growth checks. The other type—which can be characterized as high-resolution and is collected with image-processing equipment—is the spacing patterns of individual circuli of a scale. Low-resolution data are available for one of the North Sea stocks, the North Esk, which, in addition to the relationship between survival and ocean climate conditions, also shows a strong correlation between postsmolt growth and the survival-climate signal (Friedland et al. 2000). Yearly growth increments derived from back-calculated lengths are positively correlated with the observed survival rates from tagging studies for this stock (Figure 4). However, this low-resolution measure of growth reflects the entire postsmolt year or the period after the fish enters the ocean to the first winter. Though not a full year, it is a significant period of time and does not address the issue of which seasons may be more critical.

High-resolution data collected for the Girnock Burn, a small stock that neighbors the North Esk, may offer a hint about which seasons are most critical and provide a direction for future scale analyses. Circuli spacing data for the Girnock Burn collected over a span of years similar to that analyzed in the North Esk show that most of the variability in postsmolt growth can be attributed to variation in growth during the spring when fish

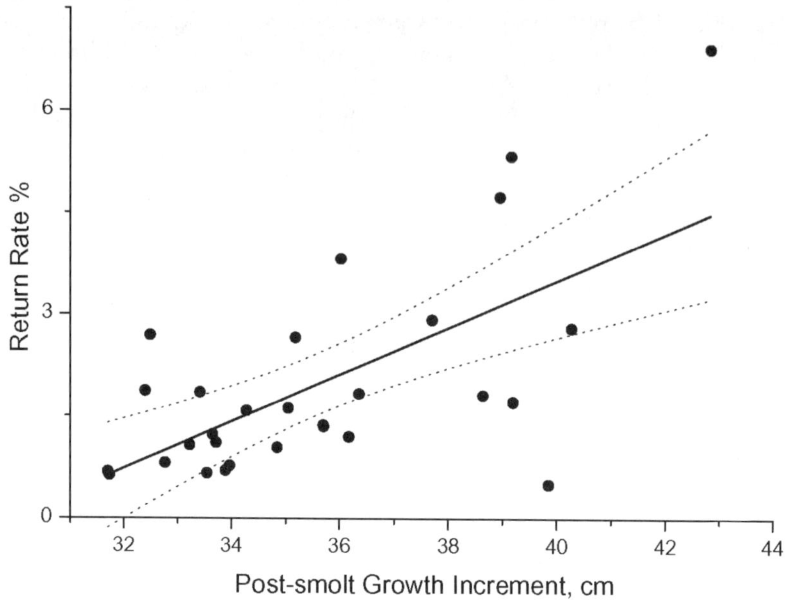

Figure 4. Return rate of 1SW salmon versus postsmolt growth increment for the North Esk stock.

first enter the marine environment (Figure 5). When Friedland et al. (1993) reported that spring temperature conditions might influence the survival of European stocks, the significance of this finding was not realized because of the limited resolution of the study. The subsequent study relating ocean conditions and survival of North Sea stocks is consistent with these earlier findings, but it goes significantly further to elucidate the mechanisms at work (Friedland et al. 1998a). The North Esk-Figgjo study is discrete over time and space, thus we can conclude that the ocean climate variation related to the survival of salmon in the North Sea area occurs in spring, when the postsmolts first enter the marine environment, and occurs in the area of the North Sea and Norwegian coast. Linking these data with the growth analysis suggests that certain hypotheses about postsmolt survival must be given greater credence than others.

Salmon postsmolts from stocks in the North Sea area enter a thermally dynamic ocean and thus experience a range of temperature conditions annually. Postsmolts are not able to outswim the ocean currents found off the British Isles and along the Norwegian coast, so for the first months at sea, they are transported northward (Jonsson et al. 1993). This would lead us to predict that postsmolts from the Figgjo and North Esk stocks would be located in the northern part of the North Sea and along the southern coast of

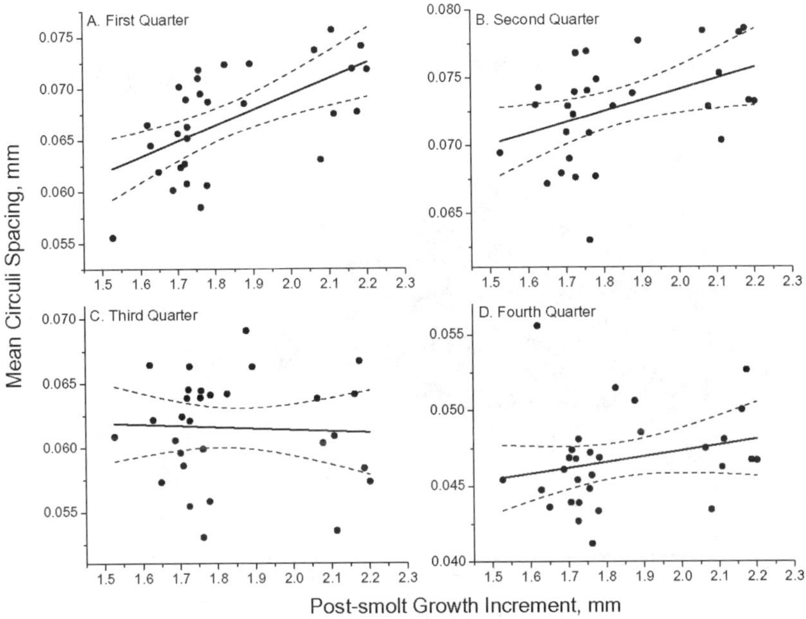

Figure 5. Mean circuli spacing, for quartered subdivisions of the postsmolt period (April–December), versus the length of the postsmolt increment for Girnock Burn salmon. Panels A–D are for quarters 1–4, respectively.

Norway during May and June, which is generally confirmed by observations of tagged fish (Jonsson et al. 1993). If we are satisfied that the variation in growth of North Esk salmon reflects the variation observed in survival of both Figgjo and North Esk fish, we should place greater weight on growth-mediated survival mechanisms such as predation. Therefore, hypotheses that are growth independent may be rejected or at least deemphasized.

Growth Signatures as Indicators of Temporal and Spatial Scale

Postsmolt growth signatures can be used to compare the dynamics of spatial distributions within and between stocks. Postsmolts collected in the Labrador Sea provide a source of annual growth signatures representative of the North American stock complex as a whole (Reddin and Short 1991). These fish were collected during three cruises over the period 1988–1991 in an area covering most of the Davis Strait (Figure 6). As reported in Friedland and Reddin (2000) and highlighted for the year 1989, growth trajectories for regionally dispersed stocks intersect after the time period

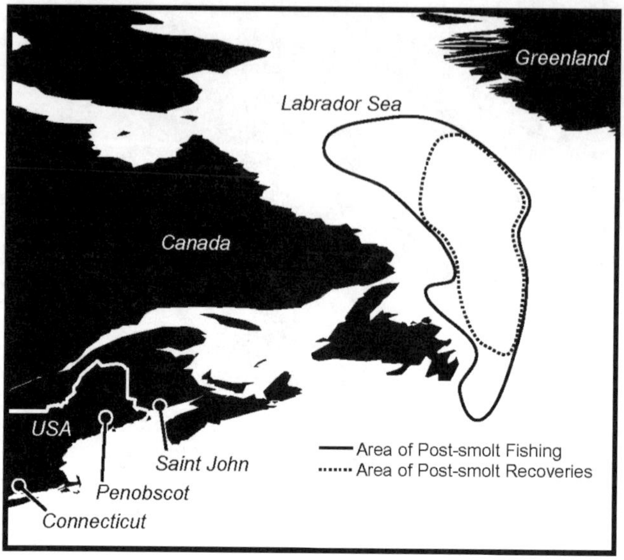

Figure 6. Map of the northwest Atlantic Ocean showing the location of postsmolt research fishing, area of postsmolt recoveries, and the locations of three southern comparison stocks.

required to deposit just four or so circuli (Figure 7). The period of time it takes to deposit a circulus changes with season. Recaptures of postsmolts used in this study were analyzed to estimate the average date of deposition for specific circuli (Friedland et al. 1993). These data suggest that circuli deposition is spaced by nearly a week during the spring and summer growth seasons and progressively slows to nearly two weeks by winter. This would suggest that Penobscot and Saint John fish do not overlap with other stocks in the North American stock complex for at least one month after their seaward migration; however, this period of regional isolation may last longer.

These data suggest a temporal and spatial scale for regional stock recruitment synchrony in North America. Regional stock groups are most likely affected by regional scale climate effects for at least the first month at sea. The temporal scale of regional synchrony also implies limits for the spatial scale of regional synchrony. A combination of swimming potential and current transport vectors will circumscribe an area relevant to a specific stock. After a few months at sea, postsmolts attain sufficient size to outswim currents and will distribute according to other behaviors such as foraging and open ocean migration. Thus, regional scale synchrony is likely limited to hundreds of kilometers in Atlantic salmon.

This assertion would appear to be supported by other observations for

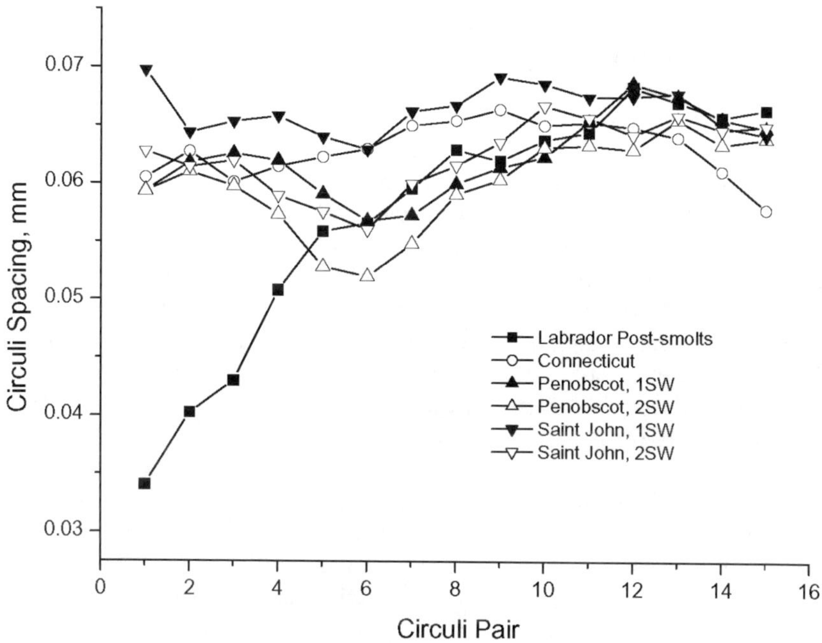

Figure 7. Circuli spacing versus circuli pair for Labrador Sea postsmolts and 1SW and 2SW age components of three comparison stocks.

North American Atlantic salmon stocks and by analyses of Pacific salmon species. A preliminary analysis of regional return data for salmon throughout Atlantic Canada shows that at least three major groups of stocks can be identified (DFO 1998). This analysis is consistent with the suggestion of a spatial scale of regional synchrony on the order of hundreds of kilometers. Patterns in regional recruitment synchrony for Pacific salmonids in the northeastern Pacific suggest that three regional groups exist as a result of ocean climate forcing (Coronado and Hilborn 1998; Peterman et al. 1998). These analyses would also suggest a scale of regional recruitment synchrony on the order of hundreds of kilometers. Taking a different approach, Myers et al. (1997) considered the problem across functional groups of fishes and produced a statistical fit suggesting that the spatial scales for freshwater and marine species are 50 and 500 kilometers, respectively, which brackets the spatial scale for anadromous species. However, there is a final point to consider. Stock groups of Pacific salmon showing regional synchrony are in negative phase to neighboring groups in response to the same broad-scale climate signal. Thus, the overall system distance for synchronous behavior is more on the order of thousands of kilometers.

Climate interactions for Pacific salmon are consistent with spatial scales of hundreds to thousands of kilometers, whereas empirical observations for Atlantic salmon in North America suggest slightly lesser distances. No single hypothesis has emerged concerning postsmolt survival of Atlantic salmon over its entire range. Relationships between climate forcing and recruitment have been described for European stocks (Friedland et al.1998a), suggesting the potential for spatial scales of synchrony on the same order as those observed for Pacific salmon. However, the same climate relationships have not been observed for North American fish. Instead, it has been suggested that the postsmolt nursery for North American salmon varies annually and is often located inshore, which supports the notion of smaller scale regional synchrony affecting these stock groups (Friedland et al. 1999). Interestingly, these same ideas about postsmolt nurseries are also being applied to Pacific salmon populations (Macdonald and Change 1993; Fukuwaka and Suzuki 1998).

Most of the information on the interaction of stocks with climate, habitat, and potentially one another was developed without the need to collect new samples. This information is based on data collected from archived scale samples, which represents one of the most important and fertile areas of research data, to address questions related to salmon in the sea. There are vast collections of stock-specific scale samples from all the salmon-producing countries in the North Atlantic, yet only a small fraction of these samples have been measured to characterize growth history of the subject stock. Much more of this type of work needs to be done because it may offer the only way to understand how regional groups of stocks respond to changes in the environment, such as those posed by global warming.

Fish as Data Collectors

New technology allowing for the miniaturization of data logging devices now allows scientists to put tags on fish to collect data describing the environments the fish have frequented. Formerly, electronic tags meant telemetry tags that were used to determine the position of fish, which in turn provided some insights on local-scale habitat utilization and behavior within environmental gradients (Quinn et al. 1989; Ogura and Ishida 1992). However, there are obvious limitations to telemetry tagging methods, including the issues of modified behavior immediately after tagging, information coming from survivors only, and data records limited to only those periods of active tracking. The first generation of data logging electronic tags was designed for large fish; they were complex devices that provided very useful data, but because of their size and cost, they could be used on only a relatively small number of animals (Lutcavage et al. 1999). Only, some researchers have begun to embrace the philosophy of "going small" by using small, cost-effective data loggers on larger numbers of fish (Walker et al. 2000). Even smaller and cheaper tags are on the horizon that may make mass releases of tagged smolts feasible. Considering the alternative approaches for describing the distribution and thermal preferences of salmon, such as ship-based surveys, data logging tags may not only be the most cost-effective way of getting these data—they may provide the only way to comprehensively describe distribution-temperature relationships for the species.

Changing Patterns of Bycatch

Atlantic salmon have been taken as bycatch in a number of fishing gears at relatively low levels; these removals have been tolerated as a minor impact on the abundance of stocks. However, concern about bycatch has been elevated by the suspected impact of recently developed pelagic fisheries in the northeastern Atlantic. These fisheries target herring and mackerel and, as we have learned through scientific survey, the distribution of these fisheries overlaps the nursery distribution of European postsmolts in both time and space (Holm et al. 2000). Though we still do not have the hard scientific evidence to characterize the impact of these fisheries on salmon stocks and in turn evaluate their role in the decline of the European stock complex, issues of circumstances continue to influence our thinking. For example, the decline of the stock complex occurred during the same time frame that the fishery developed. The high-volume gears used in the pelagic fishery target surface shoals of fish, exactly where postsmolts are codistributed with herring and mackerel. In samples from commercial and research fishing, it is clear that postsmolts will be captured by the commercial gears. The assessment community faces a monumental sampling task in estimating the bycatch in these fisheries, with the size of the problem reflecting the immense size of the fishery catch.

Farm Fish as Friend or Foe

One of the great promises of salmon aquaculture was the hope that it would relieve stress on wild populations of Atlantic salmon by providing all the fresh salmon the food markets could possibly want. The most prized of all game fish rapidly became the cow of the seas as it outperformed other species in finfish grow-out operations. Production of Atlantic salmon from sea cages in the North Atlantic now exceeds a half million metric tons, with an additional 100,000 mt coming from operations outside the North Atlantic (Figure 8). This production has so outpaced the catches coming from wild stocks that the contribution of capture fisheries to markets is now only a percent or so (Figure 9). This development has not come without cost as

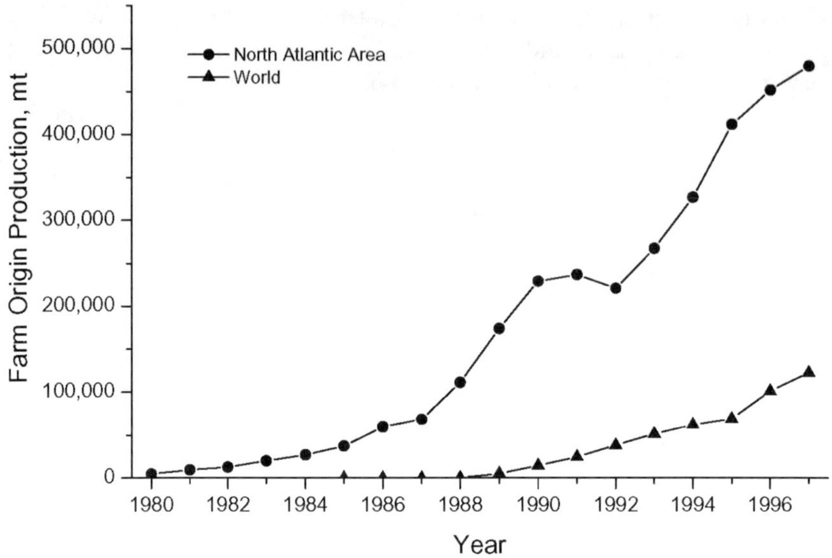

Figure 8. Production of farmed Atlantic salmon by North Atlantic area producers and producers outside the North Atlantic.

aquaculture operations pose a number of threats to wild stocks of Atlantic salmon (Jonsson 1997; Fleming et al. 2000).

The same forage fish used by wild stocks are being harvested in increasing amounts to provide the high quality meal for aquaculture operations. Is the decline in growth observed for some stocks of salmon related to the availability of food rather than climate conditions (Friedland et al. 2000)? This issue is not limited to wild and cultured Atlantic salmon. The same forage species are preyed upon by many other important commercially-harvested fish, and aquaculture operations growing other finfish place similar pressure on the forage resources.

Farmed Atlantic salmon escape cages, and concerns are growing about the spread of disease and parasites, the introgression of unwanted genes into wild populations, and competitive interactions denying wild fish access to critical habitats. Concerns about wild stock genetics are heightened by the specter of the use of transgenic salmon in aquaculture. These genetically manipulated strains attain market size in grow-out settings in a fraction of the time wild origin strains require, but are the genetic threats to wild stocks proportionally higher as well (Kapuscinski and Hallermann 1990)? This is an area of active research, but perhaps of more immediate interest is the idea that the higher growth rates of these fish open the possibility of using them in closed and completely land-based grow-out facili-

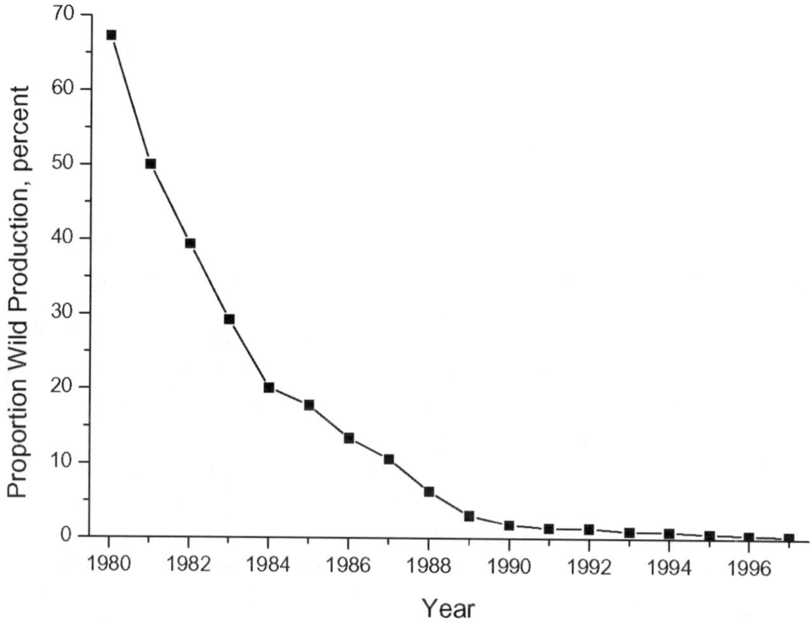

Figure 9. Percent of world Atlantic salmon production originating from harvest of Atlantic salmon stocks in capture fisheries.

ties, which might minimize the impact on wild stocks. It is not clear if consumers will accept these fish or that contained grow-out facilities will be cost effective, but it is clear that these issues will come to the fore in the near future.

Will We Be Precautionary in Our Approach?

Can we apply the precautionary approach in managing Atlantic salmon and their habitats when many think we are already being too cautious? International management of Atlantic salmon includes regulatory action aimed at controlling the harvest of mixed stock fisheries, but it also takes aim at issues related to the rational use and stewardship of salmon, including salmon aquaculture, introductions, and transfers (Windsor and Hutchinson 1994). The cornerstone of the precautionary approach is the precautionary principle, a set of principles developed in environmental management, which states where threats to a resource are serious or irreversible, the lack of data should not be used as a reason to postpone taking the action needed to protect the resource (Dovers and Handmer 1995). The precautionary approach is an adaptation of the precautionary principle deemed more appropriate for fisheries management because it incorporates economic considerations. Issues in fisheries management are usually not a threat to mankind, nor are the threats usually irreversible for the resource (FAO 1996). As reinforced in the Food and Agriculture Organization (FAO) Code of Conduct for Responsible Fisheries (FAO 1995), the precautionary approach must be applied on a stock basis, which for salmon immediately highlights populations at risk of extinction. This is clearly in conflict with the status quo in fisheries management, where actions are predicated on the average condition of the stock, thus not specifically directed to protect all stocks. Furthermore, the precautionary approach raises the status of information by establishing that the burden of proof needs to be on those whose activities may affect stock removals rather than on those charged with conserving the resource. The information demands for both biological and economic data increase significantly under a precautionary regime.

Not all advocates, however, see this as productive. Orri Vigfusson, chairman of the North Atlantic Salmon Fund, a private organization devoted to buying commercial fishing quotas, was quoted in the *Economist* (Anonymous 1998) as saying, "The basic problem is overproduction of biologists.... If governments diverted a few million dollars from studying salmon fisheries to buying out drift-netters," he argues, "the future of salmon would be safer."

But is Vigfusson's point of view precautionary? The precautionary approach is as much about understanding the resource as it is about protecting it. The assessment of risk, which is central to the precautionary approach, may examine the fuzziness of resource and environmental systems, but it does so by bringing them under intense scrutiny and examination. The decline in salmon stocks, especially in North America, may be beyond the point of requiring further protection from fishermen because the decline has continued despite significant declines in fishing. The performance of index stocks clearly shows that phenomena other than fishing are at work, and it will be impossible to make any informed decision about salmon without better information and research. The precautionary approach advocates more caution when information is uncertain, unreliable, or inadequate. Let us hope that interested parties are not too cautious about supporting the science and research aimed at developing a better understanding of what is going on with salmon.

Conclusion

We are witness to dramatic changes in salmon populations that we do not fully understand. Are we seeing a range contraction of salmon brought on by natural or anthropogenic causes? Are salmon receding into core habitats in both freshwater and the oceans that still provide tolerable thermal regimes and climate patterns? Climate appears to play a significant role in salmon survival for European stocks; our future challenge will be to better understand how climate affects North American fish and other phases of the salmon's life cycle. We must not only put our oceanographic and atmospheric databases to work on these problems, but also use archives of biological data to their fullest potential. Information on aging and growth can be of immense value in helping us understand stock survival dynamics. We need to employ new tools from other fields of technology and science, such as electronic data tags, microanalysis chemistry, and the analysis of stable isotopes, to further enhance these studies. We should also look to the commonsense principles of the precautionary approach to color the interactions between science and management because the mutual dependencies between scientists and managers run deeper then either side realizes. Future Atlantic salmon management and science will have to adapt to the realities we discover along the way.

Acknowledgements

I thank my colleagues in the salmon research community for access to samples and use of data in the many studies summarized in this paper. In particular, I thank Dave Reddin, Julian MacLean, and Lars Hansen for their help and cooperation.

References

Anderson, J. T. 1988. A review of size dependent survival during pre-recruit stages of fishes in relation to recruitment. Journal of Northwest Atlantic Fisheries Science 8:55–66.

Anonymous. 1998. End of the line: North Atlantic salmon. Economist 346(8055):80–81.

Anonymous. 1999. Report of the Working Group on North Atlantic Salmon. International Council for the Exploration of the Sea (ICES), Doc. CM 1999/ACFM:14. Québec City, Québec, Canada.

Brodeur, R. D. 1989. Neustonic feeding by juvenile salmonids in coastal waters of the Northeast Pacific. Canadian Journal of Zoology 67:1995–2007.

Brodeur, R. D., R. C. Francis, and W. G. Pearcy. 1992. Food consumption of juvenile coho (*Oncorhynchus kisutch*) and chinook salmon (*O. tshawytscha*) on the continental shelf off Washington and Oregon. Canadian Journal of Fisheries and Aquatic Sciences 49:1670–1685.

Coronado, C., and R. Hilborn. 1998. Spatial and temporal factors affecting survival in coho salmon (*Oncorhynchus kisutch*) in the Pacific Northwest. Canadian Journal of Fisheries and Aquatic Sciences 55:2067–2077.

Cunjak, R. A., T. D. Prowse, and D. L. Parrish. 1998. Atlantic salmon (*Salmo salar*) in winter: "the season of parr discontent"? Canadian Journal of Fisheries and Aquatic Sciences 55(Supplement 1):161–180.

DFO (Department of Fisheries and Oceans). 1998. Atlantic Salmon Abundance Overview for 1997. DFO Science Stock Status Report DO-02. Ottawa, Ontario.

Dickson, R. R., and W. R. Turrell. 2000. The NAO: the dominant atmospheric process affecting oceanic variability in home, middle, and distant waters of European Atlantic salmon. Pages 92–115 *in* D. Mills, editor. The ocean life of Atlantic salmon. Fishing News Books, Blackwell Scientific Publications, London.

Dovers, S. R., and J. W. Handmer. 1995. Ignorance, the precautionary principal and sustainability. Ambio 24:92–97.

Fisher, J. P., and W. G. Pearcy. 1988. Growth of juvenile coho salmon (*Oncorhynchus kisutch*) off Oregon and Washington, USA, in years of differing coastal upwelling. Canadian Journal of Fisheries and Aquatic Sciences 45:1036–1044.

FAO (Food and Agriculture Organization). 1995. Precautionary approach to fisheries part I: guidelines on the precautionary approach to capture fisheries and species introductions. FAO Fisheries Technical Paper No. 350/1 FAO, Rome.

FAO (Food and Agriculture Organization). 1996. Precautionary approach to capture fisheries and species introductions. Elaborated by the Technical Consultation on the Precautionary Approach to Capture Fisheries (including species introductions). Lysekil, Sweden, 6–13 June 1995. FAO Technical Guidelines for Responsible Fisheries. No. 2. FAO, Rome.

Friedland, K. D., J. D. Dutil, and T. Sadusky. 1999. Growth patterns in postsmolts and the nature of the marine juvenile nursery for Atlantic salmon, *Salmo salar*. Fishery Bulletin 97:472–481.

Friedland, K. D., R. R. Haas, and T. S. Sheehan. 1996. Postsmolt growth, maturation, and survival of two stocks of Atlantic salmon. Fishery Bulletin 94:654–663.

Friedland, K. D., L. P. Hansen, and D. A. Dunkley. 1998a. Marine temperatures experienced by postsmolts and the survival of Atlantic salmon (*Salmo salar* L.) in the North Sea area. Fisheries Oceanography 7:22–34.

Friedland, K. D., L. P. Hansen, D. A. Dunkley, and J. C. MacLean. 2000. Linkage between ocean climate, postsmolt growth, and survival of Atlantic salmon (*Salmo salar* L.) in the North Sea area. ICES Journal of Marine Science 57:419–429.

Friedland, K. D., and D. G. Reddin. 2000. Growth patterns of Labrador Sea Atlantic salmon postsmolts, and the temporal scale of recruitment synchrony for North American salmon stocks. Canadian Journal of Fisheries and Aquatic Sciences 57:1181–1189.

Friedland, K. D., D. G. Reddin, and J. F. Kocik. 1993. Marine survival of North American and European Atlantic salmon: effects of growth and environment. ICES Journal of Marine Science 50:481–492.

Friedland, K. D., D. G. Reddin, N. Shimizu, R. E. Haas, and A. F. Youngson. 1998b. Strontium: calcium ratios in Atlantic salmon otoliths and observations on growth and maturation. Canadian Journal of Fisheries and Aquatic Sciences 55:1158–1168.

Fukuwaka, M., and T. Suzuki. 1998. Role of a riverine plume as a nursery area for chum salmon *Oncorhynchus keta*. Marine Ecology Progress Series 173:289–297.

Furnell, D. J., and J. R. Brett. 1986. Model of monthly marine growth and natural mortality for Babine Lake sockeye salmon (*Oncorhynchus nerka*). Canadian Journal of Fisheries and Aquatic Sciences 43:999–1004.

Fleming, I. A., K. Hindar, I. B. Mjølnerød, B. Jonsson, T. Balstad, and A. Lamberg. 2000. Lifetime success, and interactions of farm salmon invading a native population. Proceedings Royal Society London B 267:1517–1523.

Healey, M. C. 1991. Diets and feeding rates of juvenile pink, chum, and sockeye salmon in Hecate Strait, British Columbia. Transactions of the American Fisheries Society 120:303–318.

Holm, M., J. C. Holst, and L. P. Hansen. 2000. Spatial and temporal distribution of postsmolts of Atlantic salmon (*Salmo salar* L.) in the Norwegian Sea, and adjacent areas. ICES Journal of Marine Science 57:955–964.

Holtby, L. B., B. C. Andersen, and R. K. Kadowaki. 1990. Importance of smolt size and early ocean growth to interannual variability in marine survival of coho salmon (*Onchorhynchus kisutch*). Canadian Journal of Fisheries and Aquatic Sciences 47:2181–2194.

Hurrell, J. W. 1995. Decadal trends in the North Atlantic Oscillation: regional temperatures and precipitation. Science 269:676–679.

Jonsson, B. 1997. A review of the ecological and behavioral interaction between cultured and wild Atlantic salmon. ICES Journal of Marine Science 54:1031–1039.

Jonsson, N., L. P. Hansen, and B. Jonsson. 1993. Migratory behavior and growth of hatchery-reared postsmolt Atlantic salmon *Salmo salar*. Journal Fish Biology 42:435–443.

Kapuscinski, A. R., and E. M. Hallermann. 1990. Anticipating environmental impacts of transgenic fish. Fisheries 15:2–11.

L'Abee-Lund, J. H., A. Langeland, B. Jonsson, and O. Ugedal. 1993. Spatial segregation by age and size in Arctic char: a trade-off between feeding possibility and risk of predation. Journal of Animal Ecology 62:160–168.

Levings, C. D. 1994. Feeding behavior of juvenile salmon and significance of habitat during estuary and early sea phase. Nordic Journal of Freshwater Research 69:7–16.

Lutcavage, M. E., R. W. Brill, G. B. Skomal, B. C. Chase, and P. W. Howey. 1999. Results of pop-up satellite tagging of spawning size-class fish in the Gulf of Maine: do North Atlantic bluefin tuna spawn in the mid-Atlantic? Canadian Journal of Fisheries and Aquatic Sciences 56:173–177.

Macdonald, J. S., and B. D. Change. 1993. Seasonal use by fish of nearshore areas in an urbanized coastal inlet in southern British Columbia. Northwest Science 67:63–77.

Martin, J. H. A., and K. A. Mitchell. 1985. Influence of sea temperature upon the numbers of grilse and multi-sea winter Atlantic salmon (*Salmo salar*) caught in the vicinity of the River Dee (Aberdeenshire). Canadian Journal of Fisheries and Aquatic Sciences 42:1513–1521.

Mathews, S. B., and Y. Ishida. 1989. Survival, ocean growth, and ocean distribution of differentially timed releases of hatchery coho salmon (*Oncorhynchus kisutch*). Canadian Journal of Fisheries and Aquatic Sciences 46:1216–1226.

Myers, R. A., G. Mertz, and J. Bridson. 1997. Spatial scales of interannual recruitment variations of marine, anadromous, and freshwater fish. Canadian Journal of Fisheries and Aquatic Sciences 54:1400–1407.

Ogura, M., and Y. Ishida. 1992. Swimming Behavior of Coho Salmon, *Oncorhynchus kisutch*, in the open sea as determined by ultrasonic telemetry. Canadian Journal of Fisheries and Aquatic Sciences 49:453–457.

Pearcy, W. G., R. D. Brodeur, J. M. Shenker, W. W. Smoker, and Y. Endo. 1988. Food habits of Pacific salmon and steelhead trout, midwater trawl catches and oceanographic conditions in the Gulf of Alaska, 1980-1985. Bulletin of the Ocean Research Institute University of Tokyo: Biology of the Subarctic Pacific (Part 2) 26:29–78.

Pepin, P. 1991. The effect of temperature and size on development, mortality and survival rates of the pelagic early life stages of marine fishes. Canadian Journal of Fisheries and Aquatic Sciences 48:503–518.

Peterman, R. M., B. J. Pyper, M. F. Lapointe, M. D. Adkison, and C. J. Walters. 1998. Patterns of covariation in survival rates of British Columbian and Alaskan sockeye salmon (*Oncorhynchus nerka*) stocks. Canadian Journal of Fisheries and Aquatic Sciences 55:2503–2517.

Peterson, I., and J. S. Wroblewski. 1984. Mortality rate of fishes in the pelagic ecosystem. Canadian Journal of Fisheries and Aquatic Sciences 41:1117–1120.

Quinn, T. P., B. A. Terhart, and C. Groot. 1989. Migratory orientation and vertical movements of homing adult sockeye salmon, *Oncorhynchus nerka*, in coastal waters. Animal Behavior 37:587–599.

Reddin, D. G. 1985. Atlantic salmon (*Salmo salar*) on and east of the Grand Bank. Journal Northwest Atlantic Fisheries Science 6:157–164.

Reddin, D. G., and P. B. Short. 1991. Postsmolt Atlantic salmon (*Salmo salar*) in the Labrador Sea. Canadian Journal of Fisheries and Aquatic Sciences 48:2–6.

Salminen, M., S. Kuikka, and E. Erkamo. 1994. Divergence in the feeding migration of Baltic salmon (*Salmo salar* L.): the significance of smolt size. Nordic Journal of Freshwater Research 69:32–42.

St. John, M. A., J. S. MacDonald, P. J. Harrison, R. J. Beamish, and E. Choromanski. 1992. The Fraser River plume: some preliminary observations on the distribution of juvenile salmon, herring, and their prey. Fisheries Oceanography 1:153–162.

Walker, R. V., K. W. Myers, N. D. Davis, K. Y. Aydin, K. D. Friedland, H. R. Carlson, G. W. Boehlert, S. Urawa, Y. Ueno, and G. Anma. 2000. Diurnal variation in thermal environment experienced by salmonids in the North Pacific as indicated by data storage tags. Fisheries Oceanography 9:171–186.

Whalen, K. G., D. L. Parrish, and M. E. Mather. 1999. Effect of ice formation on selection of habitats and winter distribution of postyoung-of-the-year Atlantic salmon parr. Canadian Journal of Fisheries and Aquatic Sciences 56:87–96.

Windsor, M. L., and P. Hutchinson. 1994. International management of Atlantic salmon, *Salmo salar* L., by the North Atlantic Salmon Conservation Organization, 1984-1994. Fisheries Management and Ecology 1:31–44.

Part II Great Lakes Salmon

Sustaining North American Salmon: Perspectives Across Regions and Disciplines, 139–154

Chapter 7

History of the Great Lakes Salmon Fishery: A Michigan Perspective

Howard A. Tanner and Wayne H. Tody

Introduction

The greatly expanded Great Lakes sport fishery of today began in Michigan with the introduction of coho salmon *Oncorhynchus kisutch* and chinook salmon *Oncorhynchus tschawytscha* in the mid-1960s. Dr. Howard A. Tanner (who acquired the name of "father of the Michigan coho") served as chief of the Michigan Department of Natural Resources (MDNR) Fisheries Division (1964–1966) when the Great Lakes salmon fisheries began, resigning to accept a position as Director of Natural Resources at Michigan State University. Dr. Wayne Tody, Tanner's assistant chief in the fisheries division during 1964–1966, served as division chief from 1966 to 1976. Together, they oversaw the early years of the salmon-stocking program that sustains today's recreational salmon fishery in Lake Michigan. The following account provides their personal insights and perspectives into the background, rationale, expectations, and outcomes of establishing this significant fishery.

Background and Early History

The surface area of the Great Lakes covers about 95,000 square miles; the Great Lakes basin spans approximately 291,000 square miles (Figure 1). With the exception of the relatively shallow, more productive Lake Erie, the lakes are deep, cold, and well oxygenated, with low to medium productivity (see Beeton et al. 1999). The Great Lakes border two nations, with

Figure 1. The Laurentian Great Lakes (from Beeton et al. 1999).

31% of their area located in Canada and 69% in the United States. In addition to the national jurisdictions, the Great Lakes are subject to the mandates of eight states, one Canadian province, and several Native American tribes (Dochoda and Jones, Chapter 11). Economic opportunity based on the vast natural resources within the Great Lakes basin fueled development, and by the end of the 20th century, more than 36 million people lived in the basin (30 million in the United States).

The Great Lakes have experienced a growing number of ecological disturbances over the centuries related to the expansion of human activities, which accelerated following the American Revolution and the War of 1812 (see Taylor and Ferreri 1999). Activities such as forest harvest, agriculture, mining, industrialization, and the resulting urbanization exposed soils to erosion, and sediments and pollutants moved easily throughout the basin's river systems, altering those arteries so vital to the Great Lakes' ecological health. The transportation networks needed for shipping, railroads, and highways further altered aquatic habitats through the construction of dams, canals, and bridges. Dams were also needed for logging, agriculture, flood control, and industrialization. By 1940, there were 110 hydropower facilities in Michigan, impounding 90% of Michigan rivers. These dams warmed waters, obliterated critical fish habitats, and restricted the access of aquatic species to remaining habitat.

Additionally, water quality throughout the basin was affected substantially by agriculture and industrial and urban development. It is estimated that almost 60%, or nearly 35 million acres of the Great Lakes coastal wetlands, have been lost since settlement (Office of the Great Lakes 2000). Nutrient levels in the Great Lakes increased, along with the concentrations of metals such as mercury and lead. Highly toxic persistent bioaccumulative organochlorines such as polychlorinated biphenyls (PCB), chlordane, and dioxins were released into the lakes. The pesticide DDT, used extensively beginning in the mid-1940s, spread throughout the Great Lakes basin, devastating bird populations and raising concerns for human health (Dempsey 2001). Pollution of the Great Lakes and the region's wildlife resources had a profound effect on the basin's residents in the 1960s and 1970s, as they became aware of the impacts of their activities and of their own vulnerability as a result.

Fisheries

Prior to the 1960s, most areas of the Great Lakes were considered the domain of commercial fishing, which was viewed as being valuable to the state of Michigan because it provided employment and economic returns. Although the Michigan Fish Commission was established in 1873 as a result of concern that the commercial fishery had overharvested Michigan's Great Lakes fisheries, the commercial fishery continued to dominate. State fisheries employees made every effort to regulate the fishery to stop overharvesting and waste, but commercial fishers guarded their interests by organizing an effective lobby to politically thwart Michigan's fishery officials. It was successful. Around 1900, the Michigan Legislature reduced the state fish management budget severely and stipulated that not "one red cent more" would be spent on the Great Lakes fishery. The legislature gradually relented and restored the budget, but the commercial fishery continued without effective regulation of gear, catch, or effort. Commercial fishermen supplied reports of catch and gear, and anyone could buy a commercial fishing license. By 1926, the approximate peak of the Great Lakes commercial fishing industry, more than 12,000 people were employed in fishing, fish processing, and fish marketing. The total Great Lakes-wide dockside value of commercial fish remained stable through the 1960s, at approximately US$12 million, although the abundance and species composition changed markedly over this time period (Brown et al. 1999).

Though Michigan had considerable management authority over the fisheries, little was exercised because of the lack of public support for managing Great Lakes waters. Importantly, in the absence of state management of the Great Lakes, the U.S. Bureau of Commercial Fisheries exercised de facto management over the U.S. portion of the Great Lakes fishery. The mission of this bureau was to assist the commercial fishing industry, and it did. It produced data on the status of Great Lakes fisheries, conducted research, and provided marketing technology to commercial fishers. This reinforced the long-standing concept that commercial fishing should be the key value for the Great Lakes fishery.

As if the Great Lakes' instability were not already adequately challenged by alterations to habitat and an extensive commercial fishery, the introduction of exotic species precipitated additional impacts. The unintended introduction of exotic species, most notably the sea lamprey *Petromyzon marinus* and the alewife *Alosa pseudoharengus*, played a major role in modifying the Great Lakes ecosystem upon which fisheries depend (Goddard, Chapter 12).

Michigan Introduces Coho and Chinook Salmon

The stage was set for the introduction of coho salmon *Oncorhynchus kisutch* by a very unusual set of circumstances—some biological, some social, and some political—that came together in Michigan in 1962–1964. In the following paragraphs, we will briefly describe the circumstances as they existed and how they contributed to the decisions to manage the Great Lakes for sport fishing and to introduce Pacific salmon.

Ecologically speaking, the time was ripe for a successful introduction of salmon. By 1960, because of a combination of overfishing, sea lamprey predation, and possibly predation by smelt on the juveniles, the native lake trout *Salvelinus namaycush*, the most important predator in the indigenous food web, was extinct in Lakes Michigan, Ontario, and Erie, nearly so in Lake Huron, and seriously depleted in Lake Superior.

The U.S. Bureau of Commercial Fisheries, under the leadership of Dr. Vernon Applegate, was producing a workable sea lamprey control program. By the early 1960s, it was reasonable to expect that a significant degree of control over the sea lamprey could be achieved.

Meanwhile, without an effective predator and with no viable commercial market, the alewife population had erupted to extraordinary abundance levels in Lakes Michigan, Huron, and Ontario. All other fish populations were at low levels. Every summer, dead alewives piled up on the beaches and frequently plugged the water intakes of Chicago. The peak of these die-offs in Lake Michigan produced what has been described as "a pile of stinking alewives one foot high and 300 mi long." Beach-oriented tourism came to a halt, resulting in serious economic impacts on lakeside communities. From another perspective, however, this glut of alewives was clearly a very large food supply upon which to build a sport fishery for salmon and trout.

Simply stated, we could try to convert a nuisance into a resource—a very large resource!

More than 60 million people live within one day's drive of the Great Lakes. By the early 1960s, with substantial leisure time, expendable incomes, and good mobility, there was clearly a large demand for fishing opportunities. The paradox was that this large sport angling population lived on the shores of the world's largest freshwater system, but this array of lakes and connecting waters offered very little fishing opportunity. With the near extinction of the lake trout and very low levels of other native species, most commercial fishers were no longer fishing. Equally important, they were nearly politically bankrupt. The need to reallocate from commercial to sport fishing was becoming abundantly clear and possible.

In 1963, Michigan adopted a new constitution that included a proviso that "natural resources shall be preeminent." For the first time, the Department of Conservation (soon to become the Department of Natural Resources) was headed by a professionally trained biologist, Dr. Ralph MacMullan. He hired Dr. Tanner in September of 1964 as chief of fisheries. At the time, Dr. Tanner was working for the Colorado Department of Fish and Game as chief of fisheries research. During his 12 years in Colorado, he had gained some experience with both kokanee *O. nerka* and coho salmon in freshwater environments. As a member of an organization then known as the Pacific Coast Fishery Biologists, he had existing contacts with salmon biologists from the Pacific Coast states. These contacts were to become extremely important.

Although prior to 1964 little attempt had been made to manage the fisheries of Michigan's Great Lakes waters, there was a major fisheries division hatchery program to raise catchable size brook trout *Salvelinus fontinalis*, brown trout *Salmo trutta*, and rainbow trout *O. mykiss*. Usually, these were released in streams "at the rod" to be caught in periods of peak fishing activity. Therefore, Michigan hatcheries had the capability to rear potential predator species for release in the Great Lakes, although capacity was limited.

During this time period in the states of Oregon and Washington, there was for the first time in many years a surplus of coho salmon and chinook salmon *O. tshawytscha* returning to the hatcheries. This surplus may well have been produced by the creation of the Oregon moist pellet. This feed came into general hatchery use in the late 1950s and is credited with producing a more vigorous smolt having better rates of survival. Until these surpluses, neither Washington nor Oregon had been willing to fill requests for salmon eggs from outside the Columbia River watershed.

These are some of the more important elements that made up the situation that existed immediately prior to policy decisions made in the fall of 1964. We in the Michigan Fisheries Division proposed that, for the first time, Michigan would manage those portions of the Great Lakes that lie within its boundaries. Further, and equally important, we proposed to manage the resource by replacing commercial fishing with sport fishing as the key value. Our director, Dr. MacMullan, and his administrative staff supported our proposal, and the Conservation Commission, the department's policy-making board, approved.

These changes in policy and management goals were very large indeed. Previously, our fisheries activities (i.e., management, hatchery production, lake and stream improvement, and research) had targeted only inland lakes and streams. Though these waters are extensive and important, they make up only about 3% of the water area of Michigan.

Now came the serious and difficult questions of *how* to manage the Great Lakes as a sport fishery. The only feasible option was to insert one or more top predators into the food web. Clearly, the physical, chemical, and biological nature of the Great Lakes dictated the top choice of predators to be trout or salmon. We quickly concluded that the native lake trout could not be the principal species employed in a program designed to replace commercial fishing with sport fishing. In precedent and by law, the lake trout was clearly established as a commercial species. Though the commercial fishing interests had been unable to block our declared intent to manage for sport fishing, their opposition was not to be ignored.

Furthermore, the U.S. Bureau of Commercial Fisheries, with headquarters for the Great Lakes in Ann Arbor, Michigan, had been assigned the roles of controlling sea lamprey and reestablishing lake trout populations. This federal agency had built and would operate the Jordan River hatchery (in Michigan) to raise lake trout. It actively opposed our shift in allocation from commercial to sport fishing. These considerations combined to make clear that the lake trout was not the species for the Fisheries Division to promote as the foundation for a sport fishery. Additionally, we could reasonably expect that the lake trout would be stocked in substantial numbers from the federal hatchery on the Jordan River in an attempt to facilitate its rehabilitation, thereby providing redundancy in the predator complex of the Great Lakes.

Our rejection of the lake trout was based on other assumptions, as well. We concluded that the lake trout was not a very interesting fish to catch. Remember that this was at a time when there were no downriggers and no fish finders. Traditional lake trout fishing techniques were usually to troll with a long metal line and heavy weights. Under those circumstances, the lake trout was not an exciting fish to catch. (As an aside, even today, in the opinion of most sport anglers, the lake trout ranks below coho salmon, chinook salmon, and steelhead *O. mykiss* as a sport fish.) Because those who would fish the big open waters have to be willing to purchase bigger and more expensive boats and gear and to invest a substantial share of their leisure time, it was our opinion that a truly exciting fishery was essential.

We started, in 1964, with a program to introduce kokanee salmon with eggs from Colorado, where Dr. Tanner had previously worked and where eggs were available. A program of stocking kokanee in a limited number of inland lakes was developed from eggs to be collected annually from Higgins Lake, a large inland lake in northern Michigan. However, the kokanee program was discontinued after a few years because of a lack of interest.

Then, in early October 1964, came the exciting and unexpected news that coho eggs might be available from Oregon and Washington. A quick review of the coho—its needs, attributes, and reputation as a sport fish—was enough to convince us that this was a species we wanted for a Great Lakes sport fishery. Most importantly, we knew that, on the west coast, it was a highly desirable sport fish and was well understood from a hatchery-rearing perspective. Additionally, in the Pacific Ocean, the coho feed on prey species not too different from alewife. Hence, we quickly decided to introduce coho salmon (Tody and Tanner 1966).

With the approval of our department's administration and commission, we requested a million coho eggs from Oregon and another million from Washington. Following an exchange of letters and phone calls, both states initially agreed to honor our request provided we follow their proven guidelines for best results. Some elements of their instructions were feed only Oregon moist pellets; rear the coho in a hatchery on a stream where we desired the fish to return for egg take and/or spawning; rear them to smolts (18 months old) before planting; and plant no fewer than 200,000 at any one location. We eagerly accepted their advice. Before the eggs could be sent, opposition developed in the state of Washington, and it was unable to deliver any coho eggs. Coho eggs were received from Washington and Alaska in subsequent years.

A total of 1 million eyed coho eggs were received from Oregon in late December 1964 and early January 1965. The eggs and subsequent fry were given the greatest attention in various Michigan hatcheries and later moved to the dirt-lined raceways of a rearing station located on the Platte River, a tributary of Lake Michigan, to be imprinted prior to release. The official release of the first coho smolts was on 2 April 1966 with a "golden bucket" ceremony near the site of our new fish hatchery on the Platte River (upstream about 10 mi from the town of Honor, Michigan). For the sake of historical completeness, we will mention that there were newspaper reports that some small number of coho smolts were released into Bear Creek, a tributary of the Big Manistee River, in late March, a few days prior to the official 2 April release. We have no personal recollection of this. Over the next few days, about 264,000 coho smolts were released into the Platte River and 394,000 into Bear Creek. State Sen. Joe Mack from Michigan's Upper Peninsula (UP) was chairman of the State Senate Appropriations Committee. To receive approval of a US$500,000 budget increase to feed the young coho in our hatcheries, we had to agree that some coho would be released into a tributary of Lake Superior. In May 1966, 160,000 coho smolts were released into the Big Huron River, a few miles northeast of the UP town of L'Anse. The salmon fisheries that would develop throughout the Great Lakes had begun.

The first coho jack salmon run in Bear Creek and the Platte River began in September 1966. More than 3,700 were caught in weirs on these streams. Anglers took an estimated 1,500 in addition. The jacks were generally 2- to 4-pound fish, with the largest weighing more than 7 pounds. Of significant importance was that 32 females were stripped of 45,000 eggs, which produced 22,000 fry. Observations indicated that natural spawning probably occurred. During this period of time, we invited the fisheries chiefs of Oregon and Washington to view the run of jack coho. Given their experience on the west coast, where the jack run of any year-class will accurately reflect the strength of the subsequent year's mature run, they told us to get ready for a very large run of big mature coho in the following year.

A favorable sequence of events brought about great public acceptance and enthusiasm for our Great Lakes salmon fishery. As described earlier, the first coho smolts were planted in the spring of 1966. The alewife die-off in Lake Michigan, in June of that year, was extensive. There was a great public outcry to solve the alewife problem. In the fall of 1966, there was a run of large jack coho weighing 2 to 7 pounds. The few hundred caught by anglers that fall raised the level of excitement and expectation. In the early summer of 1967, an even worse alewife die-off occurred. Later that summer, the first year-class of adult coho created fishing excitement that was true pandemonium. The news media were full of positive news regarding the spectacular Great Lakes salmon fishery that had been created, often referring to this event as creating coho madness.

During the summer of 1968, there were no dead alewife problems on the Lake Michigan beaches, and the public immediately concluded that the coho had controlled the alewives. What had really happened was that, by 1967, the alewife population had grown to exceed its food supply and had collapsed following the stress of spawning. Since the public "knew" the salmon had eliminated the alewife problem, the only thing we could do was accept the credit that the public insisted was due us. From that point on, the public and the legislature were enthusiastically in support of the Michigan salmon program.

As an aside, we frequently hear the Michigan salmon program described as designed to control alewives. Our intent was much more important than controlling alewives. To build the attractive sport fishery that we envisioned, we clearly needed to introduce attractive sport fish (i.e., one or more species of trout or salmon). We needed to choose species that could be expected to eat the most abundant forage species (i.e., the alewife). However, to describe our program as one designed to control alewife, which, in the process, happened to produce a great sport fishing opportunity, is demeaning. Again, and for emphasis, our goal was to turn the alewife nuisance into a very large, important, and valuable public resource.

With the success of the coho jack run, we recommended to the Natural Resources Commission (the renamed Conservation Commission) that we proceed immediately to introduce fall run chinook salmon to increase the diversity of the predator base and angling experience. By this time, the political winds had changed in the West, allowing for Michigan to receive eyed eggs of salmon populations more readily than before. The commission quickly approved our recommendation.

In this effort, Dr. Tody, who was now the chief of the MNDR Fisheries Division, developed a close working relationship with Dr. Loren Donaldson, professor at the University of Washington, who had developed a three-year strain of chinook salmon and, most importantly, endorsed our planned introductions. Cliff Millenbach, chief of the Sport Fish Division of the Washington Department of Fish and Game, volunteered as our western coordinator for planting stock, and Bud Ellis and Richard Noble, hatchery chiefs of Washington's Department of Fisheries, provided the necessary eggs—1 million eggs a year for three years. These were generously provided from stream runs most similar to Michigan's. The first plants of six-month-old chinook salmon smolts were made in the spring of 1967.

The 1967 run of coho salmon electrified not only sport fishermen, but also the general population of Michigan, and it was hailed as top news from coast to coast of the United States. Anglers took about 40,000 adult salmon averaging 12.1 lb. Our best estimate of survival to harvest was an astounding 35%. In addition, more than 1,500,000 lb was taken at Michigan Department of Natural Resource (MDNR) weirs, of which 785,000 lb were sold for human consumption before sale was halted because of contamination by DDT. This created for us a new management problem: how to dispose of millions of pounds of pesticide-contaminated salmon. At one time during this period, Michigan Gov. William Milliken asked us to consider stopping all planting of trout and salmon to avoid problems of contamination leading to human health problems. Our review quickly showed that the constituencies of those enjoying the new fishery supported environmental cleanup, and their advocacy was too valuable to lose. Therefore, we looked for other ways to protect human health while maintaining and enhancing our salmon and trout fisheries. The main thrust here was to reduce or eliminate the contaminants that bioaccumulated in the flesh of Great Lakes fish. Director MacMullan played a major role nationally in achieving the ban on DDT.

In addition to the spectacular returns of coho salmon in 1967, we were able to take 8 million eggs to continue the hatchery program. Of these, 1,100,000 were divided between Wisconsin (300,000), Ohio (200,000), Pennsylvania (300,000), New York (100,000), and the province of Ontario (200,000) to initiate the program throughout the Great Lakes. Thereafter, we continued to share eggs and worked closely with the other Great Lakes states and the province of Ontario to further the salmon program.

Establishing a population of large, highly sought after sport fishes created additional management needs as the public rushed in to utilize this new resource. Maintenance of and improvement to the fishery demanded a vast array of new facilities, both private and public. Special credit is due to Keith Wilson, chief of MDNR's Waterways Division and the Michigan Waterways Commission, for rapid development of access sites, marinas, and harbors of refuge for small craft on all of Michigan's Great Lakes. Today, even though Michigan now has more registered boats than any other state, the long lines waiting to launch at the limited facilities are gone. Boating access and boat service of all kinds are excellent. Weather warnings are more accurate and timely. Anglers are more skilled and experienced in dealing with the large, sometimes dangerous waters of the Great Lakes. The fishery has become much more sophisticated and anglers more knowledgeable. A charter boat fleet is present and busy. Boats are larger and more seaworthy, and many have been designed specifically to accommodate sport fishing. Gear has really become high-tech. Electric downriggers, radar, global positioning systems (GPS), navigational aids, fish finders, temperature probes, dipsy divers, and endless numbers of lure types are commonly present on the boats of serious Great Lakes anglers. Trolling with artificial lures is the primary technique employed when fishing from boats in the open lakes. Live, frozen, or preserved bait is rarely used. There is also a large fishing effort from piers, in the surf, and in the tributary rivers. In these modes, fresh spawn is commonly used, as well as artificial lures.

As the Great Lakes sport fisheries developed, we recognized that existing commercial fisheries posed a severe threat to the program. Wisconsin trawlers were taking millions of pounds of alewife for industrial fish meal processing plants at a price of only one to two cents a pound. The U.S. Bureau of Commercial Fisheries regional fish chief, Dr. Fent Carbine, promoted federal assistance for building additional meal plants in Michigan and aggressively renounced our predator approach to utilize the alewife as "forage." In the end, this controversy became political, and under pressure from Michigan elected officials, the administrative regional office of the bureau was removed from Michigan, as was its fishery technology branch. All other functions remained, including research and the bureau's Ann Arbor facilities that today are the Great Lakes Science Center for the Biological Resources Division of the U.S. Geological Service.

Michigan's Fisheries Division field personnel found, in an intensive 1968 survey, that commercial gill nets were killing large numbers of juvenile lake trout in northern Lake Michigan waters incidental to fishing for lake whitefish *Coregonus clupeaformis*. It became apparent that these nets posed a major threat to salmon and trout stocks as the salmonid propagation program expanded; thus, they had to be addressed. As a result, a restructuring of Michigan's commercial fishery, which had a 150-year history of aggressive fishing, was mandated. A primary premise for this restructuring was that sport and commercial fishing had to be made complementary if we were going to be successful in the rehabilitation of Great Lakes fish populations. With a policy of sport fishing being dominant over commercial fishing (which we then believed and still believe represents the best allocation of the fishery resources), a political battle was unavoidable. Sport fishermen and Great Lakes port cities organized powerful political lobbies to promote their interests. Admittedly, their interests influenced our biases. In the end, the state's fisheries regulations were rewritten, setting species, seasons, size, and catch limits for sport fishing. Large-mesh gill nets for commercial fishing were essentially banned.

Displaced commercial fishermen were reimbursed by the state's general fund for their boats and gear and were encouraged to convert to selective gear (see Lupi and Jester, Chapter 10). Michigan Fisheries Division expert Walter Crowe developed plans for limiting entry to the commercial fishery. These were adopted. Next, we recommended a plan of zone management for the combined fishery. This included sport, commercial, and restricted fishing rehabilitation zones. Plans were to gradually develop a franchise type of commercial license, wherein specific gear, catch, and seasons would be negotiated with individual fishermen. This very desirable end product was never brought to fruition because it lacked commercial fishermen support and became stalled because of new legal battles over federally-sponsored Native American treaty fishing rights.

Progress from 1968 to the Present

In the 37 years since Michigan declared its intent to shift allocation from commercial to sport fishing and to introduce salmon, many changes have occurred. Across the basin, the dockside value of commercial landings has risen sharply since the 1970s—from an estimated US$12 million to $52 million in 1990. The number of licensed commercial fishers in the Great Lakes has decreased from approximately 10,000 in 1920 to 2,200 in 1990

(Brown et al. 1999). It is not possible to compare precisely commercial and recreational fisheries values with the data available, but an indication of relative values may be made. In 1990, the basin-wide dockside value of commercial fish was US$52 million and was estimated to have a regional impact of $200 million. In 1991, U.S. recreational fisheries in the Great Lakes alone accounted for US$1.3 billion in direct angler expenditures, with an estimated total economic impact of $2 billion to $4 billion (Bence and Smith 1999). Many factors have contributed, but there is little question that there has been a basin-wide shift in values favoring sport fishing.

One result of the shift of the Great Lakes to sport fishing has been a growing advocacy group for the integrity and well-being of the Great Lakes. The popular Great Lakes sport fisheries (2 million anglers in 1996) have produced many hundreds of thousands of people who now know and appreciate the importance of the Great Lakes. These individuals represent a demonstrable voting constituency that supports stringent environmental regulations and sound scientific management of the Great Lakes. Additionally, the public's environmental awareness has increased significantly since the 1960s, resulting in the adoption of legislation at the state/province and federal levels to curb the loss of valuable habitats and bring pollution under control. As a result, many critical habitats are now protected, and efforts are underway to restore habitats that have been impaired. Much has been accomplished, but much remains to be done—particularly related to how our land use decisions affect the quality of the Great Lakes ecosystem.

Natural reproduction of fish is one indicator of improved habitat. Michigan now estimates that nearly 40% of the chinook salmon entering the fishery are from natural reproduction. Significant reproduction of coho salmon, steelhead, and brown trout is occurring in all accessible sections of rivers with good water quality. A dramatic improvement in natural reproduction was achieved when hydroelectric facilities were required to operate on a "flow of the river" regimen, rather than fluctuating the rivers by operating only at times of peak power demand. Many millions of lake trout have been stocked during the past 40 years in an effort to establish self-sustaining populations. This goal has been achieved or is clearly possible in Lake Superior and northern Lake Huron. However, it is our opinion that basic changes have occured in Lake Michigan and southern Lake Huron that may permanently preclude significant lake trout reproduction. We believe competition and predation on newly hatched lake trout by alewife and smelt are probable causes. It is our opinion that lake trout should continue to be included in the array of salmon and trout being stocked, but that we should recognize that, in Lake Michigan, southern Lake Huron, and probably Lakes Erie and Ontario, little natural reproduction can be expected.

Overall, nutrient levels have decreased in the Great Lakes, principally as a product of better sewage treatment and a ban on phosphates in detergents. Michigan was the first state to ban phosphates in detergents. Dissolved oxygen levels have increased as well. During the past 20 years, there has been a general decrease in concentrations of many toxics in the Great Lakes as a result of regulation of direct water discharges. Perhaps few remember the days when concentrations of DDT in lake trout frequently exceeded 20 ppm. There are still some warnings about the human consumption of larger trout and salmon, but concentrations of toxics found in those fish are much lower, and more fish of larger size are rated as safe for consumption.

Exotics are considered second only to loss of physical habitat in terms of severe impacts on native species, and they are a major cause of the continuing loss of biodiversity in the Great Lakes (see Mills et al. 1993). More than 144 new species have been intentionally or accidentally introduced into the Great Lakes basin since the 1800s. The greatest number of exotic species introduced coincided with the 1959 expansion of the St. Lawrence Seaway, which allowed the passage of more and larger ships into the Great Lakes. Exotic species have also found their way into the system through commercial aquaculture, the aquaria trade, and the bait industry. Many more exotics will be introduced if we do not provide an adequate protection program for the Great Lakes.

Fisheries management decision making has become much more integrative and multidisciplinary across the Great Lakes. In part, this has been a result of the efforts of the Great Lakes Fishery Commission, which has provided a forum for a more integrative and collaborative approach to fisheries management in the Great Lakes basin (see Dochoda and Jones, Chapter 11). There exists today a clear recognition among management agencies that the pursuit of individual interests is best achieved by collaborating with those other agencies with a shared interest in determining the future of a resource held in common. The focus today also moves toward management of the Great Lakes basin as an ecosystem—an approach that has become much more holistic in recognition of the fact that the Great Lakes basin habitat—physical, chemical, and biological—defines the boundaries of potential fisheries opportunities. Awareness is growing that an adaptive management approach to the future must be taken because the Great Lakes basin has changed and will continue to change into the foreseeable future.

Today's Fishery

The sport fishery in U.S. waters of the Great Lakes basin, considered insignificant in 1960, was replaced by a sport fishery that—conservatively estimated for 1996—involved 2 million anglers and 20 million days of recreation. These anglers took 17 million trips and spent US$1.4 billion on trips and equipment (U.S. Department of the Interior, Fish and Wildlife Service, and U.S. Department of Commerce, Bureau of the Census 1997). In addition to the catch by sport anglers, there is a notable commercial harvest of salmon and trout by fishermen with treaty rights.

We would be remiss if we did not recognize the web of professional and public support that has developed in support of the Great Lakes fishery programs. We realize that to name some is to risk the unintentional overlooking of others. So be it. As mentioned earlier, the Great Lakes Fishery Commission has been there since the 1950s, but its more recent success in serving as a forum for functional cooperation between all aspects of interstate and international research and management programs is an essential element for the future well-being of the resource. The Sea Grant College Program provides another set of essential services. Its service as a conveyor of useful information to all user groups of the Great Lakes fisheries is a very durable function that has been greatly appreciated by the public and the agencies. Additionally, colleges and universities are now more effectively linked to management agencies, providing much needed research and education regarding our Great Lakes and its fisheries. One such example is the Partnership in Ecosystem Research and Management unit at Michigan State University, which was organized through the leadership of the MDNR and the MSU Department of Fisheries and Wildlife. Finally, numerous public interest groups of anglers and boaters, citizen advisory panels, and elements of the boating and tackle industry all add support and strength to those responsible for policy and action programs.

Summary and Conclusions

The Great Lakes of the early 1960s were a dismal sight. Habitat—ravaged physically, chemically, and biologically—reflected the historically passive efforts of natural resource managers to maintain and nurture fishery and wildlife values.

The introduction of salmon into the Great Lakes has been a remarkable undertaking that has yielded immense benefits. It created a new, valuable, and high-profile recreational fishery. The success of the early salmon fishery served as a catalyst to encourage resource managers throughout the basin to accept the challenge of proactively managing the Great Lakes to restore and develop fishery and wildlife values.

An international array of professional fishery biologists working together during the past four decades can take pride in today's Great Lakes salmon and trout fisheries. State, tribal, federal, and international programs have, with reasonably good cooperation, produced an array of salmon and trout species that are the best means available to stabilize the inherently unstable biology of the Great Lakes. While doing so, they created an enormously valuable sport fishing industry. An international constituency of concerned and better informed citizens supports sound professional management, and existing and new problems will continue to require the best efforts of our profession. We are confident that this fishery can be sustained and improved.

References

Beeton, A. M., C. E. Sellinger, and D. F. Reid. 1999. An introduction to the Laurentian Great Lakes ecosystem. Chapter 1 *in* W. W. Taylor and C. P. Ferreri, editors. Great Lakes fisheries policy and management. Michigan State University Press, East Lansing, Michigan.

Bence, J. R., and K. D. Smith. 1999. An overview of recreational fisheries in the Great Lakes. Chapter 10 *in* W. W. Taylor and C. P. Ferreri, editors. Great Lakes fisheries policy and management. Michigan State University Press, East Lansing, Michigan.

Brown, R. W., M. Ebener, and T. Gorenflo. 1999. Great Lakes commercial fisheries: historical overview and prognosis for the future. Chapter 11 *in* W. W. Taylor and C. P. Ferreri, editors. Great Lakes fisheries policy and management. Michigan State University Press, East Lansing, Michigan.

Dempsey, D. 2001. Ruin and recovery: Michigan's rise as a conservation leader. The University of Michigan Press, Ann Arbor, Michigan.

Mills, E. L., J. H. Leach, J. T. Carlton, and C. L. Secor. 1993. Exotic species in the Great Lakes: a history of biotic crises and anthropogenic introductions. Journal of Great Lakes Research 19(1):1–54.

Office of the Great Lakes. 2000. Great Lakes trends: into the new millennium. Michigan Department of Environmental Quality, Lansing, Michigan.

Taylor, W. W., and C. P. Ferreri, editors. 1999. Great Lakes fisheries policy and management. Michigan State University Press, East Lansing, Michigan.

Tody, W. H., and H. A. Tanner. 1966. Coho salmon for the Great Lakes. Fish Management Report No. 1. Department of Conservation, Lansing, Michigan.

U.S. Department of the Interior, Fish and Wildlife Service, and U.S. Department of Commerce, Bureau of the Census. 1997. 1996 National Survey of Fishing, Hunting, and Wildlife-Associated Recreation.

Sustaining North American Salmon: Perspectives Across Regions and Disciplines, 155–180

Chapter 8

Ecological Factors Affecting the Sustainability of Chinook and Coho Salmon Populations in the Great Lakes, Especially Lake Michigan

Michael J. Hansen and Mark E. Holey

Introduction

Pacific salmon are exotic species to the Great Lakes, and their abundance has been largely driven by stocking. This leads to a different perspective on their ecology than is true elsewhere in their range (Kocik and Jones 1999). Further, salmon stocking programs have tended to focus on chinook salmon *Oncorhynchus tshawytscha* and coho salmon *Oncorhynchus kisutch*. These have led to fisheries largely dependent on these two species. Finally, salmon stocking in the Great Lakes was begun in Lake Michigan, and more salmon have been stocked in Lake Michigan than elsewhere in the Great Lakes basin. As a result, much of the study of salmon ecology has been pursued in Lake Michigan, and more of the salmon story has unfolded there than in the other lakes. For these reasons, we focus our description of salmon ecology (survival, growth, and reproduction) on chinook and coho salmon in Lake Michigan, with comparisons to the other lakes where information is available. We believe that the ecology of chinook and coho salmon in Lake Michigan provides a reasonable example for the rest of the Great Lakes basin that nicely complements the recent review by Kocik and Jones (1999).

History

The species composition of fish communities in the Great Lakes was relatively stable prior to the period of European settlement (Smith 1968). The native species complex of the Great Lakes was composed of top predators, lake trout *Salvelinus namaycush* and burbot *Lota lota*, and their prey, deepwater ciscos *Coregonus johannae*, which had evolved relatively quickly into morphologically diverse species flocks following the Wisconsin glaciation (Smith 1968). Following settlement, until 1940, the native species complex was changed by the replacement of large, old individuals by small, young individuals of the same species (Smith 1968). Stream spawning runs of lake trout and most other species were likely eliminated prior to the period of fishery records (Smith 1968).

Native fish communities in the Great Lakes changed quickly and dramatically in the mid-1900s (Smith 1968). In response to excessive fishery exploitation, lake trout stocks collapsed sequentially during 1935–1947 in Lake Huron (a decade), 1943–1949 in Lake Michigan (a half decade), and 1950–1960 in Lake Superior (a decade) (Hansen 1999). The top predator in each of the upper Great Lakes had, therefore, been removed by the midpoint of the 20th century (Smith 1968). Sea lamprey *Petromyzon marinus*, which invaded the upper Great Lakes as lake trout stocks were collapsing, turned to large deepwater ciscos as their favored prey, which in turn enabled the smallest cisco, the bloater *Coregonus hoyi*, to increase in abundance (Smith 1968).

In the absence of top predators, alewife *Alosa pseudoharengus* invaded the upper Great Lakes, and in lakes Michigan and Huron, it exploded in abundance in only a few years, drastically altering the remaining native species complex (Smith 1970). Alewife, a schooling spring-spawning pelagic planktivore, dominated fish biomass and appeared able to suppress most native species, such as bloater and yellow perch *Perca flavescens*, through predation on pelagic larvae (Smith 1970; Crowder 1980; Jude and Tesar 1985; Eck and Wells 1987). Fish species in the Great Lakes had evolved in the absence of a schooling pelagic planktivore that spawned in spring, such as the alewife, so they were ill equipped to survive in the face of such a species. Consequently, all remaining native species in lakes Huron and Michigan declined to low levels by the 1960s. By 1967, alewife composed 90% of fish biomass in Lake Michigan (Brown 1972).

Part of the rationale for introducing coho salmon *Oncorhynchus kisutch* into the Great Lakes in the 1960s was to control abundant alewife (Tody and Tanner 1966; Tanner and Tody, Chapter 7):

> It can be speculated that the Great Lakes may offer a unique environment to the coho salmon. These lakes are sufficiently large

> to offer a sea to the anadromous fish except for the dissolved mineral content. The abundance of alewife, chubs, and smelt should offer the adult fish a sufficient food supply for rapid growth.

Though it has been mostly forgotten, one of the original goals stated by Tody and Tanner (1966; Tanner and Tody, Chapter 7) was to establish Pacific salmon as self-sustaining species in the Great Lakes:

> The coho may reproduce readily in fresh water. The lack of vertebrate predators in the Great Lakes as compared with the Pacific Ocean may lead to a relatively high rate of survival. Therefore, it is certainly possible that success may be achieved in establishing the coho with just the first life cycle.

Stocking

Though salmon had been stocked as early as the 1800s, survival of these early plantings was low and failed to sustain fisheries (Parsons 1973; Kocik and Jones 1999). In contrast, survival of stocked coho salmon and chinook salmon in the Great Lakes was very high in the 1960s and fueled unprecedented growth of a sport fishery, especially in Lake Michigan (Kocik and Jones 1999). In 1972, Stanford Smith noted that ecosystem restoration must focus on improved land use, elimination of pollutants, and fish population restoration (Smith 1972):

> Measures to reduce alewives should be intensified by introduction of greater numbers of large predators such as lake trout and salmon. The most critical requirement while alewife populations are being reduced is a concurrent restoration of the small native forage species. This transition will require several decades and will require careful measurement and close regulation of the kinds and amounts of fish introduced or removed from the lakes.

The Michigan Department of Natural Resources introduced coho and chinook salmon into the Great Lakes Basin in Lake Michigan in 1966 and 1967, respectively (Keller et al. 1990). The first coho salmon were from fall runs in the lower Columbia and Cascade rivers, Oregon, and the Toutle River, Washington, and the first chinook salmon were from fall runs in the lower Columbia River (Tule strain), Oregon, and the Green River, Washington (Keller et al. 1990). Within three years, the Michigan Department of Natural Resources was supplying eggs of both coho and chinook salmon to other Great Lakes fishery management agencies, from which some developed their own stocking programs. Introductions of other strains of coho and chinook salmon have been attempted, but most hatchery production of both species in the Great Lakes basin stems from the original strains obtained by Michigan.

Chinook salmon stocking has exceeded basin-wide stockings of all other salmon since the 1960s. For example, chinook salmon stockings averaged 53% of all salmon stockings (chinook salmon, coho salmon, and steelhead *Oncorhynchus mykiss*) in all of the Great Lakes during 1983–1993 (Figure 1). Stocking programs in the Great Lakes quickly evolved to take advantage of the fact that chinook salmon smolt at a smaller size and younger age than coho salmon. Production systems could, therefore, produce many more chinook salmon than other species. Chinook salmon also reached larger sizes than other trout and salmon species, so anglers preferred them. Consequently, chinook salmon stocking grew steadily throughout the 1970s and 1980s, whereas coho salmon stocking was relatively stable after 1970.

Stocking of salmon has been substantially greater in Lake Michigan than in any other Great Lake since the 1960s. For example, average stock-

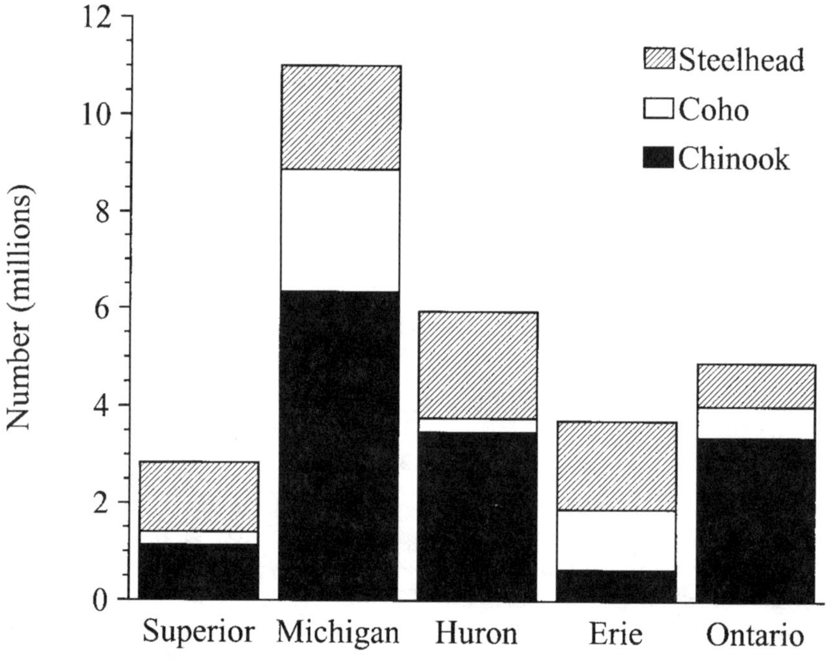

Figure 1. Average numbers of chinook salmon, coho salmon, and steelhead stocked in each Great Lake during 1983–1993 (data from Kocik and Jones 1999).

ings of salmon in Lake Michigan during 1983–1993 (11 million) were 1.8 to 3.9 times greater than in any of the other Great Lakes (Figure 1). However, stocking rates of salmon per unit of surface area were greater in Lake Ontario (250/km^2) than in Lake Michigan (190/km^2) during 1983–1993 (Kocik and Jones 1999). Stocking of salmon has been greatest in Lake Michigan, Lake Huron, and Lake Ontario, largely because of the predominance of alewife as a pelagic prey species (O'Gorman and Stewart 1999) and because nearby human population centers fueled recreational fishing effort (Bence and Smith 1999). Lake Superior was generally too cold to support high densities of alewife (O'Gorman and Stewart 1999), and Lake Erie already supported a highly productive coolwater fishery for walleye *Stizostedion vitreum* and yellow perch (Koonce et al. 1999).

Numbers of chinook salmon stocked in Lake Michigan greatly exceeded total numbers of coho salmon and steelhead stocked since 1969 because a sport fishery for chinook salmon quickly developed to utilize this species. For example, in the Wisconsin waters of Lake Michigan, harvest and stocking of chinook salmon exceeded harvest and stocking of both coho salmon and steelhead during 1980–1987 (Figure 2). Further, the ratio of annual harvest to the number stocked was much higher for chinook salmon than for either coho salmon or steelhead. The fishery quickly became known for the quality of chinook salmon fishing. Many factors were likely involved in the strength of interest in chinook salmon fishing, but the fighting ability and larger average size of chinook salmon, compared with the other species, were undoubtedly among the most important qualities that gained angler interest.

Abundance

Harvest rates of chinook salmon in Lake Michigan grew in concert with angling effort throughout the 1970s and early 1980s, while harvest rates of coho salmon varied, without trend, during the same period. For example, in Wisconsin waters of Lake Michigan, chinook salmon harvest rates increased directly with angling effort, while coho salmon harvest rates failed to increase (Figure 3). This suggests that angling effort in Lake Michigan increased with salmon abundance, which was being driven largely by chinook salmon stocking, rather than coho salmon stocking. As a result, harvest rates of chinook salmon in Lake Michigan were generally higher than for other species of trout and salmon by the mid-1970s, which suggests that chinook salmon density was higher than that of other species of trout and salmon at that time.

By 1985, the harvest of chinook salmon from Wisconsin waters of Lake Michigan exceeded that from any of the other Great Lakes (Figure 4). Fur-

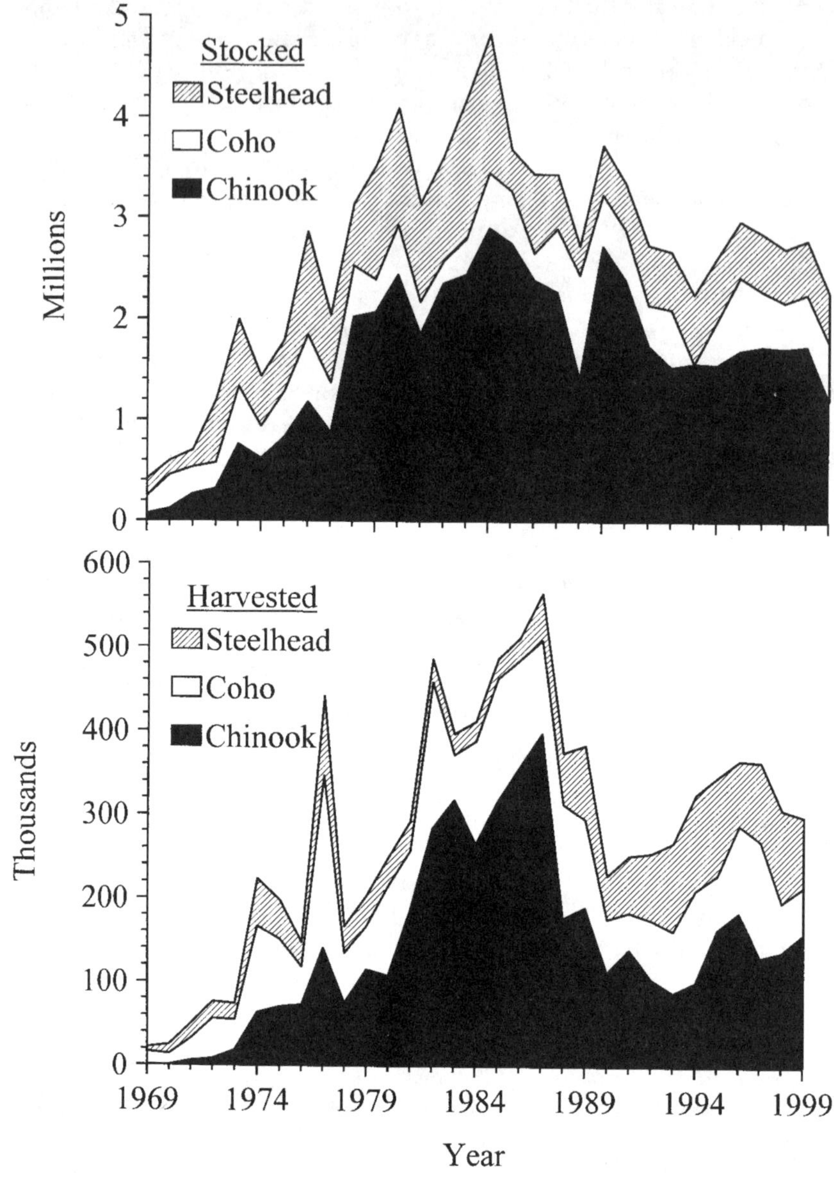

Figure 2. Numbers of chinook salmon, coho salmon, and other salmonines stocked (upper panel) and harvested (lower panel) in the Wisconsin waters of Lake Michigan during 1969–1999 (updated from Hansen et al. 1990, 1991).

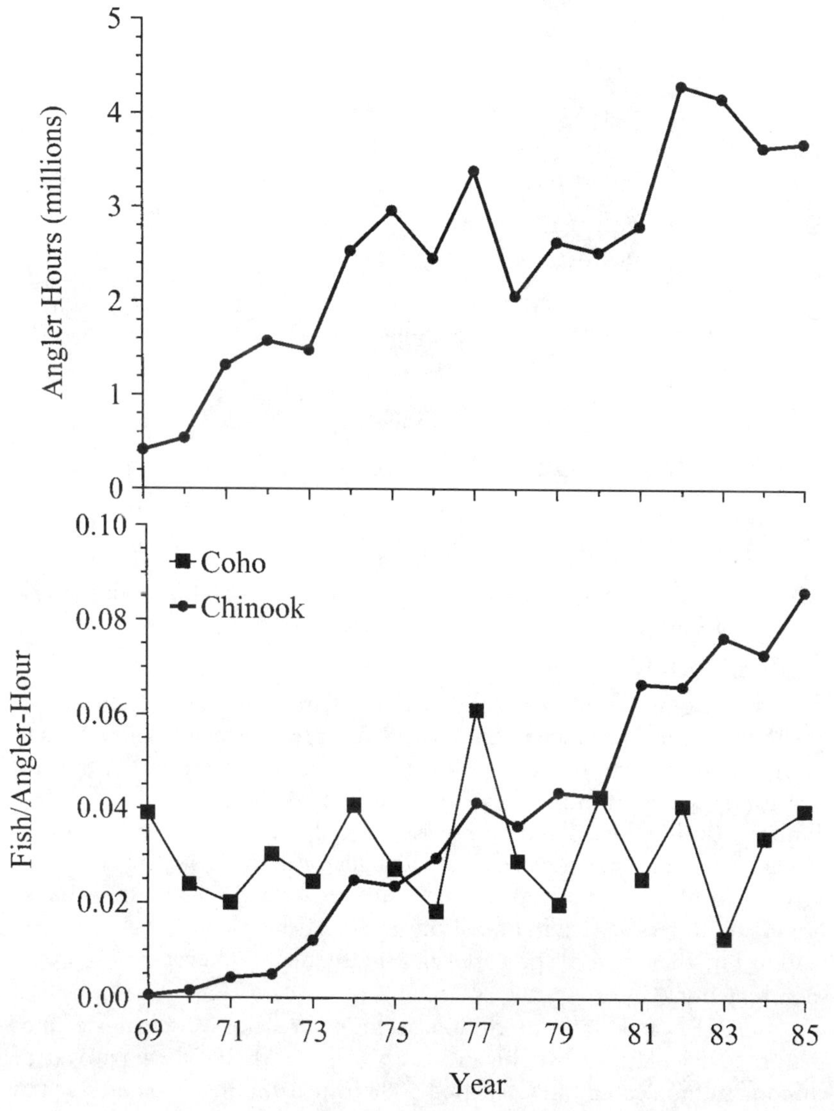

Figure 3. Angler hours (upper panel) and harvest rates (lower panel) for chinook and coho salmon in the Wisconsin waters of Lake Michigan during 1969–1985 (from Hansen et al. 1990, 1991).

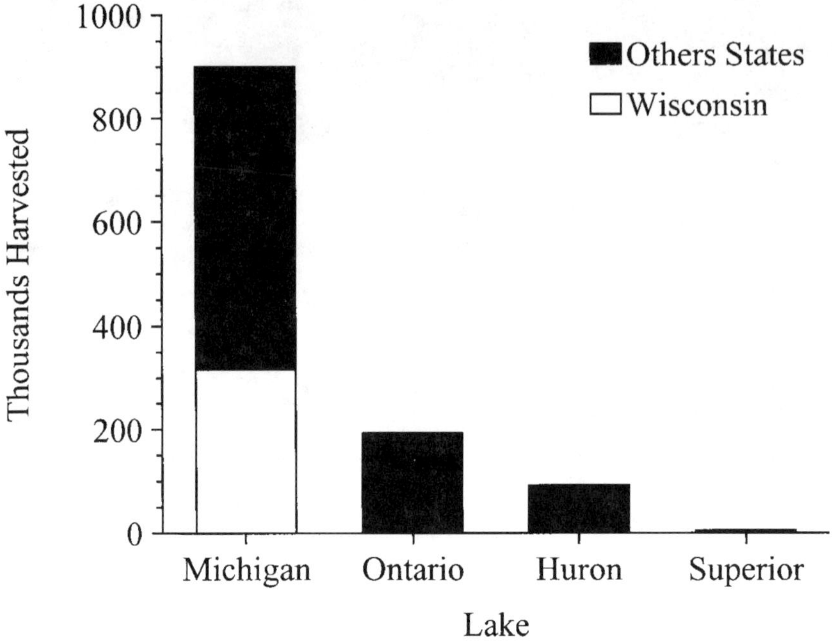

Figure 4. Numbers of chinook salmon harvested from lakes Michigan, Ontario, Huron, and Superior in 1985 (from Hansen et al. 1991).

ther, the magnitude of the chinook salmon harvest in Lake Michigan was beyond anything produced in the lake's recorded history (Figure 5). The exceptional production of chinook salmon in the 1980s seemed to indicate that this species had fulfilled the role that had been predicted by Tody and Tanner (1966). Chinook salmon seemed ideally suited to the environment of Lake Michigan that existed after the collapse of the native species, rich with an ideal prey source and free from predators (other than humans). Fishery managers basked in one of the greatest fishery management success stories, but they were ill prepared for the sequence of events that would soon play out.

In 1988, catch rates of chinook salmon in Lake Michigan declined sharply, particularly in Michigan waters (Figure 6). In Michigan waters, chinook salmon catch rates declined 50% from 1987 to 1989 and 85% by 1993. In Wisconsin waters, chinook salmon catch rates declined one year later than in Michigan, and they remained higher than in Michigan. Declining catch rates of chinook salmon in Lake Michigan, in conjunction with reduced effort caused by declining catch rates, were largely responsible for an 80% decline in yield from 1986 to 1990 (Benjamin 1998; Holey et

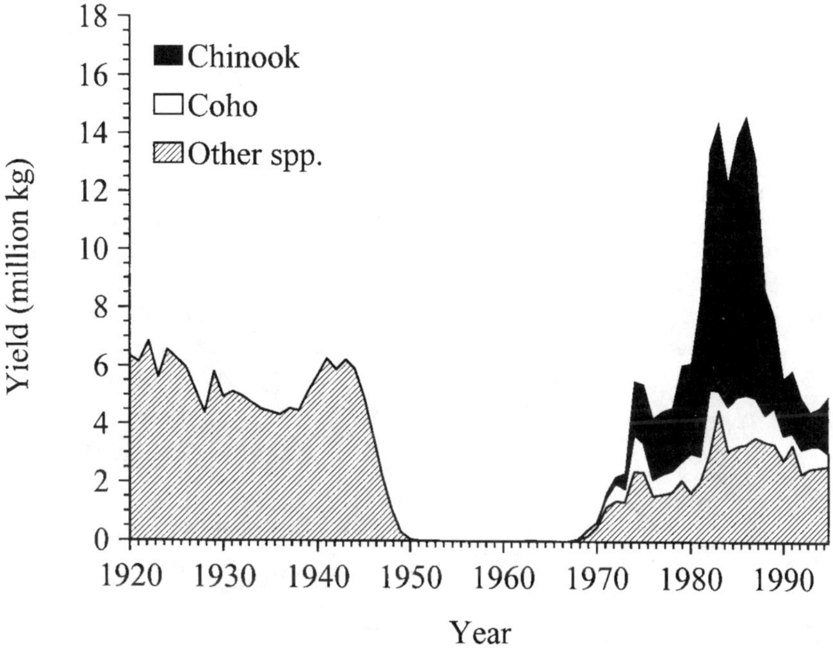

Figure 5. Harvest of chinook salmon, coho salmon, and other salmonines from Lake Michigan during 1920–1995 (from Holey et al. 1998). All harvest prior to 1950 was lake trout.

al. 1998; Bence and Smith 1999; Benjamin and Bence, in press a). The decline of chinook salmon stocks in Lake Michigan was nearly as dramatic as the collapse of lake trout stocks 40 years earlier (Figure 5).

Survival

Survival of chinook salmon from stocking to harvest was extraordinarily high and independent of stocking density in Lake Michigan during the 1970s and 1980s. For example, in Wisconsin waters of Lake Michigan, the harvest during 1969–1985 was directly proportional to the number previously stocked (Figure 7), which suggested survival was density independent and averaged 12–13% from stocking to harvest (Hansen 1989; Hansen et al. 1990; 1991). In contrast, survival of chinook salmon from stocking to harvest in most West Coast fisheries was 1% or less (Hansen 1989). The harvest of chinook salmon from Lake Michigan in 1985 reflected this extraordinary survival—more than 900,000 were harvested from 6 million stocked three years earlier, a 15% return to the creel (Hansen 1989)!

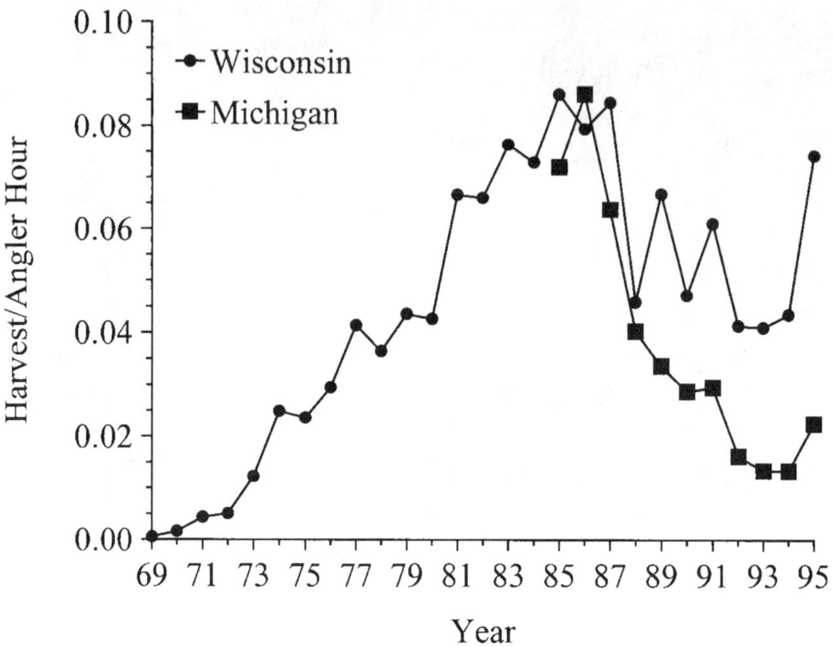

Figure 6. Numbers of chinook salmon harvested per angler-hour from the Michigan and Wisconsin waters of Lake Michigan during 1969–1995 (from Hansen et al. 1990, 1991; Holey et al. 1998).

By the late 1980s, survival changed dramatically, as dead adult chinook salmon appeared along Lake Michigan beaches (Johnson and Hnath 1991) and the fishery collapsed (Figure 5). After 1985, numbers of chinook salmon caught by angling, in relation to numbers previously stocked in Wisconsin waters of Lake Michigan, was lower and more variable than before 1985 (Figure 7). Over the entire period 1969–1995, this suggests that survival of chinook salmon was regulated by density-dependent processes in Lake Michigan. Also, changes in survival after 1985 may have been confounded with changes in movement among areas of Lake Michigan, combined with changes in stocking by jurisdictions (Benjamin 1998).

The cause of this change in survival of chinook salmon in Lake Michigan is uncertain but may relate to a change in body condition of adult fish that began in the late 1970s. For example, body condition of chinook salmon in the Wisconsin sport fishery returning to the Strawberry Creek weir declined steadily during the 1970s, then fluctuated erratically during the 1980s and early 1990s (Figure 8). The reduced body condition of adult chinook salmon in Lake Michigan may have made them susceptible to bacterial

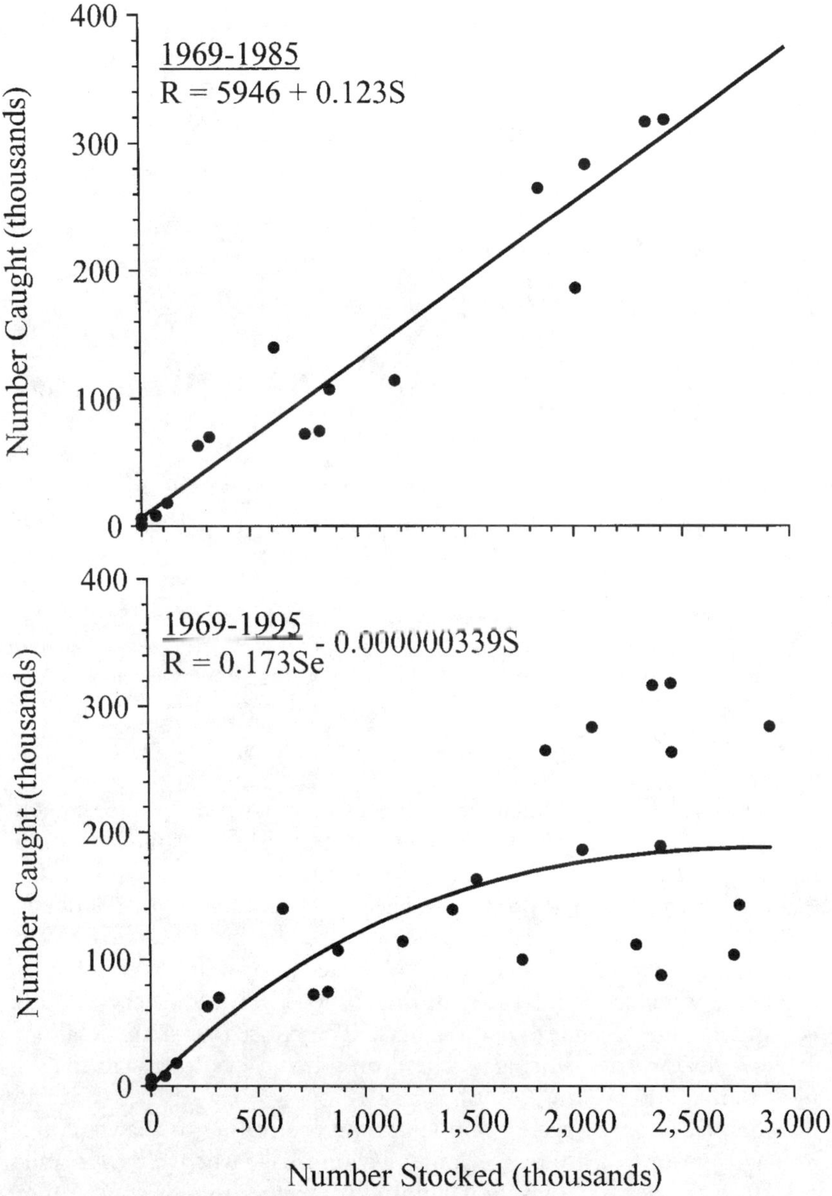

Figure 7. Number of chinook salmon caught versus the number previously stocked in the Wisconsin waters of Lake Michigan during 1969–1985 (upper panel; from Hansen 1989) and 1969–1995 (lower panel).

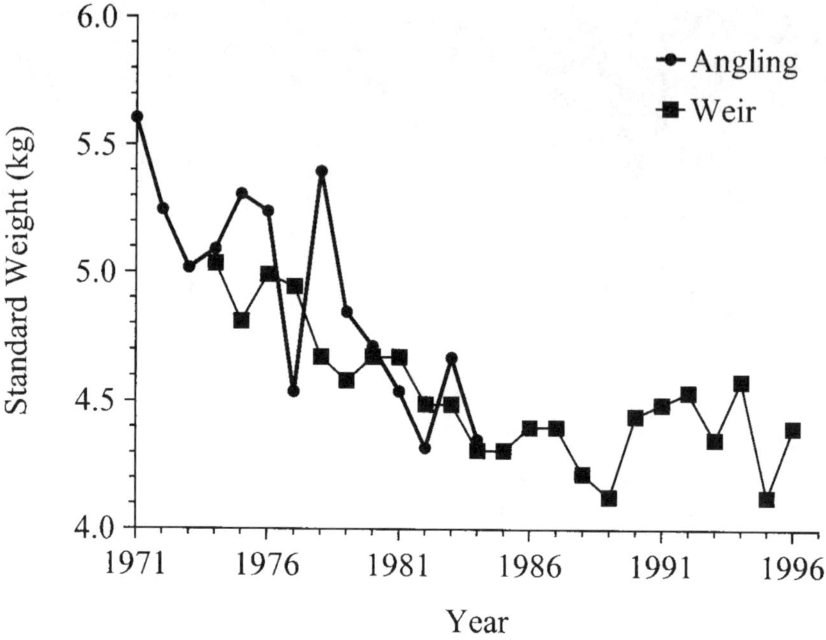

Figure 8. Predicted weight of an average length (762-mm TL) chinook salmon (standard weight) caught by anglers and returning to the Strawberry Creek weir in the Wisconsin waters of Lake Michigan during 1971–1997 (from Hansen 1986; Peeters and Royseck 1997).

kidney disease (BKD), which first appeared in spawning adults in 1986 and became prevalent thereafter. The incidence of BKD in chinook salmon returning to the Strawberry Creek weir in Wisconsin increased from 0% before 1986 to 67% in 1988, then ranged between 10% and 28% during 1989–1992 (Holey et al. 1998). In Michigan, the incidence of BKD in chinook salmon returning to the Little Manistee River weir was even higher, ranging between 53% and 100% during 1986–1991 (Holey et al. 1998). As a consequence, the rate of natural mortality increased for chinook salmon of ages 2 to 5 in Lake Michigan, based on statistical catch-at-age analysis (Benjamin and Bence, in press b).

The predominant age of chinook salmon returning to spawn shifted from age 3 prior to 1990 to age 2 after 1990 in Lake Michigan (Benjamin and Bence, in press b). Total annual mortality of ages 2 to 3 chinook salmon increased sharply during 1986–1991, largely through an increase in natural mortality, then remained high through 1995 (Benjamin and Bence, in press b). During the same period, returns of age-2 chinook salmon to the Strawberry Creek weir in Wisconsin increased after 1988 to levels more typical of

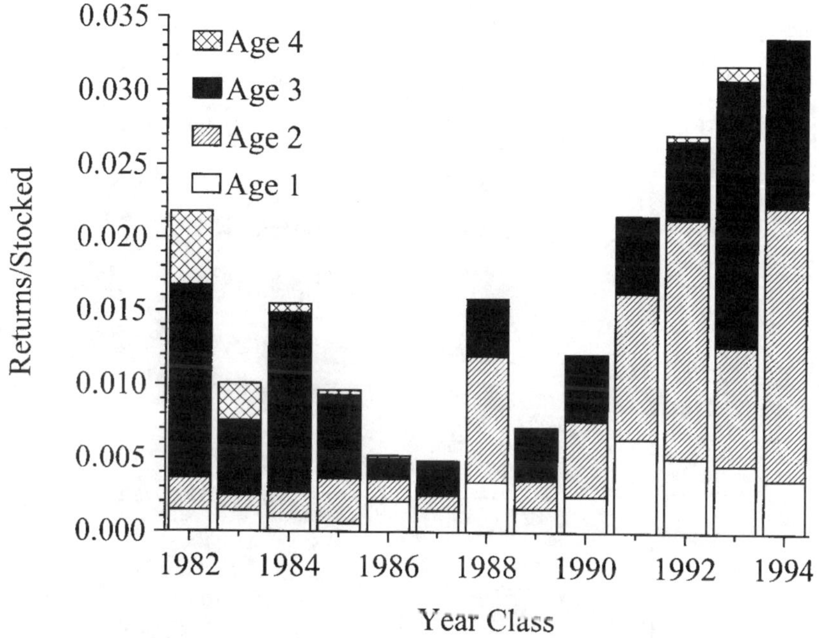

Figure 9. Rate of return of the 1982–1994 year classes of coded-wire-tagged chinook salmon stocked in and returned to the Strawberry Creek weir in the Wisconsin waters of Lake Michigan (from Peeters and Royseck 1997).

age-3 chinook salmon in previous years (Figure 9). By 1991, returns of age-1 chinook salmon increased to levels more typical of age-2 chinook salmon in previous years. By the time the 1991 year-class returned at age 1 in 1992, age 2 in 1993, and age 3 in 1994, the age structure of the spawning run had shifted sharply downward. Total returns of the 1991 year-class were 50% higher than the 1984 year-class, the last year-class to be fully recruited prior to the 1988 die-off of adults.

Growth and Diet

Size at age of chinook salmon fluctuated inversely with population density during the 1980s and 1990s in Lake Michigan (Clark 1996). For example, mean length and weight of age-2 and age-3 male chinook salmon returning to spawn in Strawberry Creek, Wisconsin, increased during 1983–1997, as catch rates in the sport fishery were declining, while age-1 fish did not increase in either mean length or weight (Figure 10). Similarly, in Michi-

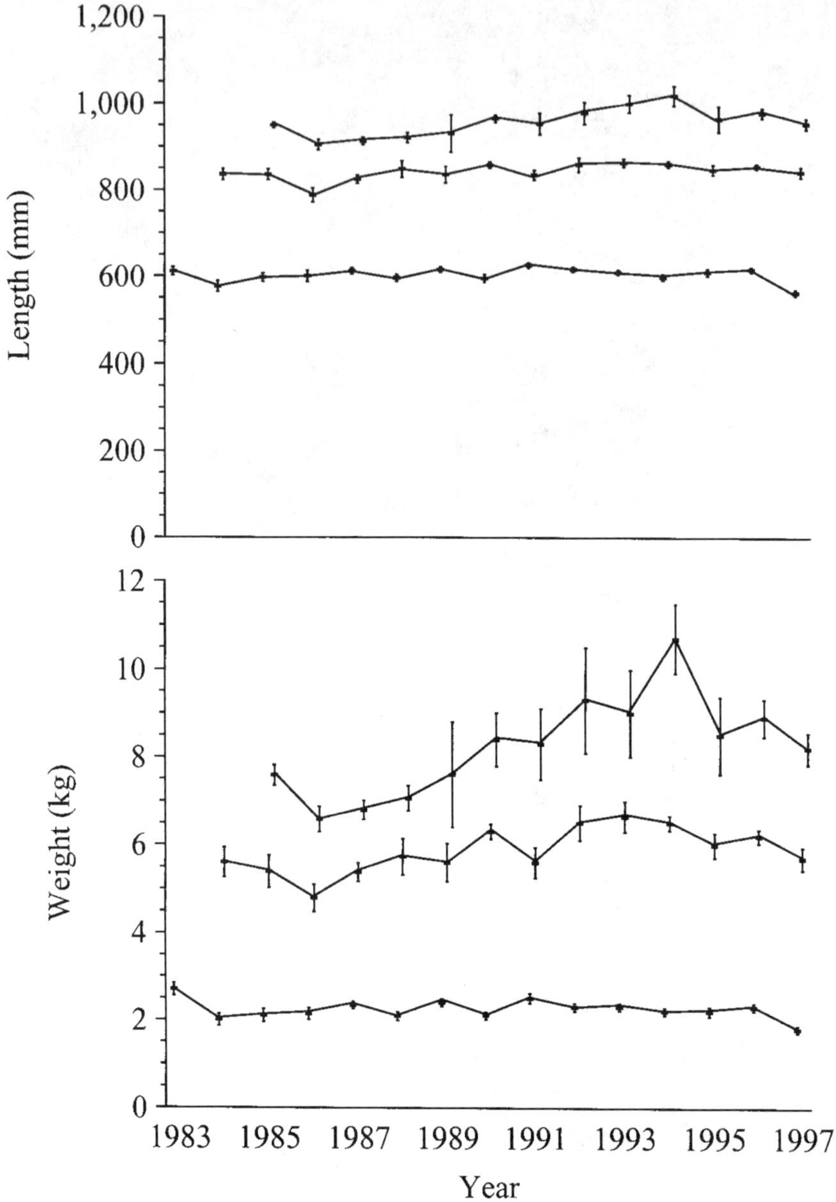

Figure 10. Mean length (upper panel) and weight (lower panel) of coded-wire-tagged male chinook salmon stocked in and returned to the Strawberry Creek weir in the Wisconsin waters of Lake Michigan during 1984–1997 (from Peeters and Royseck 1997). Error bars depict two standard errors of the mean, above and below the mean length and weight.

gan waters of Lake Michigan, mean length and weight of ages 2 to 4 chinook salmon increased from 1983 to 1986 (before BKD) to 1990–1993 (after BKD), whereas mean length and weight of age-1 chinook salmon did not change significantly between the same periods (Wesley 1996). These patterns in size at age suggest that growth of chinook salmon in Lake Michigan was strongly density dependent; adult chinook salmon were more abundant and grew slower before BKD than after BKD (Clark 1996).

Chinook salmon consume primarily adult alewife in the Great Lakes, even though other prey species have been more abundant during some periods. For example, in Lake Michigan, adult chinook salmon consumed primarily adult alewife during 1986–1995 (Elliott 1993; Rybicki and Clapp 1996; Holey et al. 1998). However, ages 2 to 3 chinook salmon consumed dramatically fewer alewife per predator in 1986, just before the BKD mortality, than during 1994–1995, several years after the BKD mortality (Figure 11). In contrast, bloaters were a significant portion of the diets of only age-2 chinook salmon during 1986 (Figure 11) and salmon 38–57 cm in

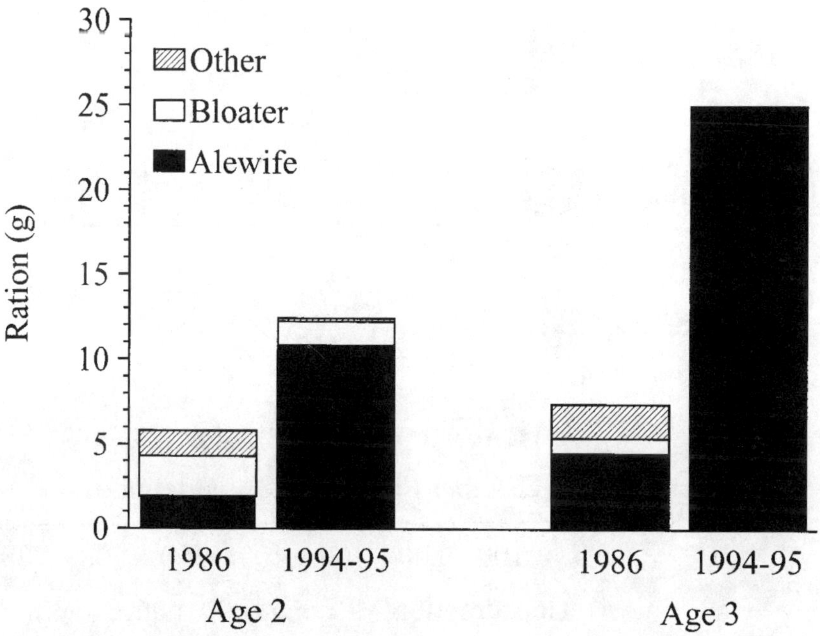

Figure 11. Index of ration of alewife and other prey types (average weight per stomach) for age-2 and age-3 chinook salmon caught by Michigan anglers in southeastern Lake Michigan in 1986 and 1994–1995 (from Holey et al. 1998).

1991–1993 (Rybicki and Clapp 1996), even though bloaters were 5–15 times more abundant than alewives at that time (Fleischer et al. 2000).

The relatively low occurrence of bloaters in chinook salmon diet, especially large salmon, in relation to bloater abundance suggests limited habitat overlap between salmon and bloater, especially adult bloater. Bloaters were a significant diet item for chinook salmon in Lake Michigan primarily when juvenile bloaters were available during the mid-1980s and early 1990s, even though chinook salmon are large enough to eat adult bloaters and adult bloaters were abundant throughout the 1980s and 1990s. Juvenile bloaters were sparse during the mid-1990s because of a decrease in recruitment (Fleischer et al. 2000). Juvenile bloaters are pelagic; adult bloaters are benthic (Wells 1966; 1968; Crowder and Crawford 1984). This favors overlap between juvenile bloaters and chinook salmon but not between adult bloaters and chinook salmon. Consequently, the high biomass of adult bloaters in Lake Michigan is largely unavailable to chinook salmon.

Chinook salmon consume more alewife per individual during their lifetime, especially adult alewife, than other trout or salmon species (Figure 12). Chinook salmon exert greater lifetime predation pressure per indi-

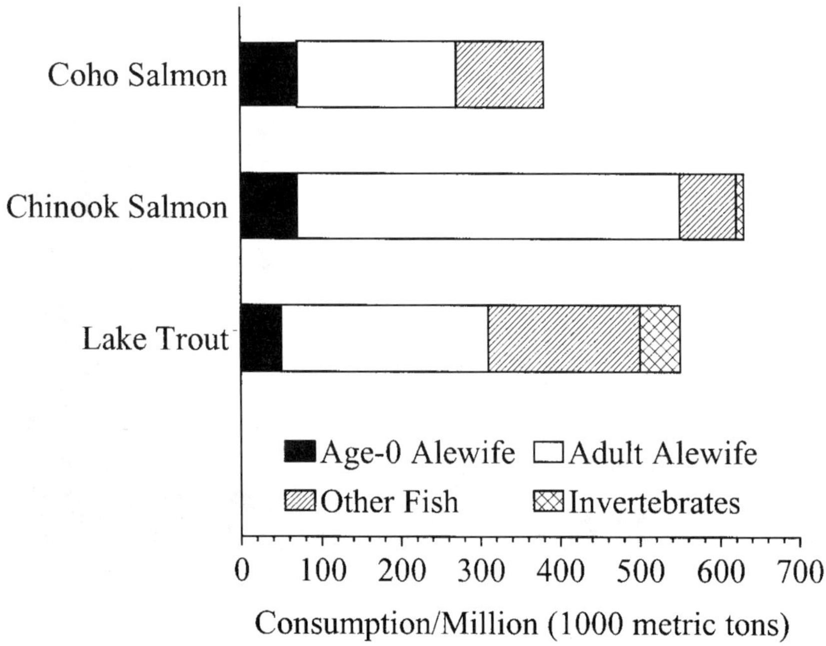

Figure 12. Estimated total consumption of various prey items (per million fish stocked) by coho salmon, chinook salmon, and lake trout in Lake Michigan (from Stewart et al. 1981).

vidual than coho salmon because chinook salmon spend one more year in the lake than coho salmon (Stewart et al. 1981). In contrast, the lifetime predation pressure exerted by chinook salmon is greater than that of lake trout because chinook salmon achieve adult size (8–10 kg) in only four years, whereas lake trout achieve a similar adult size in six to eight years (Stewart et al. 1981). Chinook salmon require more alewives per individual than other species of trout and salmon. Therefore, in condition and growth, chinook salmon are more responsive than other trout and salmon species to short-term changes in alewife abundance (Stewart and Ibarra 1991).

Predation by trout and salmon, in conjunction with severe winters in 1976–1982, suppressed alewife populations in Lake Michigan to low levels by the early 1980s (Figure 13; Stewart et al. 1981; Eck and Brown 1985; Kitchell and Crowder 1986; Eck and Wells 1987; Stewart and Ibarra 1991). After alewife declined in abundance, bloaters rapidly increased in abundance and were predominant by 1983. Abundance of age-0 bloater was inversely related to abundance of age-4+ alewife (Figure 14), perhaps through predation on larval bloater by adult alewife (Eck and Wells 1987). By 1987, salmonine predation and commercial fisheries harvested 60% of the avail-

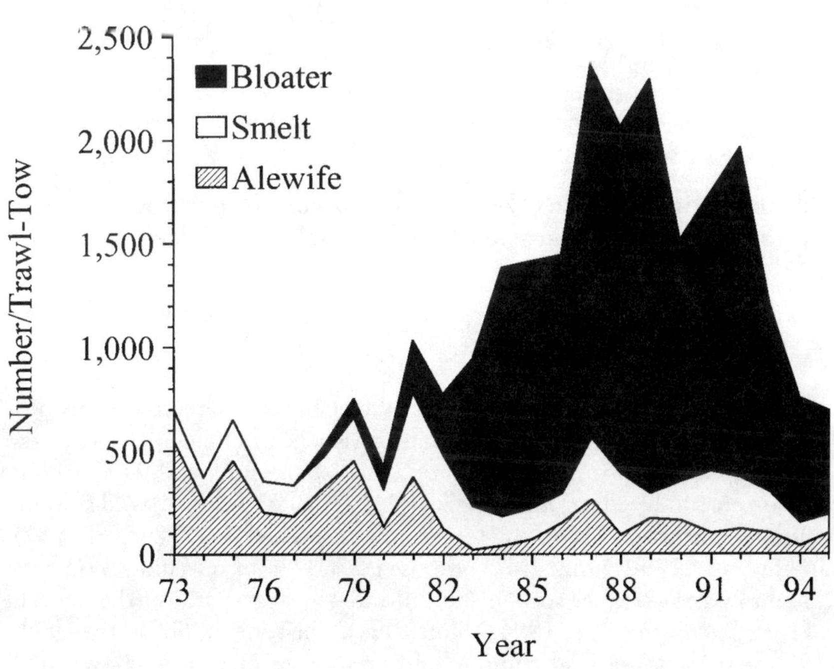

Figure 13. Number of adult alewife, rainbow smelt, and bloater per trawl tow in Lake Michigan during 1973–1995 (from Fleischer et al. 2000).

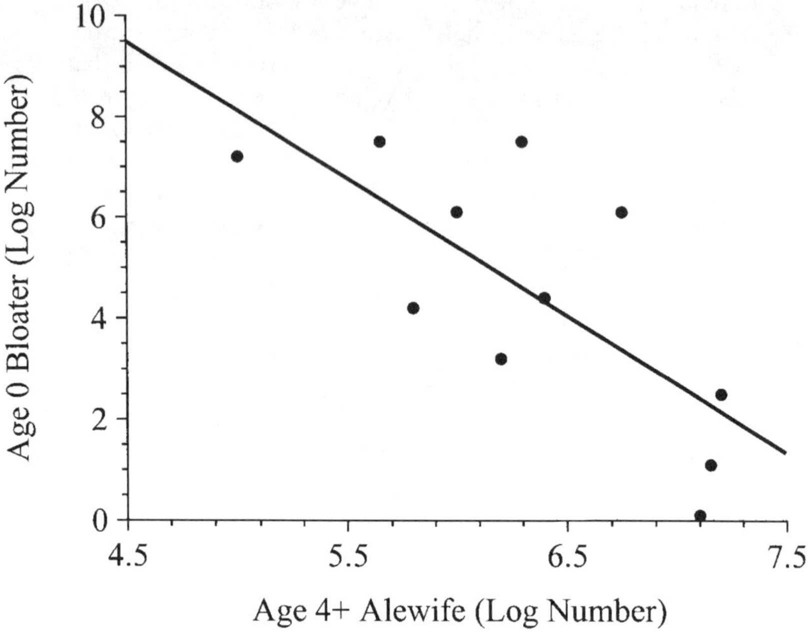

Figure 14. Estimated numbers of age-0 bloater versus estimated number of age-4+ alewife in Lake Michigan during 1974–1983 (from Eck and Wells 1987).

able alewife production in Lake Michigan and, consequently, may now regulate the alewife population at low, relatively stable levels (Stewart and Ibarra 1991).

Reproduction

Pacific salmon have reproduced in the Great Lakes wherever streams provide suitable spawning and rearing habitat (Kocik and Jones 1999). In Lake Superior, 50–90% of chinook salmon (Peck et al. 1999) and 94% of coho salmon are produced naturally (Peck 1992). In Lake Michigan, wild chinook salmon composed 13% of the total number stocked in 1972 (Carl 1982). In the early 1990s, wild fish composed 27–37% of all chinook salmon caught by anglers in the southeast basin and 45–66% in the northeast basin of Lake Michigan (Hesse 1994; Elliott 1994). The contribution of the 1992–1995 year classes of wild chinook salmon to assessment catches, when all stocked salmon of the same age were marked, ranged from 27.8% to 53.4% at age 2 (Clapp 1998). During 1996–1997, 76–82% of returning adult chinook salmon in the Manistee and Muskegon rivers were of wild origin (Rutherford et al. 1999). Increased proportions of wild chinook salmon in fisheries and spawning runs are partly caused by increased spawning success and by decreased survival of hatchery-reared fish. Reproduction by coho salmon in the same streams has been measurable (Seelbach 1985) but contributes an unknown fraction to the Lake Michigan fishery. In the other lakes, reproduction by chinook and coho salmon occurs but has not been quantified. Production of wild chinook and coho salmon could be greatly increased from current levels if prime tributary nursery habitats above dams were made accessible to adult salmon via fish ladders or dam removal.

Survival of feral salmon eggs reared in hatcheries under experimental conditions, especially of coho salmon, declined in the 1990s. From the mid-1980s to the mid-1990s, egg mortality during hatchery incubation increased from 20% to 70% for coho salmon and from 10% to 60% for chinook salmon (Marcquenski and Brown 1997; Honeyfield et al. 1998a,b). Egg survival was so poor for coho salmon in Lake Michigan that fishery managers feared there would be too few adult coho salmon returning to spawning weirs to sustain the stocking program. Consequently, management agencies adopted emergency lakewide harvest regulations in 1994 to ensure that enough adult coho salmon returned to spawning weirs.

Low survival of salmon eggs, known as early mortality syndrome (EMS), is related to low concentration of thiamine in the egg, which may be related to thiaminase concentration in predominant prey species eaten by adult salmon during egg development (Honeyfield et al. 1998b). Immersion of coho salmon eggs and fry in thiamine water baths has increased egg and fry survival to pre-EMS levels (Hornung et al. 1998), and it is now a standard practice in hatcheries that obtain eggs from feral adult salmon in Lake Michigan (Honeyfield et al. 1998b). Low thiamine levels that trigger the onset of EMS may be related to the percentage of alewife and rainbow smelt *Osmerus mordax* in the diet (Honeyfield et al. 1998b). Both alewife and rainbow smelt possess thiaminase, the thiamine-degrading enzyme (Anglesea and Jackson 1985; Ji and Adelman 1998). Alewife have been a major prey of coho and chinook salmon since their successful introduction into Lake Michigan (Elliott 1993; Stewart et al. 1981; Stewart and Ibarra 1991). The increase in EMS in Lake Michigan trout and salmon is more likely the result of a larger ecosystem change in the basin that has changed the nutritional value of alewife. Though the cause of EMS has yet to be described, its occurrence will either prevent further development of wild year classes or add increased variation in salmon spawning success in Lake Michigan and the other Great Lakes.

The Future

Chinook and coho salmon may now be functioning as important members of the fish communities in all five Great Lakes. Both species played major roles as predators in reducing the abundance of alewife in Lake Michigan, which facilitated recovery of native prey species. Sport and commercial fisheries that developed for chinook and coho salmon are popular and in high demand by the public throughout the Great Lakes basin. The ability of chinook and coho salmon to reproduce in many areas of the Great Lakes has led fishery managers to develop management goals that are intended to decrease dependence on hatchery-reared fish in the future. Fish community objectives for Lake Michigan now recognize chinook and coho salmon as naturalized members of the trout and salmon species assemblage and recommend that natural reproduction of all species be enhanced (Eshenroder et al. 1995). Subsequently, stocking objectives for chinook and coho salmon in Lake Michigan have been adopted to reduce salmon biomass, minimize BKD mortality, and promote sustainability of the fishery (Clark 1996).

Chinook and coho salmon survived best in the Great Lakes when alewife were the predominant prey fish, and they became established members of the fish community, but the future of both species is uncertain. During the 1990s in Lake Michigan, chinook salmon suffered high mortality and high prevalence of BKD when alewife no longer predominated in the prey fish community. The susceptibility of chinook salmon to BKD (Starliper et al. 1997) makes them susceptible to high mortality during low prey abundance. In lakes Huron and Ontario, reduced chinook salmon growth recently occurred in conjunction with reduced alewife abundance, and fishery managers fear a large-scale die-off in the near future in both lakes (Brandt et al. 1996; Johnson et al. 1999). EMS also threatens the sustainability of wild coho salmon in the Great Lakes because coho salmon are susceptible to EMS (Marcquenski and Brown 1997; Honeyfield et al. 1998a, 1998b) and suffer near 100% egg mortality if their eggs are not treated with thiamine. This limits the ability of coho salmon to establish self-sustaining populations in the Great Lakes.

Despite the uncertainty of the future for chinook and coho salmon in the Great Lakes, we believe both species will continue to contribute to the restoration and stability of Great Lakes fish communities in the foreseeable future, in spite of conventional wisdom that suggests introduced species do not usually fill valuable roles in ecosystem restoration and stability. Though not native to the Great Lakes, chinook and coho salmon are better adapted to controlling alewives than native Great Lakes predators, and this enhances the ecological value of chinook and coho salmon in Great Lakes ecosystems. Stocking chinook and coho salmon has been one of the most effective fishery management tools for controlling alewife abundance, and it also facilitates native species restoration and the development of sustainable predator populations. However, a return to the "good old days" of high chinook and coho salmon abundance is unlikely, even with further increases in chinook and coho salmon stocking and recruitment, because chinook salmon are so vulnerable to low alewife abundance. Species such as rainbow trout and lake trout may be needed to exert a larger portion of the predation pressure in the future should forage populations less dominated by alewife continue to develop in each of the Great Lakes. Research is needed to determine appropriate levels of chinook and coho salmon stocking for maintaining alewife populations at levels that will least affect the development of sustainable predator stocks and native species restoration. In the meantime, chinook and coho salmon will likely sustain sizeable wild populations that are supplemented by stocking because of their ecological value as biological controllers of alewives and their economic value as a sport fish.

Acknowledgements

We thank James R. Bence and two anonymous reviewers for their helpful suggestions on the manuscript. Stocking data were from the Great Lakes fish-stocking database, and diet data were provided by the U. S. Fish and Wildlife Service, Green Bay Fishery Resources Office, Green Bay, Wisconsin. This work was supported by the University of Wisconsin Sea Grant Institute under grants from the National Sea Grant College Program, National Oceanic and Atmospheric Administration, U. S. Department of Commerce.

References

Anglesea, J. D., and A. J. Jackson. 1985. Thiaminase activity in fish silage and moist fish feed. Animal Feed Science and Technology 13:39–46.

Bence, J. R., and K. D. Smith. 1999. An overview of recreational fisheries in the Great Lakes. Pages 259–306 *in* W. W. Taylor and C. P. Ferreri, editors. Great Lakes fishery policy and management: a binational perspective. Michigan State University Press, East Lansing.

Benjamin, D. M. 1998. Chinook salmon (*Oncorhynchus tshawytscha*) population dynamics in Lake Michigan, 1985-1996. Master's thesis. Michigan State University, East Lansing.

Benjamin, D. M., and J. R. Bence. In press a. Spatial and temporal changes in the Lake Michigan chinook salmon fishery, 1985-1996. Michigan Department of Natural Resources, Fisheries Research Report, Ann Arbor, Michigan.

Benjamin, D. M., and J. R. Bence. In press b. Statistical catch-at-age analysis of chinook salmon in Lake Michigan, 1985-1996. Michigan Department of Natural Resources, Fisheries Research Report, Ann Arbor, Michigan.

Brandt, S., J. DePinto, M. Hansen, K. Hartman, E. Mills, R. O'Gorman, P. Rand, H. Riessen, L. Rudstam, R. Snyder, and D. Stewart. 1996. A review of the current status of Lake Ontario's pelagic fish community: report of the 1996 Lake Ontario technical panel. Great Lakes Research Reviews 2:4–10.

Brown, E. H., Jr. 1972. Population biology of alewives, *Alosa pseudoharengus*, in Lake Michigan, 1949-1970. Journal of the Fisheries Research Board of Canada 29:477–500.

Carl, L. M. 1982. Natural reproduction of coho salmon and chinook salmon in some Michigan streams. North American Journal of Fisheries Management 2:375–380.

Clapp, D. F. 1998. Chinook salmon population dynamics in Michigan's waters of the Great Lakes. Michigan Department of Natural Resources, Federal Aid in Sport Fish Restoration Completion Report, Project F-53-R-14-471, Lansing, Michigan.

Clark, R. D., Jr. 1996. Status of chinook salmon in the upper Great Lakes, 1995. Michigan Department of Natural Resources, Fisheries Division, Administrative report, Lansing, Michigan.

Crowder, L. B., and H. L. Crawford. 1984. Ecological shifts in resources use by bloaters in Lake Michigan. Transactions of the American Fisheries Society 113:694–700.

Crowder, L. B. 1980. Alewife, rainbow smelt and native fishes in Lake Michigan: competition or predation? Environmental Biology of Fishes 5:225–233.

Eck, G. W., and E. H. Brown, Jr. 1985. Lake Michigan's capacity to support lake trout (*Salvelinus namaycush*) and other salmonines: an estimate based on the status of prey populations in the 1970s. Canadian Journal of Fisheries and Aquatic Sciences 42:449–454.

Eck, G. W., and L. Wells. 1987. Recent changes in Lake Michigan's fish community and their probable causes, with emphasis on the role of alewife (*Alosa pseudoharengus*). Canadian Journal of Fisheries and Aquatic Sciences 44(Supplement 2):53–60.

Elliott, R. F. 1993. Feeding habits of chinook salmon in eastern Lake Michigan. Master's thesis. Michigan State University, East Lansing.

Elliott, R. F. 1994. Early life history of chinook salmon in Lake Michigan. Michigan Department of Natural Resources, Federal Aid in Sport Fish Restoration Completion Report, Project F-53-R-472, Lansing, Michigan.

Eshenroder, R. L., M. E. Holey, T. K. Gorenflo, and R. D. Clark, Jr. 1995. Fish community objectives for Lake Michigan. Great Lakes Fishery Commission, Special Publication 95-3, Ann Arbor, Michigan.

Fleischer, G. W., C. P. Madenjian, T. J. DeSorcie, and J. D. Holuszko. 2000. Status of trends of prey fish populations in Lake Michigan. 1999. Great Lakes Fishery Commission, Minutes of the Annual Meeting of the Lake Michigan Committee, Ann Arbor, Michigan.

Hansen, M. J. 1986. Size and condition of trout and salmon from the Wisconsin waters of Lake Michigan, 1969-84. Wisconsin Department of Natural Resources, Fisheries Management Report 126, Madison, Wisconsin.

Hansen, M. J. 1989. Chinook salmon and Wisconsin's Lake Michigan sport fishery 1969-1985. Pages 8–17 *in* B. G. Shepherd, editor. Proceedings of the 1988 Northeast Pacific chinook and coho salmon workshop. American Fisheries Society, Northeast Pacific International Chapter, Penticton, British Columbia.

Hansen, M. J. 1999. Lake trout in the Great Lakes: basin-wide stock collapse and binational restoration. Pages 417–453 *in* W. W. Taylor and C. P. Ferreri, editors. Great Lakes fishery policy and management: a binational perspective. Michigan State University Press, East Lansing.

Hansen, M. J., P. T. Schultz, and B. A. Lasee. 1990. Changes in Wisconsin's Lake Michigan salmonid sport fishery, 1969-1985. North American Journal of Fisheries Management 10:442–457.

Hansen, M. J., P. T. Schultz, and B. A. Lasee. 1991. Wisconsin's Lake Michigan salmonid sport fishery, 1969-85. Wisconsin Department of Natural Resources, Fisheries Management Report 145, Madison, Wisconsin.

Hesse, J. A. 1994. Contribution of hatchery and natural chinook salmon to the Eastern Lake Michigan fishery, 1992-1993. Master's thesis. Michigan State University, East Lansing.

Holey, M. E., R. F. Elliott, S. V. Marcquenski, J. G. Hnath, and K. D. Smith. 1998. Chinook salmon epizootics in Lake Michigan: possible contributing factors and management implications. Journal of Aquatic Animal Health 10:202–210.

Honeyfield, D. C., J. D. Fitzsimons, S. B. Brown, S. V. Marcquenski, G. McDonald. 1998a. Introduction and overview of early life stage mortality. Pages 1–7 *in* G. McDonald, J. D. Fitzsimons, and D. C. Honeyfield, editors. Early life stage mortality syndrome in fishes of the Great Lakes and Baltic Sea. American Fisheries Society, Symposium 21, Bethesda, Maryland.

Honeyfield, D. C., J. G. Hnath, J. Copeland, K. Dabrowski, and J. H. Blom. 1998b. Correlation of nutrients and environmental contaminants in Lake Michigan coho salmon with incidence of early mortality syndrome. Pages 135–145 *in* G. McDonald, J. D. Fitzsimons, and D. C. Honeyfield, editors. Early life stage mortality syndrome in fishes of the Great Lakes and Baltic Sea. American Fisheries Society, Symposium 21, Bethesda, Maryland.

Hornung, M. W., L. Miller, R. E. Peterson, S. Marcquenski, and S. B. Brown. 1998. Efficacy of thiamine, astaxanthin, b-carotene, and thyroxine treatments in reducing early mortality syndrome in Lake Michigan salmonid embryos. Pages 124–134 *in* G. McDonald, J. D. Fitzsimons, and D. C. Honeyfield, editors. Early life stage mortality syndrome in fishes of the Great Lakes and Baltic Sea. American Fisheries Society, Symposium 21, Bethesda, Maryland.

Ji, Y. Q., and I. R. Adelman. 1998. Thiaminase activity in alewives and smelt in Lakes Huron, Michigan, and Superior. Pages 154–159 *in* G. McDonald, J. D. Fitzsimons, and D. C. Honeyfield, editors. Early life stage mortality syndrome in fishes of the Great Lakes and Baltic Sea. American Fisheries Society, Symposium 21, Bethesda, Maryland.

Johnson, D. C., and J. G. Hnath. 1991. Lake Michigan chinook salmon mortality–1988. Michigan Department of Natural Resources, Fisheries Division Technical Report 91-4, Lansing, Michigan.

Johnson, J., A. Woldt, and G. Wright. 1999. Trends in chinook salmon growth in north-central Lake Huron. Great Lakes Fishery Commission, Minutes of the Annual Meeting of the Lake Huron Committee, Ann Arbor, Michigan.

Jude, D. J., and F. J. Tesar. 1985. Recent changes in the inshore forage fish of Lake Michigan. Canadian Journal of Fisheries and Aquatic Sciences 42:1154–1157.

Keller, M., K. D. Smith, and R. W. Rybicki, editors. 1990. Review of salmon and trout management in Lake Michigan. Michigan Department of Natural Resources, Fisheries Division Special Report 14, Lansing, Michigan.

Kitchell, J. F., and L. B. Crowder. 1986. Predator-prey interactions in Lake Michigan: model predictions and recent dynamics. Environmental Biology of Fishes 16:205–211.

Kocik, J. F., and M. L. Jones. 1999. Pacific salmonines in the Great Lakes basin. Pages 455–488 *in* W. W. Taylor and C. P. Ferreri, editors. Great Lakes fishery policy and management: a binational perspective. Michigan State University Press, East Lansing.

Koonce, J. F., A. B. Locci, and R. L. Knight. 1999. Contribution of fishery management in walleye and yellow perch populations of Lake Erie. Pages 397–416 *in* W. W. Taylor and C. P. Ferreri, editors. Great Lakes fishery policy and management: a binational perspective. Michigan State University Press, East Lansing.

Marcquenski, S. V., and S. B. Brown. 1997. Early mortality syndrome in the Great Lakes. Pages 135–152 *in* R. M. Rolland, M. Gilbertson, and R. E. Peterson, editors. Chemically induced alterations in functional development and reproduction in fishes. SETAC (Society of Environmental Toxicology and Chemistry), Pensacola, Florida.

O'Gorman, R., and T. J. Stewart. 1999. Ascent, dominance, and decline of the alewife in the Great Lakes: food web interactions and management strategies. Pages 489–513 *in* W. W. Taylor and C. P. Ferreri, editors. Great Lakes fishery policy and management: a binational perspective. Michigan State University Press, East Lansing.

Parsons, J. W. 1973. History of salmon in the Great Lakes, 1850-1970. U.S. Fish and Wildlife Service, Technical Paper 68, Washington, D.C.

Peck, J. W. 1992. The sport fishery and contribution of hatchery trout and salmon in Lake Superior and tributaries at Marquette, Michigan, 1984-1987. Michigan Department of Natural Resources, Fisheries Research Report 1975, Ann Arbor, Michigan.

Peck, J. W., T. S. Jones, W. R. MacCallum, and S. T. Schram. 1999. Contribution of hatchery-reared fish to chinook salmon populations and sport fisheries in Lake Superior. North American Journal of Fisheries Management 19:155–164.

Peeters, P., and K. Royseck. 1997. Harvest, age, and size at age of chinook and coho salmon at Strawberry Creek weir and Besadny anadromous fisheries facility. Wisconsin Department of Natural Resources, Sturgeon Bay, Wisconsin.

Rutherford, E., S. T. Horrold, A. Woldt, D. Swank, and N. Golby. 1999. Abundance, production and harvest of anadromous salmonids in the Manistee, Muskegon, and Au Sable Rivers. Report to the Michigan Department of Natural Resources, Contract No. 231663, Lansing, Michigan.

Rybicki, R. W., and D. F. Clapp. 1996. Diet of chinook salmon in eastern Lake Michigan, 1991-1993. Michigan Department of Natural Resources, Research Report 2027, Lansing, Michigan.

Seelbach, P. W. 1985. Smolt migration of wild and hatchery-raised coho and chinook salmon in a tributary of northern Lake Michigan. Michigan Department of Natural Resources, Fisheries Research Report 1935, Lansing, Michigan.

Smith, S. H. 1968. Species succession and fishery exploitation in the Great Lakes. Journal of the Fisheries Research Board of Canada 25:667–693.

Smith, S. H. 1970. Species interactions of the alewife in the Great Lakes. Transactions of the American Fisheries Society 99:754–765.

Smith, S. H. 1972. Factors of ecologic succession in oligotrophic fish communities of the Laurentian Great Lakes. Journal of the Fisheries Research Board of Canada 29:717–730.

Starliper, C. E., D. R. Smith, and T. Shatzer. 1997. Virulence of *Renibacterium salmoninarum* to salmonids. Journal of Aquatic Animal Health 9:1–7.

Stewart, D. J., and M. Ibarra. 1991. Predation and production of salmonine fishes in Lake Michigan, 1978-88. Canadian Journal of Fisheries and Aquatic Sciences 48:909–922.

Stewart, D. J., J. F. Kitchell, and L. B. Crowder. 1981. Forage fishes and their salmonid predators in Lake Michigan. Transactions of the American Fisheries Society 110:751–763.

Tody, W. H., and H. A. Tanner. 1966. Coho salmon for the Great Lakes. Michigan Department of Natural Resources, Fish Management Report 1, Lansing, Michigan.

Wells, L. 1966. Seasonal and depth distribution of larval bloaters (*Coregonus hoyi*) in southeastern Lake Michigan. Transactions of the American Fisheries Society 95:388–396.

Wells, L. 1968. Seasonal depth distribution of fish in southeastern Lake Michigan. U.S. Fish and Wildlife Service Fisheries Bulletin 67(1):1–15.

Wesley, J. K. 1996. Age and growth of chinook salmon in Lake Michigan: verification, current analysis and past trends. Master's thesis. University of Michigan, Ann Arbor.

Sustaining North American Salmon: Perspectives Across Regions and Disciplines, 181–194

Chapter 9

The Many Faces of Salmon: Implications of Stakeholder Diversity in the Great Lakes

Barbara A. Knuth

Introduction

The process of managing Great Lakes salmonids occurs in a complex social milieu. Stakeholder groups in the Great Lakes region are numerous, and their views about salmon management in the region differ considerably in content and fervor. Differences in stakeholder group perspectives may be associated with the type of stakeholder group, the issue considered, and/or the fundamental values and goals seen as appropriate for Great Lakes fishery management generally. Drawing on numerous studies related to Great Lakes fisheries, I discuss in this paper these fundamental stakeholder differences—some based on empirical study of the Great Lakes salmon fishery, and others with implications for salmon management based on related issues for the Great Lakes fisheries overall.

Knuth and Nielsen (1989) argued for the importance of basing natural resource management policies on a variety of biological, physical, social, and institutional goals and constraints. The sociocultural and political components of the management environment (e.g., Krueger and Decker 1993) are among the most challenging and, in some cases, the most important. As with many natural resource management problems, Great Lakes salmon management policies based primarily on biological and ecological criteria have come under intense scrutiny because of unforeseen social, ethical, and economic consequences (Connelly et al. 1992). These concerns are often the driving force behind conflicts among stakeholder groups and associated legal and political challenges to resource management agencies.

In the Great Lakes, sociocultural differences are evident between and among those who fish for salmon, those in industries and agencies supporting the salmon fishery, and those who are concerned with the biological viability of the Great Lakes, whether or not they participate in salmon harvest activities. Brown et al. (1991) detailed the need to consider socioeconomic information in Great Lakes fishery management processes, particularly economic and social impacts, assigned and held values associated with Great Lakes fisheries, and risk assessment and management (related to the importance of chemical contaminant issues in the lakes).

Licensed anglers are but one of many important stakeholder groups. The economic potential for tourism associated with Great Lakes salmonid fisheries has been recognized for at least two decades (Brown 1977). About 587,000 anglers (29% of all Great Lakes anglers) fished the Great Lakes for salmon in 1996, spending 3,561,000 days fishing (U.S. Department of the Interior and U.S. Department of Commerce, Bureau of the Census 1996). Overall, Great Lakes anglers spent US$1,404,885,000 on fishing trips and related equipment (USDI and USDC 1996), a substantial portion of which was targeted toward salmon.

Great Lakes angling generally, however, and salmon fishing in particular, have been on the decline, at least in some areas. For example, Connelly et al. (1999) documented a substantial decline in the percent of New York licensed anglers fishing the Great Lakes from New York shores from 1988 (43%) to 1996 (35%). Gale (1987) analyzed the Great Lakes salmonid program and suggested that sustaining over time the high public catch-related expectations associated with the fishery would be a daunting task. Recent reductions in salmon stocking regimes throughout the Great Lakes and observed changes in angler participation in the fishery reaffirm Gale's (1987) assertions.

Institutionally, eight states (Illinois, Indiana, Michigan, Minnesota, New York, Ohio, Pennsylvania, and Wisconsin), one province (Ontario), and two intertribal authorities (Great Lakes Indian Fish and Wildlife Commission and Chippewa-Ottawa Fisheries Management Authority) most directly manage the Great Lakes and their fisheries. Federal agencies hold treaty-making, commerce, and conservation authority (Dochoda and Jones, Chapter 11). Broadly considered, however, activities conducted by provincial, state, and federal environmental quality agencies and public health departments also have an impact on the Great Lakes salmon fishery, yet all of these agencies strive to meet very different mandates. Several international and interstate commissions (e.g., Great Lakes Fishery Commission, Great Lakes Commission, International Joint Commission) are also involved in activities affecting fish populations and their habitats. Numerous regional nongovernmental organizations with an environmental mission are based in the Great Lakes area, as are many nongovernmental associations with angling-related interests.

This complex mix of numerous and influential stakeholder groups in the Great Lakes creates a challenging environment in which to manage salmon. In this chapter, I compare several of the major types of stakeholder groups and concerns, focusing on differences among citizen stakeholders (e.g., anglers versus environmentalists), agencies (e.g., fishery versus environmental quality), levels of government (e.g., state/provincial versus federal), sovereign entities (i.e., Canada, USA, tribes), and values potentially associated with Great Lakes salmon. Although not all of the literature from which I draw is based directly on observations associated with the Great Lakes salmon fishery, many of the issues being faced throughout the Great Lakes fisheries (e.g., chemical contaminants, debates about introduced versus native species) hold important lessons for salmon management. The many perspectives about salmon and associated management activities are summarized, and implications for future research and management are discussed.

Value Conflicts: Stakeholder Differences in Goals for Great Lakes Salmon Management

Conflicts over values characterize many natural resource management debates, and controversies about Great Lakes fisheries management are no different. Values help define the desired human relationship with Great Lakes salmon and thus identify appropriate management approaches (Knuth 1995). Basic types of values are held, assigned, and intrinsic; held and assigned values are often further categorized into social, psychological, esthetic, physiological, ecological, and educational. This section is based largely on Knuth (1995), whose analysis of values associated with wild trout pertains also to introduced salmon in the Great Lakes.

The diversity of values associated with salmon is what causes them to be viewed as desirable or useful, either relative to human perceptions and judgments or as unique organisms unto themselves (Knuth 1995). These values are part of the sociocultural context in which salmon management occurs. Stakeholder groups may differ in the importance they place on achieving goals that reflect various values. Management priorities and directions are shaped by the variety of values (and implied goals) associated with salmon, such as providing recreational opportunities for angling, stimulating local economies through tourism, fostering genetic diversity in the Great Lakes system (whether native or introduced), or reducing the extent of chemical contaminant exposure to fish consumers (e.g., by encouraging them to eat less-contaminated species or sizes of fish).

Differences in held, assigned, and intrinsic values are often the core underlying basis for many natural resources conflicts. Held values are ideas, perceptions, and attitudes about objects (Brown 1984). Held values are often translated into assigned values, which are expressed as preferences or statements about the relative importance of certain objects (Weithman 1993). Assigned values, typically represented with economic parameters, are addressed in detail elsewhere in this volume. Because various stakeholder groups may perceive Pacific salmon in the Great Lakes in very different ways, they will also hold or assign different values regarding salmon and associated management processes. Intrinsic or instrumental value is the worth that organisms have apart from people (i.e., not assigned by people and not necessarily perceived [held] by people; Rolston 1983; Callicott 1987).

Social values reflect ideas, expectations, and attitudes regarding the effect of salmon or the salmon fishery on the community and social system of concern. Such values promoted by salmon management include family or community cohesion fostered by fishing activities, salmon observation opportunities, or community festivals associated with seasonal salmon migrations. Negative social values are also possible, such as those created by the influx into a community by numerous anglers interested in the seasonal salmon run. Potential problems include increased traffic congestion, visual and odor problems caused by fish carcasses left along the shore, littering, and trespassing on private land (Connelly et al. 1992). Experiencing these impacts from having a salmon stream in their community may cause some citizens to associate primarily negative values with Great Lakes salmon.

Psychological values derive from the satisfaction, motives, or attitudes associated with use or knowledge of the fishery (Weithman 1993). These include personal satisfactions or improved self-image resulting from demonstrating skills or fishing prowess, cognitive growth from learning about salmon life history or behavior, and stress reduction from being able to escape other pressures through fishing or observing salmon (Brown and Manfredo 1987; Shaw 1987).

Esthetic values refer to the artistic and symbolic features of salmon or of the fishery (Kellert 1987). Salmon may serve as inspiration for art, music, performance, or individual appreciation. The fishery and its community environment may also contribute esthetically to people's sense of place or perception of atmosphere.

Physiological values reflect contributions to human health or functioning of the body (Brown and Manfredo 1987). These may include stress reduction through fishing activities, improved nutritional status associated with the health benefits of eating salmon, or decreased health status due to exposure to chemical contaminants from eating Great Lakes salmon.

Ecological values reflect those contributions made by Great Lakes salmon to ecological processes (Hansen and Holey, Chapter 8). These values may be manifested as positive perceptions that salmon contribute to a functioning ecosystem, providing a top-level predator in a system in which predators had been largely extirpated. Conversely, others may see negative ecological values associated with salmon in the Great Lakes as an introduced, exotic species, negatively affecting the biodiversity of the system and requiring large inputs of anthropogenic materials (e.g., hatchery-raised fish, chemical sea lamprey controls).

Educational values associated with Great Lakes salmon may be related to their contributions to helping foster greater public understanding of the ecological processes active in the Great Lakes, including understanding of ecological relationships across temporal and spatial scales. Salmon hatcheries through the Great Lakes region may be a useful mechanism for educational activities.

Education related to ethics and social norms may also be fostered through activities associated with Great Lakes salmon. Ethical issues may relate to expectations for environmental stewardship and appropriate treatment of fish and the systems in which they live. Controversies about the appropriateness of salmon snagging, for example, have provided a forum to raise general angler and public debate about appropriate relationships between humans and fish and acceptable norms for harvest-related behavior. Ethical debates stimulated by Great Lakes salmon management policies indicate that appropriate behavioral norms for harvesting a species that dies after spawning may be different from accepted norms for harvesting longer-lived fish (Connelly et al. 1992).

Value differences are reflected in various comparisons of Great Lakes salmon stakeholder groups. The sections below compare stakeholders based on national origin, level of government, types of agencies, and types of citizen stakeholders. These are broad characterizations. Other groups of stakeholders exist, and the stakeholder groups presented below as rather homogeneous on this level may be quite heterogeneous when more specific management issues are considered. The intent here is to provide an overview of stakeholder perspectives and differences and discuss their implications for Great Lakes salmon management. Although the following sections describe efforts that have reported differences in various groups' values and orientations toward fishery resource management, the reasons for those differences are not entirely clear. This would be a productive and meaningful avenue for future inquiry.

Differences Between Institutional Stakeholders

As Dochoda and Fetterolf (1987) noted, pressures from various stakeholder groups may cause management agencies to respond in different ways, depending on the particular mandates of the agency, an agency's institutional culture, and the political climate in which each is operating. They correctly noted that the agencies are social institutions and thus reflect the values of the societies that created them. For Great Lakes salmon, these institutional stakeholders are primarily federal, state, and tribal agencies. The latter are an important arm of sovereign governments in the United States; legal tribal status in Canada was still being determined at the time of this writing.

Canadian, Tribal, and U.S. Fishery Agencies

The most comprehensive comparison of various perspectives toward fishery management goals held by fishery managers in Canada, the United States, and U.S.-based tribal agencies focused on attitudes toward lake trout (Knuth et al. 1995). Nonetheless, the national differences observed are instructive for considering potential conflicts surrounding Pacific salmon management discussions between the two countries.

Knuth et al. (1995) reported that, compared with U.S. fishery managers, Canadian fishery managers supported goals associated with reestablishing native species (ecological values) more than goals associated with satisfying anglers (social and psychological values). Tribal fishery managers supported most strongly goals associated with providing edible fish (physiological values) and reestablishing native species (ecological values). Further, Canadian fishery managers were more likely to hold worldviews supporting natural systems and ecological management philosophies than were U.S. fishery managers. Tribal fishery managers' views ranged between these two groups.

Canadian fishery managers also had a broader view of the range of stakeholder groups important to consider in management decision-making processes than did their U.S. counterparts. U.S. fishery managers were more likely than Canadian fishery managers to acknowledge anglers as perhaps the most important stakeholder group. Provincial fishery programs historically were supported primarily by general revenues through Ontario, with license revenues (still distributed through the central treasury) a more recent phenomenon than in most state fishery programs, which have been supported historically primarily through direct license revenues (Dochoda and Fetterolf 1987). This difference in revenue streams is reflected in different orientations to and relationships with salmon (and other) anglers and reflects differences in the importance of assigned values.

State/Provincial versus Federal Fisheries Agencies

Differences in fisheries agencies between levels of jurisdiction were also cited by Knuth et al. (1995). Managers at the provincial/state level were more likely than fisheries staff members in federal agencies to support artificial versus natural systems and utilitarian versus ecological goals, and to emphasize anglers and economic benefits (reflecting differences in held, assigned, intrinsic, ecological, and social values). Knuth et al. (1995) suggested that these differences may predict future conflicts between agencies as resources available for fishery management programs become more scarce and the intensity of debates over native species restoration versus stocking introduced species (such as Pacific salmon) increases.

Fisheries versus Environmental Quality versus Health Agencies

A variety of government agencies beyond those charged with fisheries management influence salmon-related policies and practices in the Great Lakes. A clear example of this influence is in human health issues related to the presence of chemical contaminants in fish throughout the Great Lakes.

Each of the Great Lakes states and the province of Ontario maintains a fish consumption health advisory recommending appropriate levels of noncommercial fish consumption for various audiences (e.g., women of childbearing age, other adults, children). As in other states, public health departments and environmental quality agencies partner with fishery management agencies to monitor chemical contaminants in fish, assess potential human health risks, and develop and disseminate appropriate advice to potential fish consumers (Cunningham et al. 1994; Reinert et al. 1996).

Reflecting their differing mandates and value orientations (for example, the relative importance of ecological versus physiological values), these agencies exhibit a great deal of difference in the objectives they see as primary to achieve through fish consumption health advisory programs. Agencies differ in the importance they place on addressing health risks for various subpopulations, the priority they would assign to using health advisories to encourage support for pollution clean-up programs, and the importance they place on encouraging public enjoyment of fishery resources (Knuth and Connelly 1991). Although not unique to Great Lakes salmon management issues, these differences in approach and philosophy can be important barriers to achieving a common vision about the appropriateness of stocking and managing for a set of species whose life histories contribute to their ability to accumulate chemical contaminants and transfer them to humans who consume them.

The need to better coordinate agencies involved in Great Lakes resource management is not new. Almost two decades ago, Johnson (1980) called for greater collaboration among jurisdictions and agencies (e.g., on water quantity and quality, fisheries management, and air quality), arguing that the artificial separation of these agencies for bureaucratic ease failed to recognize the interconnectedness of the components composing the Great Lakes ecosystem and the finite capacity of the system as a whole.

Even on issues that are primarily fishery management concerns (i.e., lake trout restoration), fishery and environmental quality managers differ in perspective. Knuth et al. (1995) reported that environmental quality managers were more likely than fishery managers to support fishery management goals associated with reestablishing self-reproducing, self-sustaining populations of native species (emphasizing ecological values) and less likely to support goals associated with the economic value of fish populations (deemphasizing assigned values). These differences in agency perspectives will be increasingly important as more decision-making processes involve a range of agencies and disciplinary expertise.

Diversity of Citizen Stakeholders

Although many Great Lakes fishery management agencies describe their primary responsibility as "protecting and preserving" aquatic resources, their responsibilities also focus on "rehabilitating and enhancing" and "allocating" fishery resources (Dochoda and Fetterolf 1987). The challenge, however, is to make allocation and other management decisions while addressing their responsibilities to the public trust.

Great Lakes fishery administrators have expressed frustration with the limitations presented by "the narrow and exclusive range of interests" advocated by some stakeholder groups (Dochoda and Fetterolf 1987). Some stakeholder groups are concerned primarily with maintaining or increasing the size or quality of the salmon resource (assigned, social, and psychological values) and voice little support for "diverting" fishery management efforts into native species restoration efforts (ecological values) (Gale 1987). Others are concerned more broadly with issues of clean water, habitat protection and enhancement, restoration of native fish communities, and reduction or elimination of introduced exotic species (ecological, physiological, esthetic values). When policy is implemented through more specific regulations and other constraints, however, these stakeholder groups often splinter further into "a complex array of public groups, each with its own set of concerns and preferences" (Peyton 1987).

Anglers themselves exhibit considerable diversity in their goals and perceptions of appropriate management approaches for Great Lakes salmon (e.g., controversies about the appropriateness of snagging as a legal fishing technique) (Brown et al. 1991; Connelly et al. 1992). Siemer and Brown (1994) documented a range of motivations and satisfactions held by salmonid anglers on Lake Ontario, noting that the relative importance of these varied among tournament anglers, salmon snaggers, and highly invested salmon anglers.

Cooper (1995) noted that the largest threat facing the Great Lakes ecosystem is from the introduction of exotic species. Although management actions to control exotic species introductions are focusing mostly on new invaders (e.g., zebra mussels, ruffe *Gymnocephalus cernuus*) (Mills et al. 1993), stakeholders with strong native species desires (reflecting ecological values) may challenge the appropriateness of managing for salmonids exotic to the Great Lakes and may create an increasingly contentious issue in the future.

Stoffle et al. (1983) characterized the variety of stakeholder groups that have been involved directly in fishery management issues in local Great Lakes communities, including sports anglers, commercial fishermen, commercial charter boat captains, fishing equipment store owners, parks and recreation staff members, salmon tournament organizers, local chambers of commerce, local and regional sportfishing clubs, a variety of environmental organizations, and various commercial industry associations. That the list of potential stakeholders in Great Lakes salmon management is at least as broad emphasizes the need to develop better mechanisms to understand and respond to diverse stakeholder concerns.

Future Management Challenges and Research Needs

As Dochoda and Fetterolf (1987) noted, "Public understanding and support are essential for sustaining and increasing the benefits derived from Great Lakes fisheries." In this chapter, I have presented an overview of the variety of stakeholders holding some interest in or influence on the Great Lakes salmon fishery. Mutual recognition and understanding of these diverse stakeholder perspectives is necessary for salmon management in the Great Lakes to have a rosy future; otherwise, political, legal, and other challenges are likely. How should such recognition and understanding be fostered?

Social science research, including a variety of attitude and behavior surveys as well as social impact assessment methods, can help identify competing stakeholder interests and concerns and evaluate methods available to foster mutual understanding of these various interests (Vanderpool 1987). Improved conflict resolution, negotiation, and facilitation skills should help

Great Lakes natural resource managers work more productively with one another and with citizen stakeholders (Peyton 1987). Greater opportunities for involving diverse stakeholder groups in major decision-making processes may lead, over time, to improved public confidence in and support for decisions (e.g., Landre and Knuth 1993).

Some disagreements, however, may be rooted in basic value differences that cannot be mediated or negotiated. Better understanding of concerns underlying disagreements will help those involved identify which situations may be amenable to cooperation and for which it may be more productive to agree to disagree. Without understanding one another, however, participants in Great Lakes salmon policy issues will have difficulty distinguishing which conflicts can be addressed and which may be insurmountable.

A range of decision-making aids and processes can help the fishery manager seeking to identify and incorporate these diverse perspectives into management programs. Decker and Chase (1997) provided an overall typology of management agency and stakeholder relationships in decision processes. Most useful for situations, such as Great Lakes salmon management, that involve a wide range of stakeholder views are the approaches they term "inquisitive," "transactional," and "co-managerial" (Decker and Chase 1997). Each of these approaches emphasizes building and nurturing relationships through communication and cooperation, a critical element in addressing complex fishery management issues (Nielsen et al. 1997).

Using the inquisitive approach, fishery managers would gather ideas from various types of public meetings and scientific (human dimensions) surveys. Information gathered would help portray stakeholder values and expectations. The inquisitive approach may be most appropriate in situations in which the management agencies have a compelling reason for leading the decision-making process but seek to be comprehensive in the information considered as part of that process.

The transactional approach involves stakeholders as active participants in the decision-making process, as in multiparty task forces given authority to decide, not just advise. Great Lakes fishery issues for which decision-making authority and responsibility can be spread across a wide range of stakeholders may benefit most from this approach. This method assumes sufficient familiarity among the participants about the issues and so may require substantial information gathering and discussion to ensure all participants have a reasonably equal opportunity to understand the problems to be solved, the reasons underlying those problems, and the implications of potential alternatives.

Co-management involves stakeholders actively in many stages of the management process: information gathering, decisions, implementation, and evaluation. This approach is most appropriate for Great Lakes fishery issues that require broad community involvement in their solution—for example, local community initiatives to address the conflicts involving the charter fishing industry, birdwatchers, fish, and cormorants. Co-management by definition involves community members in the implementation and evaluation phases of management, not just in the advising or decision-making processes.

Many decisions, made at many levels of government and by many types of stakeholder groups, will determine the future of the Great Lakes salmon fishery. One thing is certain: if the diversity of stakeholders concerned with Great Lakes salmon is not acknowledged, the future of salmon management programs is controversial at best.

References

Brown, P. J. 1984. Benefits of outdoor recreation and some ideas for valuing recreation opportunities. Pages 209–220 *in* G. L. Peterson and A. Randall, editors. Valuation of wildlife resource benefits. Westview Press, Boulder, Colorado.

Brown, P. J., and M. J. Manfredo. 1987. Social values defined. Pages 12–23 *in* D. J. Decker and G. R. Goff, editors. Valuing wildlife: economic and social perspectives. Westview Press, Boulder, Colorado.

Brown, T. L. 1977. Economic impact of New York's Great Lakes sport fisheries. Journal of Great Lakes Research 3(1-2):164–166.

Brown, T. L., B. A. Knuth, and F. C. Menz. 1991. Lake Ontario's sport fisheries: socioeconomic research progress and needs. Canadian Journal of Fisheries and Aquatic Sciences 48(8):1595–1601.

Callicott, J. B. 1987. The philosophical value of wildlife. Pages 214–221 *in* D. J. Decker and G. R. Goff, editors. Valuing wildlife: economic and social perspectives. Westview Press, Boulder, Colorado.

Connelly, N. A., T. L. Brown, and B. A. Knuth. 1999. New York statewide angler survey, 1996. Report 5: insights into the future of fishing in New York. New York State Department of Environmental Conservation, Albany, New York.

Connelly, N. A., B. A. Knuth, and C. P. Dawson. 1992. The failure of success in natural resource policy: Pacific salmon harvest policy in New York State. Policy Studies Review 11:24–36.

Cooper, W. E. 1995. Risks of organochlorine contaminants to Great Lakes ecosystems are overstated. Ecological Applications 5(2):293–298.

Cunningham, P. A., S. L. Smith, J. P. Tippett, and A. Greene. 1994. A national fish consumption advisory data base: a step toward consistency. Fisheries 19(5):14–23.

Decker, D. J., and L. C. Chase. 1997. Human dimensions of living with wildlife: management challenges for the 21st century. Wildlife Society Bulletin 25:788–795.

Dochoda, M. R., and C. M. Fetterolf, Jr. 1987. Public purpose of Great Lakes fishery management: lessons from the management experience. Transactions of the American Fishery Society 116:302–308.

Gale, R. P. 1987. Resource miracles and rising expectations: a challenge to fishery managers. Fisheries 12(5):8–13.

Johnson, M. G. 1980. Great Lakes environmental protection policies from a fisheries perspective. Canadian Journal of Fisheries and Aquatic Sciences 37:1196–1204.

Kellert, S. R. 1987. The contributions of wildlife to human quality of life. Pages 222–234 *in* D. J. Decker and G. R. Goff, editors. Valuing wildlife: social and economic perspectives. Westview Press, Boulder, Colorado.

Knuth, B. A. 1995. Beyond the pocketbook: held and intrinsic values associated with wild trout. Proceedings of the Wild Trout Symposium v. Yellowstone National Park, Wyoming. 5:147–150.

Knuth, B. A., and N. A. Connelly. 1991. Objectives and evaluation criteria for Great Lakes health advisories: perspectives from fishery, health, and environmental quality agencies. Human Dimensions Research Unit Series No. 91-11. Department of Natural Resources, Cornell University, Ithaca, New York.

Knuth, B. A., S. Lerner, N. A. Connelly, and L. Gigliotti. 1995. Fishery and environmental managers' attitudes about and support for lake trout rehabilitation in the Great Lakes. Journal of Great Lakes Research 21(Supplement 1):185–197.

Knuth, B. A., and L. A. Nielsen. 1989. Social and institutional performance indicators for wildlife and fishery resource management systems. Society and Natural Resources 4:351–366.

Krueger, C. C., and D. J. Decker. 1993. The process of fisheries management. Pages 33–54 *in* C. C. Kohler and W. A. Hubert, editors. Inland Fisheries Management in North America. American Fisheries Society, Bethesda, Maryland.

Landre, B. K., and B. A. Knuth. 1993. Success of citizen advisory committees in consensus-based water resources planning in the Great Lakes Basin. Society and Natural Resources 6:229–257.

Mills, E. L., J. H. Leach, J. T. Carlton, and C. L. Secor. 1993. Exotic species in the Great Lakes; a history of biotic crises and anthropogenic introductions. Journal of Great Lakes Research 19(1):1–54.

Nielsen, L. A., B. A. Knuth, C. P. Ferreri, S. L. McMullin, R. Bruch, C. Glotfelty, W. W. Taylor, and D. A. Schenborn. 1997. The stakeholder satisfaction triangle: A model for successful management. Proceedings of the Second World Fisheries Congress 2:15–21.

Peyton, R. B. 1987. Mechanisms affecting public acceptance of resource management policies and strategies. Canadian Journal of Fisheries and Aquatic Sciences 44:306–312.

Reinert, R. E., B. A. Knuth, M. A. Kamrin, and Q. J. Stober. 1996. A review of the basic principles and assumptions used to issue fish consumption advisories. Pages 98–106 *in* L. E. Miranda and D. R. DeVries, editors. Multidimensional approaches to reservoir fisheries management. American Fisheries Society, Symposium 16, Bethesda, Maryland.

Rolston, H. III. 1983. Values gone wild. Inquiry 16:181–207.

Shaw, W. W. 1987. The recreational benefits of wildlife to people. Pages 208–213 *in* D. J. Decker and G. R. Goff, editors. Valuing wildlife: economic and social perspectives. Westview Press, Boulder, Colorado.

Siemer, W. F., and T. L. Brown. 1994. Motivations and satisfactions of Lake Ontario boating salmonid anglers. Journal of Great Lakes Research 20(2):457–470.

Stoffle, R. W., D. L. Rasch, and F. V. Jensen. 1983. Urban sports anglers and Lake Michigan fishery policies. Coastal Zone Management Journal 10(4):407–427.

U.S. Department of the Interior, Fish and Wildlife Service, and U.S. Department of Commerce, Bureau of the Census. 1996. 1996 National survey of fishing, hunting, and wildlife-associated recreation. U.S. Fish and Wildlife Service, Washington, D.C.

Vanderpool, C. K. 1987. Social impact assessment and fisheries. Transactions of the American Fisheries Society 116:479–485.

Weithman, A. S. 1993. Socioeconomic benefits of fisheries. Pages 159–177 *in* C. C. Kohler and W. A. Hubert, editors. Inland fisheries management in North America. American Fisheries Society, Bethesda, Maryland.

Sustaining North American Salmon: Perspectives Across Regions and Disciplines, 195–220

Chapter 10

Uses of Resource Economics in Managing Great Lakes Fisheries: Michigan Examples

Frank Lupi and Douglas B. Jester

Introduction

Though economics can be broadly considered the study of resource allocation (Swanson and McCollum 1991), here we focus on concepts from the field of natural resource and environmental economics. For brevity, we refer to the field as resource economics. The purpose of this chapter is to review several examples of how resource economics has contributed to the management of Great Lakes salmonid fisheries. The intent is to use salmon management examples to highlight various principles of resource economics. The chapter is not meant to be a comprehensive review of resource economics, nor is it an exact history of fishery management in the Great Lakes. Rather, it is a mix of brief case studies in which economics played a role in a salmonid or related fishery management decision. Given the experience and background of the authors, the examples focus on Michigan's waters of the Great Lakes.[1]

The chapter responds to our observation that the economics literature contains many fine expositions of economic theories and principles that

[1]Douglas Jester worked in the Michigan Department of Natural Resources (MDNR) until 1999. Most of the historical information in this chapter, as well as the discussions of the factors influencing fisheries management decisions, is based on his personal knowledge and experience during his years in the fisheries division.

bear on fisheries management yet contains comparatively few illustrations of actual cases where resource economics was applied to and affected fisheries management. More complete discussions of resource economics can be found in texts such as Tietenberg (1988), discussions that focus on fishery management include Anderson (1986) and Cunningham et al. (1985). Freeman (1993) presents methods for estimating the nonmarket values, while Swanson and McCollum (1991) provide an excellent overview of economic theories that bear on recreational fisheries management. Based on our observation that applications of resource economics to fisheries management are needed in the literature, the approach of this chapter is to highlight, by brief examples, specific cases where these well discussed economic concepts have been applied to the management of Great Lakes salmonids.

The field of resource economics is broader than can be covered in a chapter. Topics addressed range from the intertemporal allocation and sustainable use of natural resources to the valuation of these resources. Resource economic issues associated with fisheries management include optimal harvest of a renewable resource, the effects of alternative property rights regimes and the unique problems associated with open-access resources, the intended and unintended economic incentives implicit in alternative fishery management policies, safe minimum standards, and the valuation and allocation of harvest between commercial and recreational interests. Even if we wished to address all of these topics, there are simply not enough economic applications involving Great Lakes salmon to do so. Despite this, five fishery examples that to varying degrees involve Great Lakes salmon serve to illustrate a host of economic concepts that can be used to help manage salmon in a sustainable fashion. Broadly, the five examples address both commercial and recreational exploitation of salmon, where exploitation is thought of as harvest pressure, as well as other threats to salmonid populations, such as hydropower-related fish kills, chemical residues in tissues, and exotic species. Some of the economic concepts to be illustrated include the role of property rights (e.g., open-access fisheries and limited-entry programs), externalities, net economic value (consumer surplus), net present values and discounting, valuation of market goods (e.g., commercial fishery buyback programs), and the valuation of nonmarket goods (e.g., recreational fishing). The examples and concepts are germane to economically sustainable management of Great Lakes salmonids in the face of the potential for overexploitation. More generally, the economic principles addressed are applicable to a broad array of fisheries management issues.

Example 1: Limited-Entry Commercial Fishing Licenses in Michigan

This section discusses the limited-entry commercial fishery regulations adopted by Michigan in the late 1960s, which created a fixed number of transferable fishing licenses. For some time, Michigan had commercial fishing licenses, but they were not restricted in number. Thus, the Michigan waters of the Great Lakes were essentially operated under open-access fishing conditions. The basic regulatory change to a limited-entry fishery was based on an economic recommendation to create a tradable property interest in the fishery resource. Economists have long recognized the unique problems that fisheries pose for the efficient allocation of resources. Indeed, commercial fisheries management is one of the cornerstone research areas in natural resource economics.

Prior to the 1960s, the basic problem for the Great Lakes commercial fishery was one of open access, or an absence of rights to exclude others from using the resource. Open-access fisheries often are subject to overfishing and stock depletion because an open-access fishery lacks a property rights regime in which new entrants can be excluded (Anderson 1995). Indeed, evidence for overfishing was widespread in the Great Lakes by the late 1950s (Brown et al. 1999). The basic economic problem of the open-access fishery is that the inability to exclude effort from the fishery leads to overexploitation of the resource and diminished economic value (Gordon 1954). Under open access, there will be too much fishing effort, and the stock will be too small relative to economically efficient levels. Interested readers are referred to Munro and Scott (1985) or Conrad (1995) for more theoretical economic overviews of the problem.

Not only does open access lead to overexploitation of the fishery, but it also does not maximize the economic value of the fishery (Gordon 1954). One approach to solving this dilemma, suggested in an early economic analysis of the issue, would be to convert from a situation of open access to sole ownership (Scott 1955). Sole ownership would generally be infeasible for most fisheries, but the sole-ownership analysis provided key insights for fisheries management. To maximize the value of the fishery, a sole owner would exert less effort than fishers in the open-access case. With less effort, one also has larger fish stocks. This insight is the basic economic rationale for regulations aimed at limiting effort. Limited-entry license programs accomplish this through controlling the number of entrants (Rettig and Gintner 1978).

Limited licenses, by themselves, do not tend to achieve all of the efficiency of the sole-ownership solution. With limited licenses, additional regulations are usually needed to prevent effort from increasing in other ways, such as increased investments in equipment, vessels, and other forms of capital (Munro 1998; Cunningham et al. 1985)—a phenomenon known as capital stuffing. In addition to limiting entrants, making the licenses tradable can further improve the efficiency of resource utilization. Cunningham et al. (1985) discuss the potential economic efficiencies associated with license transferability; they allow high-opportunity-cost holders of licenses to sell them to low-opportunity-cost buyers. The reason for the efficiency gains is that those with lower costs of utilizing the licenses have an economic incentive to purchase the licenses from those with higher costs for using the licenses.

Another major insight provided by the Gordon (1954) analysis and the Scott (1955) solution is the recognition of the basic property rights problem associated with the open-access fishery. This recognition of the primary importance of property rights is embodied in the economics of modern fishery management regimes, such as individual transferable quotas (ITQs; Anderson 1995). Not long after their exposition and acceptance, these economic insights found their way into the fisheries management practices on the Great Lakes. In the late 1960s, Anthony Scott, then an economics professor at the University of British Columbia, made recommendations for a limited-entry fishery in the Great Lakes. About this same time, the Wisconsin waters of Lake Superior also began a limited-entry fishery (see Bishop et al. 1978 for a discussion of the Wisconsin experience).

Michigan had discussed the imposition of commercial fishing regulations from the 1880s through the 1920s, and comprehensive regulations were adopted in Michigan in 1929 (Talhelm 1978) and about this same time elsewhere in the Great Lakes. The 1929 regulations contained a licensing requirement and instituted gear and season regulations. In addition, the regulations began the systematic reporting and collection of data on catch. Principal commercial species at the time were lake trout *Salvelinus namaycush*, whitefish, and ciscos (lake herring) *Coregonus artedi* in the upper lakes and walleye and perch in the lower lakes and Saginaw Bay. Because there was no limit on the number of commercial fishing licenses, however, the fishery was still characterized by open-access property rights.

The effects that the open-access conditions were having on the fishery were brought to the fore by events in the 1950s. About this time, sea lamprey *Petromyson marinus* began to take a severe toll on the commercial fishery of the Great Lakes. Following the impacts of lamprey and commercial fishing pressures on the lake trout fishery, there was a shift of effort to whitefish. A similar decline in whitefish led to another shift of effort to cisco (and chubs). The biggest and shallowest stocks were exploited first, and then effort shifted to deeper stocks (bloater chubs *C. hoyi*). Eventually, the three largest species of cisco became extinct (Coon 1999). On Lake

Michigan, a brief period in the late 1940s and early 1950s saw the collapse of the lake trout, whitefish, and cisco stocks. A similar shift in species, followed by collapse, occurred on Lake Huron, albeit more gradually (Brown et al. 1999). Most of the native species were heavily fished in the commercial fishery, and several, including lake trout, were pushed over the edge by sea lamprey mortality. Also, in the mid-1950s, alewife had become established in the Great Lakes. By the 1960s, major commercial fishery targets were gone, and Lake Michigan was dominated by alewife *Alosa pseudoharengus*. Large spring die-offs of alewife resulted in unsightly beaches and other problems.

In the early 1960s, the governor of Michigan charged the fisheries division of the Michigan Department of Natural Resources (MDNR) with examining the fisheries situation in the Great Lakes. About this time, the fisheries division chief retired, and Jim McFayden, then a professor at the University of Michigan, was asked to serve as division chief. He agreed and undertook a strategic planning process for the division. The strategic plan included two novel components. First, the plan recognized the potential of the Great Lakes as a sport fishery. Second, the plan pointed out the need for new commercial fishery regulations. After McFayden's service, Howard Tanner became chief of the MDNR fisheries division. Working with Wayne Tody, also of the fisheries division, Tanner initiated the introduction of salmon to the Great Lakes (Tanner and Tody, Chapter 7) and pursued a program of limited entry for commercial fishing in Michigan. On the recommendation of the MDNR, the Michigan Legislature adopted a comprehensive revision of Michigan commercial fishing laws. Following some necessary administrative rule making, the new laws were adopted in late 1967.

These new rules and regulations involved several notable policy shifts. First, the commercial harvest and sale of salmon and lake trout were banned in 1967 (Brown et al. 1999). This ban on commercial harvest and sale of salmonids was, in part, a recognition that the value of these species to the recreational fishery was greater than it was to the commercial sector (Talhelm 1978). Second, large-mesh gill nets were banned, forcing a shift to impoundment gear. Impoundment gear reduced mortality for nontarget salmonid species that were valuable to the recreational fishery. Third, state money was allocated to compensate for some of the costs of the shift in gear type. Fourth, a limited-entry fishery was established by banning further issuance of licenses. Fifth, license renewals were practically assured. Sixth, licenses were now made transferable. The latter two policy shifts created an ongoing property interest in the existing commercial fishing licenses.

One of the new restrictions on the licenses was spatial. With the new rules, license holders were limited to fishing within 50 mi of a prescribed home port. An administrative change restricted the licenses to the area where the most recent fishing had been occurring. In addition, licenses were now written with quantitative limits on either fishing gear or harvest. In the Lake Superior lake herring fishery, and in a few other situations, harvest limits were also set, thereby effectively creating a total allowable catch for these fisheries. In terms of gear restrictions, most licenses specified the number of nets that could be fished and a named vessel that would be used. Furthermore, zones were created where commercial fishing was to be excluded, thereby serving as refuges for recovering stocks such as lake trout.

License allocation occurred as follows: those existing license holders that had not caught some amount of fish in recent years were told they would no longer have licenses. Thus, very small fishing operations were eliminated. A fixed pool of money was allocated by the Michigan Legislature for compensation for the eliminated licenses, and the money was divided on the basis of historical catch. Individuals who had been fishing gill nets were offered compensation for the costs of converting to impoundment gear, and they were given the option of taking the compensation and converting or taking the compensation and leaving the fishery.

The effects of the rules and regulations began to take hold in the early 1970s. One effect was a dramatic reduction in licenses. Prior to the changes in licensing, there were more than 1,000 commercial licenses, but by the late 1970s, the number of commercial licenses had fallen below 200. In the mid-1980s, the number of license holders fell to about 160. However, the reduction in licenses is confounded by an increase in tribal fishing effort, which is not subject to state licensing requirements.

The new state licensing requirements and regulations were contributing factors in the reassertion of tribal fishing rights in the 1970s in Michigan treaty waters of the Great Lakes. Since the 1985 consent decree settling the treaty dispute between Michigan and several Indian tribes, tribal commercial fishing is no longer subject to state licensing requirements. The 1985 consent decree created separate fishing zones for Michigan's state-licensed commercial fishing and tribal commercial fishing. Because a large segment of the commercial fishing activity—that conducted by the tribes—is not state licensed, it is difficult to fully assess the success of the Michigan limited-entry program. Nevertheless, evidence does suggest that limited-entry licensing and related regulations have been successful in preventing overexploitation by the state-licensed commercial fishery. Since 1985, state-licensed commercial fishing effort has primarily targeted whitefish, and the whitefish stocks have recovered in Lakes Michigan and Huron since the mid-1970s (Brown et al. 1999). These stock improvements are attributable in part to fishing regulations, limited-entry licenses, and suppression of sea lamprey.

In sum, the Michigan limited-entry, transferable license program was based on the recommendation of an economist to create an excludable and tradable property interest in the resource. That the license applies to a specific vessel and a specific number of nets, combined with other gear restrictions, serves to limit partially some of the overcapitalization problems often associated with limited-entry programs (Munro 1998; Cunningham et al. 1985). Finally, even though there have been relatively few license transfers, the fact that licenses are transferable allows for the licenses to be allocated in an economically efficient manner.

One final point that should be made about these fishery regulations concerns the external effects of commercial fishing on the recreational fishery. External effects, or externalities, are a major concept in resource economics. An externality exists whenever the well-being of some economic agent (either a firm or an individual) depends not only on his or her activities but also on the activities under the control of some other agent (Tietenberg 1988). As a classic textbook example, pollution caused by a firm may directly affect the health of others, or it may diminish the quality of a fishery. Either of these are examples of production externalities because the production practice imposes external costs on others. One economic problem with externalities such as pollution is that, to the extent that pollution is a cost that is external to the producer's decision making, too much of the externality will occur. In a sense, unregulated pollution is like a production input for which no price is paid. Thus, the firm will emit more pollution than is socially desirable. One approach for dealing with pollution externalities that is advocated in the economics literature is to charge a price for pollution (for example, a per unit discharge fee). The price will then induce the firm to include the amount of pollution emitted in its profit calculations. If the fee is set at the point where it equals the marginal economic value of the damages associated with the externality-generating activity, it is considered economically efficient. Faced with this fee, the externality will be internalized (i.e., the external costs will be fully accounted for in the decision making of the firm). Real-world examples of such marginal social cost pricing are rare, but an example will be provided in the third case of this chapter. More commonly than social cost pricing schemes, laws and other social institutions are often structured so as to reduce external effects, as in the case of commercial fishing.

The effects of commercial fishing on recreational fishing—direct effects due to harvest of a recreational species or indirect effects such as gear-induced mortality on nontarget recreational species—are examples of production externalities. The general trend in Great Lakes commercial fishery management, especially in Michigan, has been to recognize the larger relative values of the recreational fishery (Samples and Bishop 1982) and manage the commercial fishery to reduce the external effects on the recreational fishery. For example, commercial fishing for most recreational species has been banned, and many gear regulations, such as requirements to use impoundment gear, are aimed at reducing mortality to nontarget species (especially for salmonids such as lake trout and chinook salmon).

Example 2: Commercial Fishing License Buyback Efforts

This case discusses a buyback of commercial fishing licenses and the associated efforts to establish the economic value of fishing licenses. As a part of the negotiations between Michigan and tribal fishing interests that led to the 1985 consent decree, Michigan agreed to set aside certain zones in Lakes Michigan, Huron, and Superior for tribal fishing. The state agreed to remove state-licensed commercial fishers from these zones. To accomplish this, Michigan sought to buy back the licenses of the state-licensed commercial fishing operations within the newly created tribal zones. Targeted license holders were offered the larger of the market value of their assets or the present value of the income generated by their licenses. The acquired fishing assets were transferred to tribal fishermen to encourage conversion from gill nets to trap nets, which reduce mortality on nontarget species such as lake trout. Because the fishing licenses were not purchased under a voluntary buyback program, the state had to develop a means of arriving at compensation for the licenses. This effort required the determination of the value of each business as an ongoing concern. The state used a third-party accounting firm, Price-Waterhouse, to establish these license values. Assets were valued "but for" the glut in the (local) market due to the buyback.

As a part of the legal and political proceedings associated with the buyback, there was some confusion and controversy over how to value a business. We note that the general approach for valuing a business is well established in the accounting, finance, and economics literature. The basic idea is to calculate the profits or income stream associated with the fishing enterprise over time and then translate that into a single dollar value. That single dollar value represents the value today, or the present value of the entire stream of income that could be produced by the business. The process of discounting and computing net present values is one of the key economic concepts of this chapter.

Determining the present value is akin to finding the amount of money invested today that would generate that stream of income. Put differently, one seeks to determine the capitalized value of residual cash flows associated with holding a license. If the discount rate is r, the discount factor for year t is $d_t = (1 + r)^{-t}$. The present value of the benefit, B_t, that occurs in year t is $B_t \times d_t = B_t/(1 + r)^t$. Thus, as the interest rate becomes larger, the net present value of a benefit or cost that occurs in year *t* becomes lower. The specific formula employed to calculate net present value is

$$\text{net present value} = \sum_{t=0}^{T} \frac{1}{(1 + \text{rate})^t} \times \{\text{benefits}(t) - \text{costs}(t)\},$$

where *t* indicates time periods and "rate" refers to the discount rate. Figure 1 illustrates how discount factors decrease over time. Figure 1 shows the discount factor applied to dollar values for years along the horizontal axis at different discount rates. For example, at a 10% discount rate, a dollar in 30 years is worth 6 cents today; yet at a 3% discount rate, a dollar in 30 years is worth 41 cents today. The reason for this is that, if one invested 6 cents today at a 10% interest rate, it would grow to one dollar in year 30, whereas with the lower 3% interest rate, one would have to invest more, 41 cents, for the value to reach one dollar in year 30. Therefore, as the rate of return that can be earned on alternative investments becomes higher, the weight that gets placed on future benefits and costs in the net present value calculation decreases.

Returning to the buyback issues, fishing licenses can be thought of as an asset that enables the fishing enterprise to generate income from fishing. The economic value of that asset then depends on the present value of the

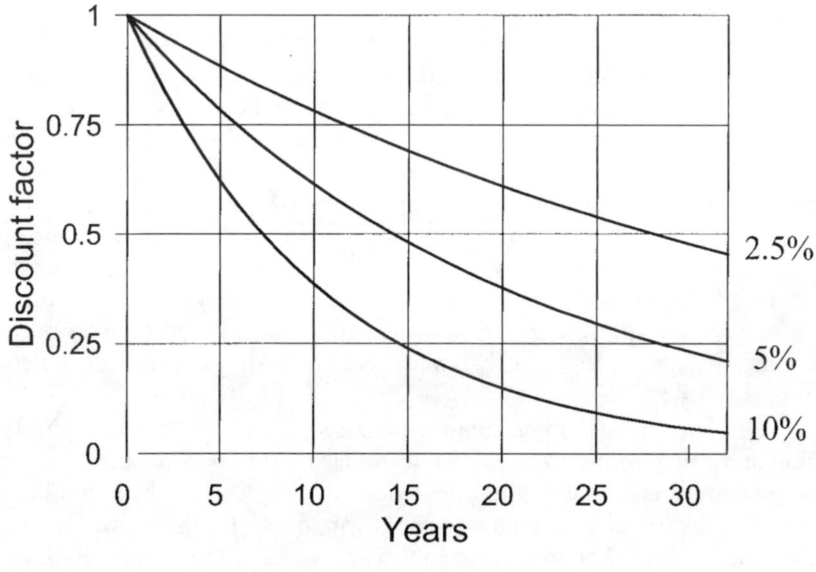

Figure 1. Decline in discount factors over time, illustrating how higher interest rate results in less weight being placed on future benefits.

stream of income that can be generated by that asset. Similar to the net present value formula presented above, this asset value for an income stream can be computed as follows:

$$\text{license value} = \text{present value of income stream}$$
$$= \sum_{t=1}^{T} \frac{1}{(1+r)^t} \times \text{income}_t \qquad ,$$

where r is the discount rate (interest rate from alternative investment options) and t represents time periods. To calculate the income stream that could be provided by a fishing enterprise, one must determine the revenues and costs of the operation for the present and the future. If it is possible to assume that income and the discount rate do not change over time and that the planning horizon is infinite, then the present value formula simplifies and is given by the income divided by the discount rate.

In the above formula, income refers to profits—that is, the difference between revenues and costs (which is directly analogous to benefits minus costs in the net present value calculation described above). Revenues refer to the value of business sales or the gross value of landings in the case of a commercial fishing enterprise. In other words, revenues are the sum of daily harvest multiplied by price. Among other things, costs consist of officer compensation, crew compensation (wages), operating expenses such as fuel, finance charges, and depreciation on capital such as boats and nets.

In the policy discussions and the court cases associated with the buyback offers, the commercial fishermen, their attorneys, and some policy-makers expressed the opinion that the approach taken by the state was very controversial, and this notion persists to this day. We point this out because different methods of calculating costs or revenues affect the calculated value of the business, but they do not invalidate the basic formula. In practice, there are many serious issues to deal with when performing a valuation. For example, the choice of discount rate typically would need to account for the degree of risk within the industry as well as the liquidity of the operations. As one might expect, it was the assumptions about the components of the business valuation formula that were actually in dispute, not the general approach itself.

One such disagreement stemmed from some confusion between inflation and discounting. In the above formulae, when revenues and costs are based on current dollars, the calculation of net present values should be based on a real discount rate adjusted for inflation. To compare dollar amounts in different periods, economists often make adjustments for inflation based on the consumer or producer price index. The process of discounting and calculating present values of future dollar amounts is sometimes confused with methods of accounting for inflation. Inflation is an entirely different phenomenon. Inflation refers to a rise in price levels such that it costs more money to acquire a fixed set of goods. Inflation, per se, has nothing to do with the time value of money. However, when making calculations regarding future benefits and costs, economists usually deal in real or inflation-adjusted terms so that benefits and costs are expressed in the real dollars at some point in time, and inflation-free rates of return are used in the calculations. Unlike inflation, discounting has to do with the time value of money, not differences in price levels over time. Distinguishing between inflation and discounting became a point of disagreement in the 1985 buyback. Eventually, because the Price-Waterhouse income computations were based on current dollars, it was agreed to use a real, inflation-adjusted discount rate for computing net present values. The resulting license values were larger than initially estimated by the state, but this was the correct decision from the standpoint of economic theory.

Another disputed component of the commercial fishing business valuation was the treatment of managerial time. In the valuation formula, managerial time is typically valued at some opportunity cost, such as a wage rate that could be earned in alternate employment. If a zero wage is used, the analysis will overstate that value of the enterprise—after all, losing the license does not prevent one from other employment. Alternatively, choosing a wage rate for managerial time that exceeds the wage that could be earned elsewhere will understate the value of the enterprise. Again, determining what wage rate to use was another issue over which the state and the fishermen disagreed.

The discussion of these areas of disagreement is not meant to trivialize the assumptions that enter into the valuation formula but rather to stress that these details, not the general approach, cause controversy. The general formula for the value of a business is not controversial. Resource economists devote a significant amount of time and effort explaining nonmarket valuation of resources to noneconomists. One of the interesting lessons of the buyback was how the public and policy-makers lacked a general understanding of market valuation principles. For example, the value of the firm must consider operating costs—it is not simply based on gross receipts. There is also broad confusion over stocks and flows (present values or capitalized values versus income in any year). In light of potential misunderstandings surrounding business valuations, it is not surprising that some confusion continues over nonmarket valuation of the recreational and ecosystem services provided by fisheries.

Example 3: Natural Resource Damage Assessment and Externality Pricing

This section discusses the role of economics in a court settlement regarding natural resource damages associated with the operation of a hydroelectric facility. The case discussed is noteworthy for several reasons. First, it is an example of a natural resource damage assessment for fish kills in the Great Lakes. Second, the values at issue were for recreational fishing resources and were established through the use of a nonmarket valuation approach referred to as the travel cost method. Third, the fish kills illustrate the economic concept of external effects (externalities), and the settlement internalizes these effects through future payments equal to the economic value of the damages caused by this hydropower production. Before discussing the case in detail, we turn to a brief discussion and definition of some of the economic concepts that will be addressed.

Net Economic Value (Consumer Surplus)

The economic concept of value is driven by the notion that trade-offs are associated with any choice. When a choice is made, that which is forgone is referred to as the opportunity cost of that choice. This notion figures into the economic concept of value because economic value is based on people's willingness to give something up to obtain something else. Economic value could be translated into any standardized units, but money serves as a convenient scale. For this reason, economists measure the economic value of a good or service by individuals' willingness to pay for it—that is, their willingness to give up money to obtain it. Thus, economic values are not prescribed by economists, planners, or managers—they are inferred from the choices made by individuals.

The net economic value of a good or service is typically calculated using the concept of consumer surplus. Consumer surplus reflects the amount of money, above and beyond what is currently spent, that an individual would be willing to pay for a good. That is, consumer surplus is the difference between willingness to pay and what is actually paid; it is not measured by the cost (or price) of a good. Consumer surplus is the appropriate measure of the net economic value associated with an activity such as recreational fishing. An estimate of consumer surplus is the theoretically desired measure of economic benefits for use in a cost-benefit analysis (Freeman 1993). Moreover, reductions in consumer surplus are recognized as the relevant measure of damages in natural resource damage assessment cases (Ward and Duffield 1992). Below, the concept of consumer surplus is further illustrated in the discussion of the travel cost method for measuring the net economic value of recreational fishing.

Valuing Nonmarket Goods

Many goods are traded in markets and have prices associated with them. The market is a convenient mechanism for revealing economic values because the prices provide signals that can be used to infer economic values. However, many goods and services, such as the recreational use of a fishery, are not traded on markets and do not have market prices. The fact that something does not have a market price does not mean that it does not have economic value. Economists have developed a suite of valuation techniques for such nonmarket goods or services. These techniques include the travel cost method and the contingent valuation method (the latter is explained in example 4), both of which have been used to influence salmon management in the Great Lakes.

The essence of the travel cost method is to determine statistically the relationship between price (travel costs) and quantity (the number of visits to some site); this relationship is referred to as the demand curve (Bockstael et al. 1991). Figure 2 presents a graph of a hypothetical recreational demand curve for trips to a recreation site. The price per trip is on the vertical axis, and the horizontal axis shows the number of trips. As price decreases, we expect anglers to take more trips; therefore, demand curves slope downward. The travel cost demand curve shows the number of trips an individual will make at any given price. In Figure 2, when the price of a trip to the site is TC, the number of trips taken by anglers is *N*. In addition, the demand curve shows the maximum amount that individuals would be willing to pay to take each trip. For the initial trips, the amount anglers would be willing to pay exceeds the price, TC. Anglers will take trips until their benefit (willingness to pay) for an additional trip just equals their cost—point *B* on the demand curve. The total willingness to pay for *N* trips is given by the area 0ABN (all of the shaded areas in the figure). The total amount that anglers actually pay for *N* trips is *N* × TC (the area labeled "expenditures"). What is actually spent is less than the total willingness to pay. The accumulated difference between what individuals would pay and what they actually pay is referred to as consumer surplus (area TCAB). Consumer surplus is the appropriate measure of the net economic value or benefits of a recreational fishing site (Freeman 1993; Ward and Duffield 1992).

The Ludington Pumped Storage Plant

This case study involves the Ludington Pumped Storage Plant (LPSP), located on the eastern shore of Lake Michigan in Ludington, Michigan. This hydroelectric plant was designed to supply electricity during periods of peak electrical demand. The LPSP sits on a hill and pumps Lake Michigan water

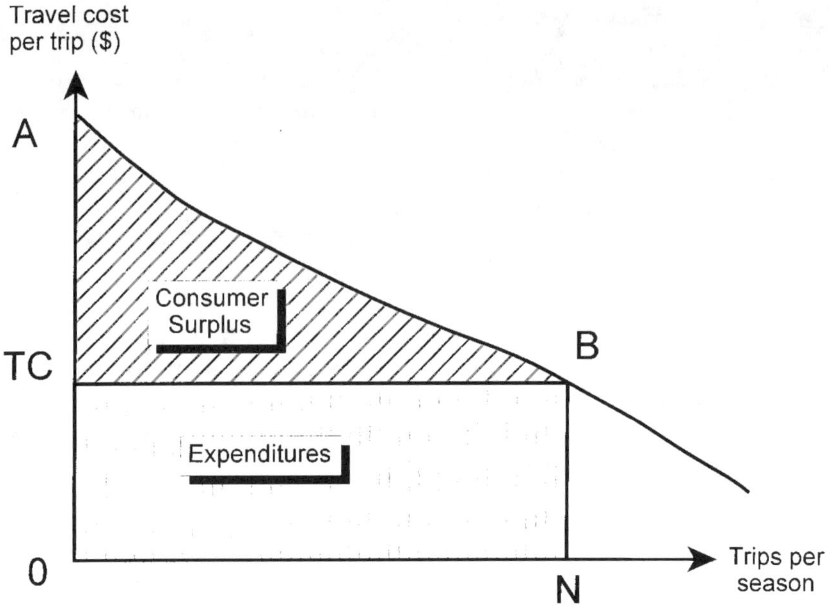

Figure 2. Demand curve for a recreation site, illustrating consumer surplus.

up to a storage reservoir 360 ft above lake level. During times of low electrical demand, base load power is used to pump water up to the reservoir, and during times of peak electrical demand, water is released through turbines. The LPSP is the largest facility of its kind in the United States. The facility is capable of generating 1.8 million kilowatts of energy when operated at full capacity (although overall, the facility consumes more energy than it generates). In a form of arbitrage, the plant is economically viable because it takes advantage of the difference in the peak value and the off-peak value of electricity: it pumps water when electrical demand (and price) is low and generates electricity when demand (and price) is high.

The operation of the LPSP results in an economic externality in the form of fish kills. Fish are entrained in the water that is pumped out of Lake Michigan. In the process of being pumped up to the reservoir and then returned to Lake Michigan, a significant portion of the fish are killed or suffer mortal injuries (GLFT 2000). As a result of the damages to the fishery caused by the LPSP, several conservation organizations filed actions with the Federal Energy Regulatory Commission (FERC), which licenses the facility, and also sought regulation of the facility under the Clean Water Act. The state of Michigan also filed for compensation for fish kills (Civil Action No. 86–7075-CE, Michigan circuit court) and entered the FERC proceedings seeking the installation of barriers to reduce future damages to Great Lakes resources. The conservation groups, the MDNR, the federal government, and several Indian tribes with damage claims joined in a comprehensive settlement agreement with the utility (GLFT 2000).

A trust, the Great Lakes Fishery Trust (GLFT), was established to handle the assets created by the settlement. The GLFT is a nonprofit organization governed by members of organizations that were party to the settlement. Because the parties to the lawsuit included state and, indirectly, federal interests, as well as tribal and private interests, this trust was devised as a means to facilitate the settlement. The parties generally did not want to give up control over the assets that would be awarded under the damage assessment, but all agreed that the funds should "benefit the Great Lakes fishery by providing funding for enhancement, propagation, protection and replacement of Great Lakes fishery resources" (GLFT 2000). Thus, the trust was an innovative institutional response that solved the economic and political problems of allocating the settlement funds among the parties.

As a part of the settlement, the utility must maintain a seasonal barrier net to restrict the number of fish entering the facility, which minimizes future fish losses. The barrier net is in place from April to October of each year. In addition, the utility makes annual compensation payments that are based on the effectiveness of the barrier net. In compensation for past damages, the utility also provided substantial land holdings to Michigan and to the GLFT, in addition to undertaking numerous projects to improve angler access on utility-owned properties.

The compensation claims associated with the case were based on estimates of the natural resource damages due to the fish kills. The natural resource damages are the market and nonmarket economic values that were lost because of the fish kills. The nonmarket values for the fish kills served two purposes in the negotiations. First, they were used to set damage amounts for damages prior to installation of the barrier net. Second, they were used to establish the fees for fish kills that would be paid on future fish losses. These fees are tied to net effectiveness, as well as the volume of water pumped. The fees are discussed more fully following a discussion of how the economic damages were calculated.

The case for damages was based on losses of four types of fishes: sturgeon, whitefish and bloaters, trout and salmon, and forage fish. Each of these served as a category for which economic damages or forgone net economic values were calculated. For sturgeon, there is no commercial fishery, but there is a very small and highly restricted recreational fishery. Because of the diminished nature of sturgeon stocks in Lake Michigan (a native species of special interest in Michigan), as well as the sturgeon's unique characteristics as a large, long-lived "dinosaur" of the Great Lakes, it was considered likely that there would be significant nonuse values associated with sturgeon. However, it was considered feasible at that time to replace the sturgeon losses through additional stocking, so sturgeon was valued at the cost of replacing the fish. These costs were estimated on the basis of internal MDNR estimates of sturgeon stocking costs.

Whitefish and bloaters are targeted by commercial fisheries in Lake Michigan. The LPSP caused long-term injury to these stocks but only in a local area around the plant because these stocks do not generally move over large areas. Because of the limited movement of these fish and the limited-entry licensing in effect, the affected whitefish and bloater fisheries were treated as a single, local stock with one fishery. The losses to the local commercial fishery were then valued using the profit margin from the Price-Waterhouse records from the 1985 license buyback (discussed in example 2).

For recreational fishing, the operating assumption was that if the plant had not killed fish, recreational effort would have killed them (directly as trout and salmon catch or indirectly as food for trout and salmon that are then caught by recreational effort). Forage fish availability was considered to be a limiting factor to the population of trout and salmon in Lake Michigan. Thus, any reduction of forage fish would result in a loss of trout and salmon, and an agency could not simply stock trout and salmon to replace them because the availability of food in the form of forage fish was limiting.

The benefits for trout and salmon losses, both direct and due to forage losses, were estimated using the travel cost model developed by Jones and Sung (1993). This was a multiple-site model that related the recreation demand at a Great Lakes fishing site to the trout and salmon catch rates at the site. Hence, the model could be used to estimate economic value to recreational anglers due to a change in catch rate at a site. The method used to estimate values is based on measuring changes in the consumer surplus (net economic value) of recreational fishing that are associated with the changes in catch rates. We discuss these values in three parts.

First, an estimate was made of the benefits that would accrue because of increased catch rates that would occur if more forage fish were available. This is essentially a production function or damage function approach to valuation (Freeman 1993), in which the forage fish are inputs into the production of the recreational harvest of salmonids. The resulting changes in Lake Michigan trout and salmon catch rates were valued using the Jones and Sung (1993) model. The costs of stocking the trout and salmon that would have eaten these forage fish were subtracted from these benefit estimates because it was assumed the MDNR would have stocked additional trout and salmon had more forage fish been available. This value for changes in salmonid catch rates as a result of forage abundance was just under half the annual damages.

Second, there were some losses in freshly stocked trout and salmon smolts entrained and killed by the plant. These losses were valued using replacement costs based on the stocking cost to offset this decline. This was a small part of the settlement. Third, an estimate was made for the losses associated with trout and salmon entrained by the LPSP. Any adult trout and salmon that were entrained in the LPSP were directly lost from the pool of catchable fish, so they were directly valued via a change in catch rates in the Jones and Sung model. This value for changes in salmonid catch rates due to entrainment of adult salmonids was also just under half the annual damages.

The annual damage amount was about US$6 million. Almost all of the damages were the result of the direct and indirect trout and salmon recreational losses. The damage amounts for the losses in sturgeon replacement costs, commercial fishing for whitefish and bloaters, and stocking costs for entrained smolts were a smaller portion of overall damages. Most of these (about $270,000) were due to the sturgeon stocking costs, with about $25,000 in annual damages for the losses in whitefish and bloaters.

Three issues with important economic aspects were addressed in the settlement negotiations: past fish kills, the use of barrier nets as prevention of fish entrainment, and payments for ongoing fish kills that depend on net effectiveness as well as the degree to which the plant is used. To move forward toward a settlement, the parties agreed that the above fish values would be used for valuing all future fish losses, but they would bargain and compromise about past values. This approach was adopted to ensure that in future plant-operating decisions, the utility would face appropriate *incentives* for operating the plant and investing in net effectiveness. The parties also agreed to structure future payments so they could be built into the costs and rate structure of the utility.

The mesh on the barrier net to prevent fish from being entrained was greater than four inches, but most of the forage fish (about half the annual value) are smaller than that. Therefore, the settlement required that the effectiveness of the barrier net be measured and the amounts paid for annual fish kills be reduced on the basis of barrier net effectiveness. The net effectiveness is monitored by measuring the diversity of fish inside and outside the barrier net.

Fish mortality data and economic values were arranged in tables that present, on a monthly basis, the economic value of fish expected to be killed per volume of pumping. These tables essentially provide information on the value of fish killed per kilowatt hour of electricity generated by the plant. The calculation tables separate the values for fish that could, in principle, be protected by the net because of their size and those that could not be protected by the net. The value can then be adjusted for barrier net effectiveness by changing the percentage of the first component that enters into the damage fee calculation. The dollar value of damage per kilowatt hour is calculated by multiplying the net ineffectiveness factor by the value of the fish that can be protected by the net and adding to that the value of fish that cannot be protected by the net.

In the tabulations, the fish values and the net effectiveness vary by months to account for seasonal variations in fish abundance in the vicinity of the plant. Thus, the annual payments made by the utility for fish kills can be altered by changing the timing of plant operation, as well as by changing the amount of electrical generation. This provides an economic incentive to the utility to use the plant less often during times of peak fish abundance in the vicinity of the plant. Consequently, the utility will use the LPSP less and other sources more when the externality charges for the LPSP are highest. Internal cost estimates were made by the utility, and the operating schedule was adjusted in response to the externality charges for the fish kills. As a result, the cost to the utility is less than US$6 million per year because of the net effectiveness and the changes in operating amounts and timing. About half the annual future damages were mitigated by the installation of the barrier net. In addition, another 10% of the future losses were mitigated by changing the amount and timing of electrical generation.

The LPSP settlement illustrates several aspects of natural resource economics. First, the case illustrates the use of nonmarket valuation in a major natural resource damage assessment. In this case, the nonmarket valuation technique is a modern variant of the travel cost method for estimating the demand and value of a recreational resource, and the bulk of the economic damages resulted from losses in recreational fishing opportunities for salmonids. Second, the settlement led to the establishment of the GLFT, which has become a significant financial resource that provides funding for research to assure the sustainability of the fishery. The trust itself represents a unique institutional innovation that solved the interjurisdictional problem of controlling the settlement funds. Third, the settlement established a forward-looking mechanism for compensation for future fish kills. Pricing the effects of the plant operation on the fishery internalized the external damages of the plant operation. The utility now has an economic incentive to make investments and adjust production amounts and timing to reduce these charges. Externality charges such as the fish-kill payment act as prices that affect marginal decisions by a firm. Though this type of externality pricing is widely advocated in theoretical literature in resource economics, externality pricing (basically a pollution tax in this case) is relatively rare in practice.

Example 4: Contingent Valuation for the Value of Information in Fish Consumption Advisories

Many species of fish in the Great Lakes, including salmonids, have been found to contain chemical residues. In many situations, the residues are at levels thought to pose health risks to individuals who might consume the fish. Thus, the Great Lakes states and Ontario issue public health advisories (fish consumption advisories) that provide the public with information about the potential risks of consuming fish with chemical residues. The fish consumption advisories are generally specific to certain sites and differ by species and size of fish. Because of their position in the food chain and their fat content (in which chemicals accumulate), salmonids are frequently listed in the consumption advisories. As pointed out by Krieger and Hoehn (1998), though the advisories generally indicate the sites where fish have been found to contain chemical residues, few of the advisories indicate the sites where fish have been tested and found to be safe. Moreover, because agency budgets are limited, not all fishing sites have been tested. This case concerns the amount of testing that is conducted to monitor fish for contaminants, and the example illustrates how the contingent valuation method of estimating economic value has been applied to a nonmarket good.

Like the travel cost method discussed above, the contingent valuation method is another technique that economists have developed to measure economic values for nonmarket goods (Freeman 1993). Contingent valuation measures economic values by directly asking individuals to state a preference, or in some cases to state their valuation, for a carefully defined good or service. In many contingent valuation studies, a referendum approach is used to present people with the choice of a good or service at a fixed price. The offered price is then varied across a sample of individuals, and econometric methods are used to estimate how the discrete "yes" or "no" responses vary with the price. This process yields a model of people's willingness to pay for the good or service. The method can be applied to market or nonmarket goods, and it is even capable of estimating values that are not directly related to an individual's use of a resource (Freeman 1993).

Krieger and Hoehn (1998) used a contingent valuation study to examine Michigan anglers' willingness to pay for additional testing of sites under the state's fish contaminant monitoring program. The Krieger and Hoehn study administered the contingent valuation approach with a mailed questionnaire using the referendum elicitation format. As a part of the study design, Krieger and Hoehn also tested the effect that the provision of information about testing results has on the willingness to pay for additional testing. Krieger and Hoehn examined two information strategies: full and partial disclosure of tested sites. Under a partial-disclosure program, anglers were informed about sites where fish were tested and found to contain chemical residue levels that would trigger an advisory, but they were not informed of the sites tested and found to be safe. Under a full-disclosure program, anglers were informed about advisory sites, and they were also offered information about the sites where fish had been tested but no advisory was necessary (i.e., where fish were found to be safe).

The Krieger and Hoehn results indicated that Michigan anglers were willing to pay little for continuation or expansion of a testing program that does not inform them of safe sites. In contrast, anglers were willing to pay a substantial amount for a program that tells them about relatively safe sites. With partial disclosure, willingness to pay for testing of additional sites each year was not significantly different from zero. However, with full disclosure, the value of testing additional sites was positive and significantly different from zero. With full disclosure, the economic benefits to anglers of testing an additional 110 sites per year were estimated to be US$6.14 per year per angler. The per-angler cost of testing an additional 110 sites per year was estimated to be in the range of $1.10 to $2.20 per year. Hence, additional testing with full information would result in a net economic value of about $4 to $5 per year per angler.

This case demonstrates one of the many ways in which the contingent valuation method of estimating the value of nonmarket goods can be applied to the evaluation of fisheries management actions. In addition, the application emphasizes that information can be an output of fisheries management, and the value of a program can depend on the information available to anglers. In partial response to these results, the Michigan Department of Community Health has since altered Michigan's advisories to include information on how anglers could obtain tables showing the test results for each lake tested.

Example 5: Sea Lamprey Treatments on the St. Mary's River

Sea lamprey are exotic to the Great Lakes and prey on lake trout and other Great Lake salmonids. Ongoing sea lamprey control efforts have permitted the recovery of lake trout stocks in Lakes Superior and Michigan. However, rehabilitation of lake trout in Lake Huron has been hampered by large populations of lamprey believed to be originating from the St. Mary's River, the large channel system connecting Lakes Superior and Huron. This case reports on research that estimates some of the economic benefits and costs associated with several new options for controlling sea lamprey in the St. Mary's River. All treatment options were shown to have positive net present value, even though only part of the economic benefits has been measured.

Efforts to restore lake trout in Lake Huron and northern Lake Michigan have been stalled by the large numbers of lamprey that spawn in the St. Mary's River. Because of the large flow volume and size of the St. Mary's River, traditional control methods such as the lampricide TFM would require substantial funds or would be of limited effectiveness. However, newly developed, lower cost treatments offered the potential to control lamprey populations. Even so, treating the St. Mary's River would require a substantial increase in the lamprey treatment budget for the Great Lakes. Funding the lamprey treatments in the St. Mary's River by reducing treatments in other tributaries would compromise gains in lamprey suppression in other areas of the Great Lakes. Recently, Lupi et al. (1999) examined the economic benefits and costs of treating sea lamprey in the St. Mary's River. The Lupi, Hoehn, and Christie analysis addressed the following question: was the magnitude of the economic benefits of reducing sea lamprey populations large enough to justify the additional cost?

Lupi, Hoehn, and Christie examined three St. Mary's River treatment options that consisted of combinations of two treatments: sterile male release and trapping (SMRT) and granular bayluscide applications (GB). The sterile male release and trapping program involves trapping lamprey, sterilizing the males, and releasing the sterile males. Granular bayluscide is a chemical treatment that can be applied by helicopter to habitats where larval lamprey are concentrated. Most of the larval lamprey in the St. Mary's River are concentrated in a relatively small area, so this method is far more effective than a typical riverwide TFM treatment. The granules sink to the riverbed where larval lamprey are located. The three sea lamprey treatment options considered by Lupi, Hoehn, and Christie were as follows: annual sterile male release and trapping (SMRT only), annual SMRT along with applications of granular bayluscide every five years (SMRT + GB), and annual SMRT along with a *one-time* application of granular bayluscide (SMRT + GB1). In cost, granular bayluscide is much more expensive than sterile male release. Applications of granular bayluscide cost just less than US$5 million dollars per application. The sterile male release and trapping program costs about US$300,000 a year.

Lupi, Hoehn, and Christie applied a travel cost model developed by Lupi and Hoehn (1998) to estimate Michigan anglers' demand for recreational fishing on the Great Lakes. The model related anglers' choice of recreational fishing site to salmonid catch rates. By using data on anglers' trade-offs between travel costs and catch rates, the model can estimate anglers' willingness to pay for changes in catch rates of salmonids. The model was used to evaluate some of the benefits of changes in lake trout populations that resulted from control of sea lamprey. As with the Ludington example mentioned earlier, benefits were measured using the concept of consumer surplus or net economic value. Recall that consumer surplus measures the difference between anglers' maximum willingness to pay for changes in environmental quality and what they actually pay (willingness to pay minus expenditures).

Lupi, Hoehn, and Christie reported that the economic benefits associated with increases in fishing success for lake trout that would accrue to Michigan's resident recreational anglers outweighed the cost of treating lamprey on the St. Mary's River for all of the treatment alternatives examined. At discount rates often used for federal water resource projects (3.5%), the annualized angling benefits, reduced by the treatment costs, ranged from about US$3 millon to $6 million per year, depending on the treatment alternative. The results are consistent with earlier research that found that the benefits of sea lamprey control in the Great Lakes outweigh the costs (Talhelm and Bishop 1980).

Note that this is a marginal analysis in the terms of Anderson (1994; see also Swanson and McCullom 1991 or Peterson and Cordell 1991), in that the change in the economic value of the fishery that results from the additional lamprey treatments is compared with the increased cost of the treatments. The total overall economic value of the fishery, though an often requested statistic, has no bearing on the management decisions at play here. The reason for this is that elimination of the fishery is not a realistic management action. In contrast, changes in the quality of the fishery are much more relevant to fisheries management.

In the case of treating lamprey on the St. Mary's River, there was a disparity between when the control costs are incurred and when the benefits are realized. The benefits of lamprey control accrue in the future because it takes time for the lake trout populations to recover. However, the costs of control begin to be incurred in the present. In addition, the benefits are not constant through time—they grow as the stock of fish grows. This raises a fundamental economic question of how to compare policy options where the costs and benefits differ in timing as well as in their magnitude over time. As mentioned in example 2, in such situations, economists often argue for comparing the net present value of the alternative investment options.

Among the treatment alternatives examined, Lupi, Hoehn, and Christie found that the option with the largest measured net present value depended on the discount rate used. For example, at low discount rates, the treatment involving both continued granular bayluscide applications and sterile male trapping and release was preferred to the SMRT alone. However, at higher discount rates—greater than 6%—the converse was true. With the combined treatments, lake trout populations grow the fastest and achieve a larger long-run population, but this option also has large recurring costs. Alternatively, SMRT alone has the lowest costs, but it also has the slowest growth in lake trout populations. As the discount rate rises, relatively less weight is placed on the larger future lake trout population, and relatively more weight is placed on treatment costs that are incurred in early years, prior to the growth of lake trout populations.

The only values that were quantified in the St. Mary's River study were those associated with recreational fishing by Michigan resident anglers. These values outweighed the full costs of treating the St. Mary's River. In this case, the benefits that were quantified outweighed all of the costs. This illustrates the fact that cost-benefit analysis does not always require one to estimate all possible benefits associated with some management action to yield results that can be used for fisheries management decisions. In the Lupi, Hoehn, and Christie example, it is likely that there are recreational fishing benefits for users in other jurisdictions, and it is possible that there are substantial nonuse values associated with restoration of populations of lake trout (a native salmonid species) and reduced populations of sea lamprey (an invasive species). However, in the case at hand, it was not necessary to quantify those benefits. In the lake trout case, the use values to a portion of the beneficiaries were large enough to justify the treatments, even though nonuse values associated with lake trout restoration were not measured. The Lupi, Hoehn, and Christie result provides an example of the one-way benefits test described by Krutilla and Fisher (1985).

Like the fish contamination monitoring example, the St. Mary's River sea lamprey treatment case provides another example of cost-benefit analysis, in which a nonmarket valuation technique was used to estimate economic values. In this case, the travel cost method was employed. This case also illustrates the importance of discounting and accounting for benefits and costs that are unevenly distributed over time, a common feature of natural resource management. Finally, this case illustrates that all possible benefits do not necessarily need to be quantified for an analysis to provide useful information.

Conclusions

This chapter presents several examples in which resource economics has contributed to the management of salmonids in the Great Lakes. The first case highlighted the limited-entry, transferable commercial fishery licenses by Michigan. The second case illustrated economic principles of asset valuation for market goods and their use in a fishing license buyback effort. The third case discussed a major natural resource damage assessment settlement, as well as an innovative externality pricing approach that was adopted as a part of the settlement. Cases 4 and 5 illustrated applications of cost-benefit analysis to fisheries management, and cases 3, 4, and 5 highlighted various ways in which nonmarket valuation techniques have influenced management. These efforts have contributed to the success and sustainability of salmonids in the Great Lakes by reducing commercial exploitation pressure, achieving compensation for environmental damage and installing incentives to reduce this damage, improving information for anglers and consumers about fish contamination, and demonstrating the benefits of controlling sea lamprey, thereby helping to reduce damages caused by exotics.

As these examples illustrate, the use of economics in the management of Great Lakes fisheries has great potential. We have presented a few success stories, but fishery management decisions in the Great Lakes continue to be made almost exclusively by biologists and politicians. Examples of thorough economic analyses of alternative management decisions are the exception, rather than the rule. In fact, with a few exceptions, most Great Lakes fishery management agencies do not currently employ or have never employed economists. Ultimately, effective utilization of economic concepts and information for sustaining Great Lakes salmon fisheries is more likely to occur when economics is brought to bear in the early stages of any decision, rather than after the fact.

References

Anderson, L. G. 1986. The economics of fisheries management (revised and enlarged edition). Johns Hopkins University Press, Baltimore, Maryland.

Anderson, L. G. 1994. An introduction to economic valuation principles for fisheries management. Special Publication 94-2. Great Lakes Fishery Commission, Ann Arbor, Michigan.

Anderson, L. G. 1995. Privatizing open access fisheries: individual transferable quotas. Pages 453–474 *in* D. W. Bromley, editor. The handbook of environmental economics. Blackwell Scientific Publications Publishers, Cambridge.

Bishop, R. C., G. V. Johnson, and K. Samples. 1978. Wisconsin's limited entry experience. Pages 317–332 *in* R. B. Rettig and J. J. C. Gintner, editors. Limited entry as a fishery management tool. University of Washington Press, Seattle.

Bockstael, N. E., K. E. McConnell, and I. E. Strand, Jr. 1991. Recreation. Pages 227–355 *in* J. B. Braden and C. D. Kolstad, editors. Measuring the demand for environmental quality. North Holland Publishing, New York.

Brown, R. W., M. Ebener, and T. Gorenflo. 1999. Great Lakes commercial fisheries: historical overview and prognosis for the future. Pages 307–354 *in* W. W. Taylor and C. P. Ferreri, editors. Great Lakes fisheries policy and management: a binational perspective. Michigan State University Press, East Lansing, Michigan.

Conrad, J. M. 1995. Bioeconomic models of the fishery. Pages 405–432 *in* D. W. Bromley, editor. The handbook of environmental economics. Blackwell Scientific Publications Publishers, Cambridge.

Coon, T. 1999. Ichthyofauna of the Great Lakes Basin. Pages 55–71 in W. W. Taylor and C. P. Ferreri, editors. Great Lakes fisheries policy and management: a binational perspective. Michigan State University Press, East Lansing, Michigan.

Cunningham, S., M. Dunn, and D. Whitmarsh. 1985. Fisheries economics: an introduction. St. Martin's Press, New York.

Freeman, A. M, III. 1993. The measurement of environmental and resource values: theory and methods. Resources for the Future, Washington, D.C.

GLFT (Great Lakes Fishery Trust). 2000. Website and documents for the Great Lakes Fisheries Trust: *<http://www.glft.org>*.

Gordon, H. S. 1954. The economic theory of a common property resource, the fishery. Journal of Political Economy 62:124–142.

Jones, C. A., and Y. D. Sung. 1993. Valuation of environmental quality at Michigan recreational sites: methodological issues and policy applications. Final report: EPA Contract No. CR-816247–01–2. U.S. Environmental Protection Agency, Washington, D.C.

Krieger, D. J., and J. P. Hoehn. 1998. Improving health risk information for sport anglers: the value of information about chemical residues. North American Journal of Fisheries Management 18:411–421.

Krutilla, J. V., and A. C. Fisher. 1985. The economics of natural environments: studies in the valuation of commodity and amenity resources (second edition). Resources for the Future, Washington, D.C.

Lupi, F., and J. P. Hoehn. 1998. A partial benefit-cost analysis of sea lamprey treatment options on the St. Marys River. Report to the Great Lakes Fishery Commission, Ann Arbor, Michigan.

Lupi, F., J. P. Hoehn, and G. C. Christie. 1999. Valuing non-indigenous species control and native species restoration in Lake Huron. In W. D. Shaw, editor. Benefits and costs in natural resource planning. Western Regional Research Publication, Reno, Nevada.

Munro, G. R. 1998. The economics of overcapitalization and fishery resource management: a review. Discussion paper 98-21. Department of Economics, University of British Columbia.

Munro, G. R., and A. D. Scott. 1985. The economics of fisheries management. Pages 623–674 *in* Handbook of natural resource and energy economics. Volume 2. North Holland, Amsterdam.

Peterson, G. L., and H. K. Cordell. 1991. Estimating recreation demand functions: experience with the public area recreation visitor survey. Pages 316–327 *in* D. Guthrie, J. M. Hoenig, M. Holliday, C. M. Jones, M. J. Mills, S. A. Moberly, K. H. Pollock, and D. R. Talhelm, editors. Creel and angler survey methods in fisheries management. American Fisheries Society, Symposium 12, Bethesda, Maryland.

Rettig, R. B., and J. J. C. Gintner, editors. 1978. Limited entry as a fishery management tool. University of Washington Press, Seattle.

Samples, K. C., and R. C. Bishop. 1982. An economic analysis of integrated fisheries management: the case of the Lake Michigan alewife and salmonid fisheries. Journal of Great Lakes Research 8:593–602.

Scott, A. D. 1955. The fishery: the objective of sole ownership. Journal of Political Economy 63:116–124.

Swanson, C. S., and D. W. McCollum. 1991. Application of economics to recreational fisheries management: an overview. Pages in 299–315 *in* D. Guthrie, J. M. Hoenig, M. Holliday, C. M. Jones, M. J. Mills, S. A. Moberly, K. H. Pollock, and D. R. Talhelm, editors. Creel and angler survey methods in fisheries management. American Fisheries Society, Symposium 12, Bethesda, Maryland.

Talhelm, D. R. 1978. Limited entry in Michigan fisheries. Pages 300–316 *in* R. Rettig and J. J. C. Gintner, editors. Limited entry as a fishery management tool. University of Washington Press, Seattle.

Talhelm, D. R., and R. C. Bishop. 1980. Benefits and costs of sea lamprey (*Petromyzon marinus*) control in the Great Lakes: some preliminary results. Canadian Journal of Fisheries and Aquatic Sciences 37:2169–2174.

Tietenberg, T. 1988. Environmental and natural resource economics. Scott, Foresman and Company, Glenview, Illinois.

Ward, K. M., and J. W. Duffield. 1992. Natural resource damages: law and economics. Wiley, New York.

Sustaining North American Salmon: Perspectives Across Regions and Disciplines, 221–242

Chapter 11

Managing Great Lakes Fisheries Under Multiple and Diverse Authorities

Margaret R. Dochoda and Michael L. Jones

Whoever devised the Great Lakes Fisheries Convention and the American statute enforcing the same could have spared Solomon many an anxious moment as he proceeded to judgment upon babies and other equally indivisible entities.

Henry Reiff (1959, cited in Piper 1967)

Introduction

The Great Lakes make up a vast and complex ecosystem, and the fisheries they support have a colorful and frequently troubled history of use, abuse, and rehabilitation by the human inhabitants of its basin. No less complex than the lakes themselves are the many institutions that share responsibility for, or at least a substantive interest in, the management of Great Lakes fisheries. No fewer than two federal governments, two intertribal authorities representing 15 Indian tribes, one provincial government, eight state governments, and one international commission play key roles in the process of making fishery management decisions on the Great Lakes. Further, this classification does not include the many other stakeholder interest groups, divisions or departments with governments, and other transjurisdictional institutions whose actions can affect Great Lakes fish. In this chapter, we examine these institutional arrangements, focusing on those aspects of the current arrangements and the history of their development that explain why Great Lakes fishery management has been relatively free of conflict, at least when compared with the other salmon management case histories considered in this book. To tell this story it is impossible to separate Great Lakes salmon management from other fish species and stocks of concern in the basin, so our discussion is not specifically about salmon but rather about institutional arrangements that bear on the management of all economically valuable fish stocks in the Great Lakes.

Why examine how Great Lakes fishery agencies cooperatively manage the fishery? The many, different authorities responsible for Great Lakes fish often revisit—and revise—the processes they employ to better respond to new challenges and opportunities. For example, in 1997, the agencies revised their joint strategic plan (GLFC 1997, see below), and in 2000, the Great Lakes Fishery Commission reorganized and renamed the Habitat Advisory Board to better meet its strategic plan commitments to state, provincial, and tribal fishery agencies. Moreover, comparisons of cooperative institutional arrangements suggest that there is much to be learned from others' experiences (Miller 1984; Gunderson et al. 1995; Kerr and Ryder 1997). This is a dialogue that has only recently been joined by fishery managers from the Great Lakes.

In this chapter, we will focus on two agreements fundamental to coordinating fish management in the Great Lakes: the Canada-United States Convention on Great Lakes Fisheries (1954), which created the Great Lakes Fishery Commission; and the Joint Strategic Plan for Management of Great Lakes Fisheries (GLFC 1997). The constitutional framework and factors such as geography will be examined for their influence on coordinated management. Finally, we will briefly touch on the interplay between the commission and the fishery agencies in implementing the two agreements.

Authorities for Managing Great Lakes Fish

Management arrangements for the Great Lakes fishery are ultimately shaped by political and constitutional realities. There are six categories of constitutional or treaty authority of interest in Great Lakes fish management decisions (Table 1). In Canada, authority is constitutionally divided between Canada, which preserves and conserves, and Ontario, which "owns" the fish and determines access. In the United States, the states have responsibility for both conservation and regulation of access, except as legislated by the Congress under the interstate commerce or treaty-making clauses of the constitution. In implementation, the two countries differ in that Canada has tended to delegate its preservation responsibilities to Ontario—for example, in Canada-Ontario Agreements. In the U.S. waters of the Great Lakes, federal override of state authority requires federal legislation such as

Table 1. Agencies with interests in Great Lakes fishery management and signatory to A Joint Strategic Plan for Management of Great Lakes Fisheries, 1981 (as revised in 1997).

Chippewa-Ottawa Resource Authority	New York Department of Environmental Conservation
Fisheries and Oceans Canada	Ohio Department of Natural Resources
Great Lakes Indian Fish and Wildlife Commission	Ontario Ministry of Natural Resources
Illinois Department of Natural Resources	Pennsylvania Fish and Boat Commission
Indiana Department of Natural Resources	U.S. Fish and Wildlife Service
Michigan Department of Natural Resources	U.S. Geological Survey
Minnesota Department of Natural Resources	Wisconsin Department of Natural Resources
National Marine Fisheries Service (U.S.)	

the Endangered Species Act of 1973 or a treaty such as the Convention on Great Lakes Fisheries. Native fishing rights have emerged in both countries as important considerations in fish management. Implementation of fishing rights guaranteed in U.S. Indian treaties has brought two intertribal agencies and their combined 15 member tribes into the decision-making process in Lakes Superior, Michigan, and Huron. In Canada, the nature of aboriginal rights guaranteed by section 35 of the 1982 Canadian constitution is still being determined in the courts (Dochoda 1999).

As one might expect, treaty and constitutional realities have greatly influenced selection of strategies to coordinate fish management in the Great Lakes. Because of the international border, a United States-Canada treaty (the 1954 Convention on Great Lakes Fisheries), not a U.S.- or Canada-only instrument, was the preferred umbrella for coordination. As a treaty, the convention imparts a certain level of formality into coordinated activities conducted under its umbrella. Because U.S. treaties, including Indian treaties, and interstate commerce legislation can preempt state authority, states tend to be alert for what they might perceive as encroachments on their respective authorities. The ever-alert Great Lakes states always attend questions of process, especially those related to the 1954 convention or the 1997 strategic plan. This attention to process has been critically important to any success the Great Lakes agencies have enjoyed in coordinating their management. Because there are so many agencies with fish management interests (Table 2), the process that has developed over the years focuses on shared objectives for fish communities for each of the Great Lakes, and it relies on mutually derived, science-based knowledge of the structure and function of those communities as the basis for recommending and assessing management actions.

Table 2. Nature of constitutional or treaty authority or interest in Great Lakes fish management decisions.

	Canada	United States
Federal	1. Treaty-making 2. Conserve & preserve (often delegated to Ontario)	1. Treaty-making 2. Interstate commerce (must be legislated)
Ontario/states	Proprietary (i.e., regulate access)	Preserve and regulate
Natives	Aboriginal rights (being defined in courts)	Treaty fishing rights

History of Management Arrangements

Many of the key elements of today's binational fishery management process in the Great Lakes owe their origin to the 1954 (U.S.-Canada) convention on Great Lakes fisheries, which led to the creation of the Great Lakes Fishery Commission. Prior to the 1950s, however, attempts to create institutional arrangements that enabled binational fishery management were largely unsuccessful (Fetterolf 1980). According to Fetterolf (1980), international agreements were prepared in 1908 (treaty between the United States and Britain—fisheries in U.S. and Canadian waters) and 1946 (convention between the United States and Canada for the development, protection, and conservation of fisheries of the Great Lakes), but they failed to achieve their purpose because the U.S. Congress either failed to approve consequent regulations (1908 treaty) or failed to ratify the agreement (1946 convention).

Why were these agreements unsuccessful, and what was different about the 1954 convention? The earlier agreements failed largely because they included provisions to allow an international body to impose regulations on Great Lakes fisheries. Such provisions were deemed unacceptable because they would lead to an erosion of authority for the existing regulatory agencies, particularly U.S. state governments. It now seems clear that the 1954 convention was successful for two reasons. First, as stated by Fetterolf (1980, p. 1590), "The delegates [responsible for drafting the treaty] had past experience working on their side. They knew that granting regulatory power to a commission would be touchy." As a result, Article IV of the convention assigned the following duties to the Great Lakes Fishery Commission:

(a) To formulate a research program or programs designed to determine the need for measures to make possible the maximum sustained productivity of any stock of fish in the convention area which, in the opinion of the commission, is of common concern to the fisheries of the

United States of America and Canada and to determine what measures are best adapted for such purpose;

(b) To coordinate research made pursuant to such programs and, if necessary, to undertake such research itself;

(c) To recommend appropriate measures to the Contracting Parties on the basis of the findings of such research programs;

(d) To formulate and implement a comprehensive program for the purpose of eradicating or minimizing the sea lamprey populations in the Convention Area; and

(e) To publish or authorize the publication of scientific and other information obtained by the commission in the performance of its duties.

Of critical significance is the fact that none of these duties implies any formal regulatory authority for the commission over Great Lakes fisheries (except possibly as pertains to sea lamprey). Second, the preparation and signing of the convention was motivated primarily by the universal recognition of the urgent need to develop an international program aimed at controlling the exotic sea lamprey, the commission's only management responsibility. Concerned over the serious damage to some fisheries and the continuing threat to others (Convention on Great Lakes Fisheries 1954), Great Lakes fishery managers regarded the sea lamprey as the principal impediment to the recovery of these stocks. The parties to the convention had a clearly recognized common enemy, the sea lamprey, and were prepared—indeed eager—to work together to rehabilitate an economically and ecologically decimated Great Lakes fishery.

Nonfederal entities such as the states and Ontario had many—if not most—of the management responsibilities for the Great Lakes, and the commission was given no rule-making authority, so this question arose early: how could the commission be effective in advising and coordinating its activities with nonfederal authorities? The convention called for governor-nominated representatives (representing state agencies, users, and the public at-large) to advise U.S. commissioners. These advisors continue to serve the commission today, and in 1999, the commission also established Canadian advisors to serve a similar role. As well, U.S. enabling legislation (the Great Lakes Fisheries Act of 1956) directed the secretary of state to copy the governor of each Great Lakes state on any conservation measures recommended by the commission, plus "his comments thereon." (As deemed appropriate, the secretary could also inform other public agencies.) While U.S. agency advisors were formally provided for in the convention, two other provisions eventually supported development of even more constructive relationships with fish management agencies: as part of its "study and advise" mandate, the commission was encouraged to coordinate as well as to undertake research, and it could work with any public or private person or agency to do so. These provisions created wide discretionary power for the commission to pursue virtually any issue deemed relevant to Great Lakes fisheries. Ultimately, it was this that led to the development of the Joint Strategic Plan for Management of Great Lakes Fisheries, which we discuss in detail later in this chapter.

Though initially engaged in developing its program for successfully controlling the predacious sea lamprey, the GLFC soon turned its attention to its other duty under the convention, studying and advising Canada and the United States on issues and measures to ensure the productivity of the fishery. The early initiatives were well received by the states and the province, without whose support the commission's efforts would have been much less successful. Examples of early initiatives include lake trout rehabilitation efforts; stemming the decline of Lake Erie walleye; and broader reviews of Great Lakes fishery issues, information gaps, and needed measures. In 1958, the commission, alarmed at the decrease in numbers of Lake Superior lake trout, convened a special committee on lake trout rehabilitation in which all agencies in the upper three lakes were invited to coordinate and assess measures for the preservation of broodstock and the disposition of hatchery stocks (GLFC 1958). In 1960, representatives of Michigan, Minnesota, Ontario, and Wisconsin went so far as to request commission guidance on harvesting, precisely the regulatory authority that states defended so carefully in discussions leading to the formation of the commission. As a consequence, an ad hoc committee on regulation of the Lake Superior lake trout fishery recommended that the states and province adopt an annual catch limit recommended by the commission for distribution among the four jurisdictions (GLFC 1973a). In 1962, Ohio requested that the commission consider the decline of Lake Erie walleye and recommend appropriate management measures (GLFC 1973a). This request eventually led to a science-based, interjurisdictional process that resulted in coordinated harvest restrictions in the wake of the 1970 closure of the fishery because of mercury contamination (Hatch et al. 1987). In 1964, the commission completed the Prospectus for Investigations of the Great Lakes Fishery, in which agency researchers cooperated with the commission in setting out, in general terms, their views on fishery problems and necessary investigations with recommendations both general and specific to the lakes (GLFC 1964a), most of which were being pursued a decade later (GLFC 1973a).

The commission was also effective in strengthening existing institutional arrangements. At its 1964 annual meeting, the commission asked its cooperators from both national sections (i.e., U.S. and Canadian biologists and managers) how the commission might encourage the development of programs from the prospectus. Ken Loftus, one of Ontario's most respected fisheries scientists and administrators, responded to this request:

> Mr. Loftus indicated that the Research Branch of the Ontario Department of Lands and Forests has been pleased with the informal association of the Commission with lake committees established before the commission was organized. It felt that a more formal association of the Commission would be even better now. He suggested that committees sponsored by the Commission be established for each lake. These committees would meet annually to plan and construct cooperative investigations and coordinate management on the lakes (GLFC 1964b, p. 11).

Beginning the following year, fishery agencies abandoned existing arrangements and joined lake committees organized under the umbrella of the commission (GLFC 1965).

In 1973, the commission established a Great Lakes fish disease control committee (now the Fish Health Committee) at the recommendation of the Lake Michigan committee (GLFC 1973a), whose study group had surveyed fishery agencies throughout the Great Lakes and found only three with specific regulations for controlling the importation of fish diseases in the Great Lakes basin (GLFC 1973b). (With the sea lamprey predation on lake trout and the overabundance of the exotic prey fish alewife, agencies had come to rely on lake trout and salmon hatcheries and thus needed a basinwide approach for disease management.) The new disease committee was charged with studying and proposing measures for the control and eradication of fish disease. The disease committee's principles and programs (Hnath 1993), though formally proposed to the commission, came to be followed carefully by the fishery agencies represented on the committee.

A Joint Strategic Plan for Management of Great Lakes Fisheries

Satisfied with progress in commission lake committees, the states declined an opportunity to form a U.S. regional fishery management council under the Magnuson Fishery Conservation Act of 1976 (Dochoda 1999). Instead, in 1981, directors of eight state, one provincial, and two U.S. and one Canadian federal fishery agencies negotiated and signed the Joint Strategic Plan for Management of Great Lakes Fisheries, which reiterated and further developed their commitment to the lake committee process. Two U.S. intertribal agencies and the U.S. Geological Survey—the latter had been assigned the Great Lakes Science Center from the U.S. Fish and Wildlife Service—were invited to sign later, in 1989 and 1997, respectively. The commission is not a signatory but was crucial in facilitating the plan's development and implementation, and it has twice passed resolutions of support. The plan's "action arms" are the lake committees and other management committees of the Great Lakes Fishery Commission. The plan was updated in 1986 and 1997.

The joint strategic plan has four parts: an overarching goal statement, a list of general issues related to the goal, four strategies for working together, and strategic procedures that specify working arrangements and processes. The goal statement orients fish agencies, through good stewardship of self-sustaining fishes and some level of stocking, to work together to secure fish communities to meet needs identified by society (GLFC 1997, p. 9):

> To secure fish communities based on foundations of stable, self-sustaining stocks, supplemented by judicious plantings of hatchery-reared fish, and to provide from these communities an optimum contribution of fish, fishing opportunities, and associated benefits to meet needs identified by society for
>
> - wholesome food,
> - recreation,
> - cultural heritage,
> - employment and income, and
> - a healthy aquatic ecosystem.

The plan identifies and discusses five broad issue areas: lost fishing opportunities, instability of fish communities, inadequate environmental quality, competition and conflicts among users of the fishery resources, and new emerging issues. Within the area of community instability, the plan gives specific attention to sea lamprey and overharvesting, as well as invasions and introductions of other nonnative organisms. Land use concerns, water use concerns, and atmospheric inputs are highlighted in the environmental quality area. The user conflicts section singles out allocation among jurisdictions, commercial versus sport fishing, native people versus other users, and access to the resource. Finally, the emerging issues section raises concerns about the possible effects of future changes in global climate on Great Lakes fisheries.

Signatories to the plan have subscribed to four strategies for coordinating their management: consensus, accountability, ecosystem management, and information to support management:

1) *Consensus must be achieved when management will significantly influence the interests of more than one jurisdiction.*

The authors of the plan again recognized the importance of assuring the signatory agencies that the plan would not usurp the individual agencies' authority to manage their fisheries. Nevertheless, they argued strongly for the mutual benefits of cooperative agency action, claiming that such consensus "would not only strengthen all of them [individual agencies] in their individual requests for support of needed management initiatives but would also provide a good deal of incentive to act in accordance with the group interest and intent." Among other matters, consensus is required to establish fish community objectives for each of the Great Lakes, as discussed below.

2) Fishery management agencies must be openly accountable for their performance.

For consensus management to be effective, the cooperating parties must have access to information on the management plans and performance of each partner agency. The plan argues that mutual evaluation of management proposals is important not only to ensure that each party's interests are being maintained, but also to afford opportunities for "integrated operational programming employing the best available fisheries science and technology."

3) The parties must exercise their full authority and influence in every available arena to meet the biological, chemical, and physical needs of desired fish communities.

The plan advocates an ecosystem approach to Great Lakes fishery management. An obvious implication of such an approach is that issues beyond the traditional scope of fishery management bear on the success of management programs. Most of these issues—for example, water quality, exotic invasions—broaden the institutional scope for management well beyond the mandates of the agencies signatory to the plan. As a result, this strategy exhorts the parties to broaden their activities—for example to "engage in a process of structured, multi-level, ongoing dialogue with federal, state, provincial, and tribal environmental agencies...." (This strategy was strengthened to one of advocacy in 1997. In 1981, fishery agencies had adopted the milder mode of a supplicant: "Fishery agencies *shall endeavor to obtain full consideration* by the Great Lakes environmental agencies of the potential impacts of their activities and decisions on fishery needs and objectives.")

4) Fishery agencies must cooperatively develop means of measuring and predicting the effects of fishery and environmental management decisions.

Finally, the plan recognizes the central importance of high quality information to support management. The particular emphasis of this strategy is to seek means of implementing common approaches to data collection and developing information systems, such as databases, that are readily accessible to all the parties.

To realize the goal and to implement the strategies, the plan specifies 18 strategic procedures detailing processes for achieving consensus, ecosystem management, information sharing, accountability, and plan governance. Principal among the strategic procedures is the requirement to establish "fish community objectives," which calls for agencies on each lake to agree upon the desired structure of their respective fish community and to develop a means of measuring progress toward its achievement. There are, however, 20 strategic procedures addressing such matters as conflict resolution, habitat advocacy, data standards, and reporting arrangements, including a state-of-the-lake conference held in rotation every five years for each lake. The complete set of strategic procedures is listed in Table 3.

Dispute Resolution under a Joint Strategic Plan

On first exposure to the joint strategic plan, the question most frequently asked is "What happens when Great Lakes fish managers disagree?" The

Table 3. Strategic procedures from A Joint Strategic Plan for Management of Great Lakes Fisheries (GLFC 1997, p. 21–24).

Consensus

1. ***Fish Community Objectives***— The Great Lakes Fishery Commission will maintain lake committees, which will define objectives for the structure of each of the Great Lakes fish communities and develop a means of measuring progress toward their achievement. Upon request of a lake committee, the commission will support development of fish community objectives that are comprehensive and principle-based.
2. ***Operational Plans***— Each fishery agency should identify its plans, which may include joint lakewide fishery management plans, for achieving the fish community objectives identified by the lake committees.
3. ***Changes in Practice***— Each fishery agency should submit all substantive changes from existing practice to the appropriate lake committee before implementation.
4. ***Consensus on Changes***— Any agency proposal for change which other agencies believe will influence their interests may become the subject of negotiations within lake committees until consensus of affected agencies is achieved.
5. ***Conflict Resolution***— If consensus cannot be achieved, a party may
    - Request the commission to arrange/facilitate an information exchange forum;
    - Seek advice of existing plan committees;
    - Ask the commission to arrange third-party mediation with any resolution being endorsed through normal plan procedures; and/or
    - Ask the commission to arrange a process involving a mutually acceptable third-party intermediary to make a nonbinding recommendation.

Table 3. continued.

Ecosystem Management

6. ***Environmental Issues***— The lake committees will identify environmental issues which relate to or may impede achievement of their fish community objectives and will work with other ecosystem initiatives, such as LaMPs, in developing and furthering plans for achieving, refining, and assessing progress on environmental and fish community objectives.
7. ***Coordination with GLWQA/LaMPs***— Lake committees will work with LaMPs to develop joint proposals to the Great Lakes Fishery Commission or other organizations to identify environmental needs relative to fish community objectives, related management plans and assessments, and to thereby provide feedback on fish community objectives.
8. ***Environmental Issue Resolution***— Unresolved or emerging environmental issues may be referred by lake committees to the parties, the Council of Great Lakes Fishery Agency, or the Great Lakes Fishery Commission to represent fishery interests in these issues to the most appropriate body or process.
9. ***Habitat Advisory Board***—The Great Lakes Fishery Commission will maintain an expert Habitat Advisory Board (HAB) to assist each lake committee, the Great Lakes Fishery Commission, and the parties to develop ecosystem objectives and identify critical habitats essential to achieving fish community objectives.
10. ***Exotic Species***— Fishery agencies will collectively identify and promote procedures that will protect aquatic resources from unauthorized introductions of nonnative species (e.g., via aquaculture and shipping).

Information Sharing

11. ***Data Standards***— The Great Lakes Fishery Commission will coordinate development and implementation of standards for recording and maintaining fishery management and assessment data, to ensure compatibility among the parties and other agencies.
12. ***Models***— The Great Lakes Fishery Commission and the parties will coordinate development and implementation of models for common use by the parties and other agencies
13. ***Information Access***— The parties are encouraged to maintain databases on the Internet. The Great Lakes Fishery Commission will maintain Internet links to party and others' databases to facilitate access, including a catalog of Great Lakes fishery assessment and research programs, which are planned or in progress.
14. ***Data Sharing***— The agencies are encouraged to provide their data to other agencies upon request if the collecting agency has had reasonable time to verify and interpret the data (such time should not normally exceed one year for assessment data and three years for research data) and to collectively develop shared information services under the umbrella of the Great Lakes Fishery Commission.

Table 3. continued.

Accountability

15. ***Decision Record***— Consensus decisions will be recorded appropriately (e.g., meeting minutes).
16. ***Agency Reports***— The fishery agencies, separately or jointly, should measure and make annual reports to the lake committees of the progress made toward achieving their mutual committee objectives.
17. ***Lake Committee Reports—*** Each lake committee will prepare an annual progress report and make recommendations to both the agencies and the Great Lakes Fishery Commission. In addition to annual reporting, on a rotational basis once every five years, each lake committee will convene a special conference focused on the state of the lake. The product of this conference will be a formal report capable of serving as a 5-year report card on progress toward a comprehensive set of fish community objectives and related measures and indicators.
18. ***Commission Report—*** The Great Lakes Fishery Commission's annual report to the governments and the public will include a summary of the lake committee reports and recommendations regarding fishery and environmental objectives, programs, and activities needing specific attention.

Plan Governance

19. ***Plan Changes—*** All parties must approve changes in or the addition of new parties to the plan.
20. ***Council of Great Lakes Fishery Agencies—*** The parties hereby establish the Council of Great Lakes Fishery Agencies. Each party will have representation on the council. The council will establish terms of reference. The terms of reference should reflect decision-making by consensus of the parties to

    - Ensure mutual accountability of the parties;
    - Ensure accountability for the implementation and periodic review of the plan;
    - Provide guidance and support to the plan's institutional arrangements;
    - Ensure timely and effective information exchange between law enforcement and fishery managers;
    - Ensure that environmental objectives are articulated and reciprocal strategies with environmental agencies are developed;
    - Provide for non-party participation by Environment Canada, U.S. Environmental Protection Agency, the International Joint Commission, and other agencies as appropriate; and
    - Inform and educate on a basin-wide perspective, including the development and implementation of a strategic communications framework that details the roles and responsibilities of the parties and the commission.

The council and the commission will cooperatively determine the appropriate mechanisms for commission support of the council.

joint strategic plan provides for dispute resolution when consensus cannot be reached on fish community objectives or other management disagreements (e.g., a change in management practice perceived to affect another jurisdiction). In such instances, an unsatisfied party may ask the Great Lakes Fishery Commission to provide nonbinding arbitration or other mediation services. It is a testament to productive relationships that independent-minded states, whose concerns delayed for 60 years the formation of the commission (Fetterolf 1980), joined with Ontario and federal fishery agencies to call on that same commission some 25 years later to provide nonbinding arbitration upon request.

Nonbinding arbitration works at several levels (Dochoda and Koonce 1994). Initially, a request for commission arbitration can prompt a reluctant partner to engage in negotiations with the concerned party. Frequently, the specter of arbitration is used by agencies to goad themselves to serious negotiation; referral to the commission is seen as an admission of defeat, and the agencies would clearly prefer not to leave decisions to a third party. Nonbinding arbitration is typically seen as a last resort where parties to a disagreement allow their positions to be subjected to the climate of persuasion. Though arbitration by the commission is nonbinding, it is not without weight. Managers acting against a public recommendation could be risking reputations or careers if eventually proven wrong.

Since the strategic plan was first signed in 1981, the commission has twice been asked to arbitrate. Most recently (August 1992), the Ohio Department of Natural Resources petitioned the commission to arbitrate an arrangement for allocation of yellow perch in Lake Erie, identifying the province of Ontario as counteragency. Arbitration procedures proposed by the commission were acceptable to both Ohio and Ontario, testimony was prepared, and arrangements were made for a hearing. However, Canadian federal officials grew concerned when they learned that the commission chairman, then an employee of Fisheries and Oceans Canada and potentially a representative to one of the parties, might possibly have to break a tie between Canadian and U.S. commissioners on an allocation issue. As recommended by Canadian officials, the arbitration process became the subject of discussion between the U.S. State Department and its Canadian counterparts. With the process under debate, the two parties, Ontario and Ohio, in March 1993 negotiated the first of two agreements that would over the next decade phase in a sharing formula based on surface area of the lake. Because of the delay in providing for an arbitration process to act on Ohio's request for arbitration, the 1997 revision of the joint strategic plan added mediation services to nonbinding arbitration as dispute resolution mechanisms that may be requested of the commission.

The other issue concerned decisions about stocking juvenile lake trout survivors of an unusual mortality then attributed to epitheliocystis (GLFC

1986a). The Lake Superior and Michigan committees referred the question to the Great Lakes Fish Disease Control Committee, which replied that not enough was yet known about the epizootiology of the disease to assess risk and referred agencies to its policy to prevent the release of fish suffering excessive mortalities from epizootics, recommending that agencies must act on their own perceptions of risk until more definitive information on the disease was available. Michigan DNR hatchery manager John Driver wrote the commission to state his belief that the infected fish should not be stocked. His recommendation was referred to the assembled lake committees, which could not reach consensus, and deferred to the commission. Uncertainty and conflicting advice persisted during discussions at the commission's 1986 annual meeting, and the commission replied to the Council of Lake Committees that it "...remained committed to the objectives and guidance of the Model Fish Disease Control Program. The burden of proof regarding a proposed action's compatibility with the program lies with the originator of the action. In this instance, the commission concurred that the evidence did not justify destruction of stocks" (GLFC 1986b).

By 1 July 1988, scientists had identified a new herpes virus, epizootic epitheliotropic disease (EED) virus, which was found to be the pathogen responsible for the large-scale lake trout mortalities described above. The Council of Lake Committees agreed to abide by the Fish Disease Control Committee's subsequent recommendation that "no fish from a facility at which EED has been demonstrated shall be planted into waters of the Great Lakes basin, or to stations or waters with no previous history of EED. Following disinfection, subsequent year classes may be stocked into previously affected waters. Following disinfection and two years without EED, transfers into waters with no history of EED will be permitted" (GLFC 1988a). Millions of lake trout and salmon were killed as USFWS, Wisconsin, and Michigan hatcheries were disinfected. Even today, progeny of broodstock from potentially infected lakes are quarantined for two years before being allowed into agency hatcheries. In addition to relying upon the commission to arbitrate the 1986 issue, the commission-sponsored Fish Disease Control Committee helped the agencies to show remarkable resolve and adherence to science in identifying and eliminating the risks posed by the EED virus. As of 2001, EED had never been detected in the Great Lakes.

Legal Status of the Strategic Plan

What is the legal status of the joint strategic plan? It could be described as a nonbinding "gentleman's agreement," rooted in the 1954 convention. In addition to U.S. and Canadian federal fishery agencies, the strategic plan is signed by nonfederal entities—fishery agencies of the eight Great Lakes states, Ontario, and two U.S. intertribal groups. The actual signatories include a provincial minister of the crown, a U.S. assistant secretary, a Canadian deputy minister, a DNR director, and a fish chief, among others.

In 1997, the U.S. State Department dismissed as "firmly...without merit" a claim by Ohio commercial fishermen that participation of Ohio fishery managers in the commission's Lake Erie committee process under the strategic plan—in particular recommending harvest allocations in concert with Ontario—"represents an inappropriate, and perhaps even unconstitutional exercise of foreign affairs power entrusted to the U.S. Federal government under the Constitution" (28 February 1997 internal State Department memo from David A. Balton, Office of the Legal Advisor, to David Herman, Office of Marine Conservation). Apparently "Article 1, Section 10 [of the U.S. Constitution] does not...prohibit U.S. states from coordinating their legitimate regulatory authorities in a nonbinding manner with those of a neighboring nation or province, especially where such coordination does not impinge on the authority or the foreign relations of the United States." Indeed, in 1981, counsel for the Wisconsin Department of Natural Resources conditioned that state's execution of the memorandum of acceptance for the strategic plan, saying that its acceptance "shall be construed in a manner which recognizes the administrative rulemaking processes of the states, nations, or provinces signatory to this Memorandum and shall support full compliance with such processes when a course of action of a party in furtherance of the Joint Strategic Plan is interpreted as an administrative rule by their sovereign" (Appendix 1.B, GLFC 1997; 1980 version).

Moreover, according to State Department legal advisors, state participation in the Lake Erie committee is expressly endorsed by the federal government, consistent with federal law: "Both the Convention and the Great Lakes Fisheries Act of 1956 envision the participation of the relevant U.S. States and their officials in furtherance of the Convention. The 198(1) Plan, also adopted in furtherance of the Convention, provided more specifically for the participation of officials from the relevant U.S. States and from the Province of Ontario in this effort" (Great Lakes Fishery Commission correspondence, on file).

Implementation of the Strategic Plan

How exactly is the joint strategic plan implemented? The Great Lakes Fishery Commission implemented the creation of formal lake committees for each of the five Great Lakes, whose membership consists of managers from each of the state, provincial, and tribal agencies with fishery management interests on a given lake. Though signatory to the strategic plan, Fisheries and Oceans Canada, the U.S. Fish and Wildlife Service, the U.S. Geological Survey, and the U.S. National Oceanographic and Atmospheric Administration are not members of lake committees—the first three, however, provide critical input in technical subcommittees. These committees are specified in the plan as its "action arms." A key task for each lake committee is the development of common objectives for its respective fish communities. This focus encourages a long-term perspective by the fishery agencies represented on the lake committees and draws attention to in-lake results. That is, the objectives provide a means for evaluating the performance of management relative to some common expectation. They also encourage agencies to examine a range of management options and to be tolerant of other jurisdictions' short-term deviations, provided they continue to work toward compliance with the agreed-upon strategy over the longer term. Further, fish community objectives can dictate the script for collective advocacy on behalf of the Great Lakes ecosystem.

The development of fish community objectives (FCOs) and the evaluation of progress toward their achievement has been a major activity of all five lake committees. The Lake Superior committee (LSC) is perhaps most advanced, having developed its first set of objectives in 1990 (Busiahn 1990). Draft revised objectives are currently being reviewed. As well, the LSC has published two "state of the lake" reports (1989, 1992) and will be presenting a third in 2002. The Lake Michigan committee published its FCOs in 1995 (Eshenroder et al. 1995), used their publication to foster development of a lakewide assessment plan (Schneeberger et al., unpublished), and reported on the state of Lake Michigan at the 2000 Lake Michigan committee meeting. Lake Huron and Ontario committees have also published FCO documents (Huron: DesJardine et al. 1995; Ontario: Stewart et al. 1999) and have completed "state of the lake" assessments (Huron: 1990, 1995, 2001; Ontario 1989, 1996). The Lake Erie committee has completed a draft set of FCOs and, in 1999, contributed to an ecosystem-wide review of the state of Lake Erie (Munawar et al. 1999).

Important contributors to lake committee success are their technical subcommittees, consisting of agency biologists who collectively assess the resource, advise lake committees on management options and progress, and write reports for the lake committees (Dochoda and Koonce 1994). Recognizing the risk of sociopolitical pressures compromising biology-based recommendations (Fraidenburg and Lincoln 1985), agency biologists operate in a culture where they are encouraged—and, indeed, exhort one another—to put aside agency considerations or "agency hats" when working in the technical committees. A few factors that Holder (1988) believed contributed to the success of the more effective technical committees were a clear demand for the product from within the management community; clear terms of reference endorsed by the official lake committee representatives; enthusiastic, qualified participants; sufficient available science and support data; and continued (at-home) support from the originators.

What is remarkable to the authors, having closely watched the process for more than two decades, is how the system captures and preserves good ideas, remaining receptive to new ideas as understanding grows. Innovations are imported from other lake committees, and negative experiences are not repeated. Progress may not always be as rapid as one might wish, but, in the long term, progress toward realization of shared objectives for the community is sustained.

There are several major successes in Great Lakes fishery management for which credit is due to the lake committees, the GLFC, and the joint strategic plan. These include the recovery of decimated species such as Lake Superior lake trout (GLFC 1996) and Lake Erie walleye (Hatch et al. 1987). Lake Huron's lake trout rehabilitation plan (developed by the Lake Huron Technical Committee) became the basis for a state-tribal judicial agreement (Dochoda and Koonce 1994). With the support of the Great Lakes Fish Health Committee, diseases such as salmon's infectious hematopoietic necrosis have been denied entry, and others such as the lake trout disease previously discussed (epizootic epitheliotropic disease) have been contained—EED virus is now known to exist only in a freezer (Joe Marcino, Minnesota Department of Natural Resources, personal communication). As described earlier, jurisdictions have formulated agreements for total catch and its allocation, particularly in Lake Erie. As illustrated above for Lake Ontario, species-based stocking ceilings are negotiated and renegotiated as phosphorus reduction and zebra mussels influence predator-prey balance. As water quality agencies met and seemed ready to manage phosphorus loading below targets of the Great Lakes Water Quality Agreement, a concerned Lake Erie committee called for a scientific understanding of how such changes will influence the composition and productivity of fish communities (GLFC 1998). There are protocols for coordinated data collection and an increasing number of databases. Jurisdictions routinely consult on proposed new introductions (Council of Lake Committees 1992), although recently the proposals under consideration have concerned new strains and sources of eggs, rather than new species. In 1988, the Lake Superior committee, newly invaded by Eurasian ruffe, called for action to prevent the transfer of aquatic organisms in ballast water (GLFC 1988), sparking a series of events that resulted in the 1989 Canadian ballast water exchange guidelines and the 1990 U.S. National Aquatic Nuisance Species Prevention and Control Act.

Frances Westley (1995) characterizes environmental group processes as learning-led, vision-led, and planning-led group processes, each with its characteristic strengths and weaknesses. The strategic plan process can be characterized as learning-led. Typical of learning-led processes, issues are defined incrementally under the strategic plan. Lake committees' fish community objectives are updated as understanding of the system improves, as conditions change, and as opportunities emerge. In learning-led processes, there is precommitment to act. Indeed, the prospect of coordinating their management actions is the primary reason agencies devote so much time to commission committee activities. Resources are often limiting for learning-led group processes—long-term survival of structures is frequently in peril, and resources often come with strings attached. The strategic plan has somewhat sidestepped this peril through its association with the commission, which has established management committees and supports the plan—in large part to satisfy its research-coordinate-and-advise mandate. Member agencies, of course, devote much staff support to committee activities. And, increasingly, the United States seems to be supporting committee activities through such vehicles as the 1998 amendment to the 1990 Great Lakes Fish and Wildlife Restoration Act. Still, the learning-led, nonfederal lake committees under the strategic plan tend to be perceived as low profile relative to their planning-led environmental counterparts: the Lakewide Management Plans (LaMPs) under the Great Lakes Water Quality Agreement (GLWQA).

With five Great Lakes, we have an opportunity to make comparisons regarding the role of ecology, geography, politics, and other features in implementation of coordinated management (Table 4). No two lakes have exactly the same set of features, in kind or quantity, and thus the challenges vary from lake to lake, as do the strategies used by the participating agencies. First, we will compare Lakes Superior and Erie, followed by Lakes Michigan and Ontario. Finally, we will consider unusual features of the fifth lake, Lake Huron.

Lakes Superior and Erie are both bisected by the U.S.-Canadian border. U.S. Indian treaties address fishing rights on Lake Superior but not on Lake Erie. Both the international border and Indian treaties seem to engender a more formal approach to management coordination, as does the relatively large number of state, Ontario, and intertribal fishery agencies that

Table 4. Features influencing coordination of fish management in the five Great Lakes.

Feature	Superior	Michigan	Huron	Erie	Ontario
International border	Yes	No	Yes	Yes	Yes
Indian treaty	1836[a], 1842[b]	1836[a]	1836[a]	No	No
No. agencies on lake committee	6	5	3	5	2
Wide/narrow lake	wide	narrow	wide	narrow	narrow
Pelagic fishery	native	put & take	mixed	native	put & take

[a] Treaty of Washington, March 28, 1836, 7 Stat. 491.
[b] Treaty with the Chippewa, October 4, 1842, 7 Stat. 591.

serve on the lake committees for these lakes. Fisheries on both lakes tend to focus on native species, though not exclusively. The lake committees on both lakes have developed strong technical subcommittees to support management in identifying management options and assessing progress, perhaps because of their focus on managing and rehabilitating native species as well as their relatively large pools of agency biologists. The most obvious difference between the two lakes is in scale: Superior is broad and deep, whereas Erie is shallow and relatively narrow. The narrowness means that fish managers on Lake Erie never entertain the perception that fish management on one side of a border is little felt on the other. In Lake Superior, the connection between the two sides of the lake is less compelling, although lakewide programs exist (e.g., sea lamprey control).

Lakes Michigan and Ontario have many similar features. Both lakes are narrow enough that fish managers appreciate their respective capacity to influence one another's interests. They are the two lakes most dependent on put-and-take salmonine stocking, although efforts are ongoing to rehabilitate native fish communities. Consequently, predator-prey management and perceived conflicts between salmon stocking and lake trout rehabilitation have fostered a constituency eager to apply political pressure. The lake committees for these two lakes tend to be a little less formal and less structured than those for the other lakes. Lake Ontario is bisected by the U.S.-Canadian border, which tends to engender a degree of formality, but there are only two fish management agencies (New York and Ontario) on the lake committee. As a result, they tend to be familiar with each other's activities and have perceived less need for structure and formal integration—at least in the form of joint reports to the lake committee. Lake Michigan is the only Great Lake not bisected by the U.S.-Canadian border, although the exercise of treaty fishing rights in the 1980s and the subsequent appointment of the Chippewa-Ottawa Resource Authority to the Lake Michigan committee engendered a more formal tone than had been evident previously. When only states were members of the committee, it may have been that there were more common interests, assumptions, and issues among committee members. Addition of Canadian or intertribal representatives tends to diversify the interests. On the other hand, Lake Michigan has sufficient fishery agencies to staff an active technical committee. The challenge, however, is to utilize technical capacity effectively in managing a frequently contentious fishery in which complaints quickly reach senior officials.

In ecology, geography, and political features, Lake Huron truly is the lake in the middle. Indian fishing rights are of key importance in Lake Huron, the U.S.-Canadian border bisects this lake, and the salmonine stocking is an important component of management; but it is a very large lake with only three players—Michigan, Ontario, and the Chippewa-Ottawa

Resource Authority. For the most part, management coordination on this lake has been moderately formal but without the inducements to action (intensively managed fisheries, a critical mass of support biologists, or proximity) that have encouraged formality on other lakes. In the future, however, control of sea lamprey—still a serious problem in Lake Huron—and the avowed objective of Michigan, Ontario, and the Chippewa-Ottawas to rehabilitate lake trout will no doubt prompt more intense coordination of fish management.

In summary, major forces that shaped coordinated fish management in the Great Lakes are the federal/national interests related to a shared border, treaties with Indian tribes, and state/provincial interests related to a long history of natural resource management at the local level. Two types of institutions, the convention's Great Lakes Fishery Commission and the agencies represented on commission management committees (under guidance of the Joint Strategic Plan for Management of Great Lakes Fisheries), operate cooperatively in meeting each other's needs: the commission's mandate to coordinate, research, and advise, and the agencies' need for an umbrella that provides access across the border and administrative and other support for coordinated fish management. Both the convention and the strategic plan act to entrain resources for shared fish community objectives, and this focus tends to reduce involvement in less productive disputes. The system tends to be self-directed, voluntary, and based on shared information. Limitations are in resources to support information gathering and analysis and planning. As elsewhere, the rate of progress depends on the participation of persons with "clarity of vision and strength of purpose" willing to make use of the available process. Fortunately, the system tends to preserve policy gains achieved by strong leaders using sound science.

References

Busiahn, T., editor. 1990. Fish community objectives for Lake Superior. Great Lakes Fishery Commission Special Publication 90-1. Ann Arbor, Michigan.

Council of Lake Committees. 1992. Introductions in the Great Lakes basin: procedures for consultation. In minutes of 1992 annual meeting of the Council of Lake Committees. Great Lakes Fishery Commission, Ann Arbor, Michigan.

DesJardine, R. L., T. K. Gorenflo, R. N. Payne, and J. D. Schrouder. 1995. Fish community objectives for Lake Huron. Great Lakes Fishery Commission Special Publication 95-1. Ann Arbor, Michigan.

Dochoda, M. R., and J. F. Koonce. 1994. A perspective on progress and challenges under a joint strategic plan for management of Great Lakes fisheries. University of Toledo Law Review 25:425–442.

Dochoda, M. R. 1999. Authorities, responsibilities, and arrangements for managing fish and fisheries in the Great Lakes ecosystem. Pages 93–220 *in* W. W. Taylor and

C. P. Ferreri, editors. Great Lakes fisheries policy and management. Michigan State University Press, East Lansing, Michigan.

Eshenroder, R. L., M. E. Holey, T. K. Gorenflo, and R. D. Clark, Jr. 1995. Fish community objectives for Lake Michigan. Great Lakes Fishery Commission Special Publication 95-3. Ann Arbor, Michigan.

Fetterolf, C. M. 1980. Why a Great Lakes fishery commission and why a sea lamprey international symposium? Canadian Journal of Fisheries and Aquatic Sciences 37:1588–1593.

Fraidenburg, M. E., and R. H. Lincoln. 1985. Wild chinook salmon management: an international conservation challenge. North American Journal of Fisheries Management 5:311–329.

GLFC (Great Lakes Fishery Commission). 1958. Minutes of special committee on lake trout rehabilitation. Great Lakes Fishery Commission, Ann Arbor, Michigan.

GLFC (Great Lakes Fishery Commission). 1964a. A prospectus for investigations of the Great Lakes fishery. Doc. 64-10. Great Lakes Fishery Commission, Ann Arbor, Michigan.

GLFC (Great Lakes Fishery Commission). 1964b. Minutes of the annual meeting. Great Lakes Fishery Commission, Ann Arbor, Michigan.

GLFC (Great Lakes Fishery Commission). 1965. Minutes of Lake Erie committee. Great Lakes Fishery Commission, Ann Arbor, Michigan.

GLFC (Great Lakes Fishery Commission). 1973a. Annual report for the year 1973. Great Lakes Fishery Commission, Ann Arbor, Michigan.

GLFC (Great Lakes Fishery Commission). 1973b. Minutes of the Lake Michigan committee. Great Lakes Fishery Commission, Ann Arbor, Michigan.

GLFC (Great Lakes Fishery Commission). 1986a. Minutes of the annual meeting. Great Lakes Fishery Commission, Ann Arbor, Michigan.

GLFC (Great Lakes Fishery Commission). 1986b. Minutes of the 7–8 May 1986 executive meeting. Great Lakes Fishery Commission, Ann Arbor, Michigan.

GLFC (Great Lakes Fishery Commission). 1988. Minutes of Lake Superior committee. Great Lakes Fishery Commission, Ann Arbor, Michigan.

GLFC (Great Lakes Fishery Commission). 1996. Minutes of Lake Superior committee. Great Lakes Fishery Commission, Ann Arbor, Michigan.

GLFC (Great Lakes Fishery Commission). 1997. A joint strategic plan for management of Great Lakes fisheries. Great Lakes Fishery Commission, Ann Arbor, Michigan. (Supersedes 1980 and 1994 versions). *<http://www.glfc.org/fishmgmt/sglfmp97.htm>*.

GLFC (Great Lakes Fishery Commission). 1998. Minutes of Lake Erie committee. Great Lakes Fishery Commission, Ann Arbor, Michigan.

Gunderson, L. H., C. S. Holling, and S. S. Light, editors. 1995. Barriers and bridges to the renewal of ecosystems and institutions. Columbia University Press, New York.

Hatch, R. W., S. J. Nepszy, K. M. Muth, and C.T Baker. 1987. Dynamics of the recovery of the western Lake Erie walleye (*Stizostedion vitreum vitreum)* stock. Canadian Journal of Fisheries and Aquatic Sciences 44(Supplement 2):15–22.

Hnath, J. G., editor. 1993. Great Lakes fish disease control policy and model program. Great Lakes Fishery Commission Special Publication 93(1):1–38. Ann Arbor, Michigan.

Holder, A. S. 1988. Technical committees of the lake committees—their strengths and weaknesses. Pages 59–61 *in* M. R. Dochoda, editor. Committee of the whole workshop on implementation of the Joint Strategic Plan for Management of Great Lakes Fisheries (reports and recommendations from the 18–20 February and 5–6 May 1986 meetings). Great Lakes Fishery Commission Special Publication 88-1. Ann Arbor, Michigan.

Kerr, S. R., and R. A. Ryder. 1997. The Laurentian Great Lakes experience: a prognosis for the fisheries of Atlantic Canada. Canadian Journal of Fisheries and Aquatic Sciences 54:1190–1197.

Miller, A. 1984. Professional collaboration in environmental management: the effectiveness of expert groups. Journal of Environmental Management 16:365–388.

Munawar, M., T. Edsall, and I. F. Munawar. 1999. State of Lake Erie: past, present and future. Backhuys, Leiden, Netherlands.

Piper, D. C. 1967. The international law of the Great Lakes. Duke University Press, Durham, North Carolina.

Reiff, H. 1959. The United States and the treaty law of the sea. University of Minnesota Press, Minneapolis, Minnesota.

Stewart, T. J., R. E. Lange, S. D. Orsatti, C. P. Schneider, A. Mathers, and M. E. Daniels. 1999. Fish community objectives for Lake Ontario. Great Lakes Fishery Commission Special Publication 99-1. Ann Arbor, Michigan.

Westley, F. 1995. Governing design: the management of social systems and ecosystems management. Pages 391–427 *in* L. H. Gunderson, C. S. Holling, and S. S. Light, editors. Barriers and bridges to the renewal of ecosystems and institutions. Columbia University Press, New York.

Sustaining North American Salmon: Perspectives Across Regions and Disciplines, 243–258

Chapter 12

The Future of Pacific Salmon in the Great Lakes

C. I. Goddard

Introduction

The introduction of chinook *Oncorhynchus tshawytscha* and coho salmon *Oncorhynchus kisutch* into the Laurentian Great Lakes was a recent event in terms of ecological time. With the reality of niche overlap and a finite forage base, all salmonid species in the Great Lakes exist at some expense to one another (Negus 1995), and the ultimate ecological niche of Pacific salmon (chinook and coho) in these lakes is difficult to predict. Moreover, the Great Lakes ecosystem is still subject to widespread structural and functional changes, and the nature and sustainability of future fish communities is an open question (Steedman and Regier 1987). Nevertheless, on the basis of the science and discussions with fisheries managers, I believe that it is reasonable to predict that Pacific salmon will continue to play a major role as pelagic piscivores in the Great Lakes but at lower abundance than in the past. To support this prediction, I will review the role of Pacific salmon from ecological and management perspectives.

Changes in the Fish Community

At the time of the European settlement of the Great Lakes basin, the fish community was self-sustaining and unaffected by large-scale human activity (Christie et al. 1987). Productivity was low in the cold, deep, oligotrophic waters that make up most of the Great Lakes, and the lake trout *Salvelinus namaycush* was the dominant piscivore and likely structured the fish community (Edwards et al. 1990). The lake trout was dominant because of its large size, abundance, and wide distribution (Smith 1972). The burbot *Lota lota*, a second piscivore in the offshore community, was more benthic and less abundant than lake trout (Smith 1968). The Atlantic salmon *Salmo salar*, an obligate shallow-water pelagic piscivore, was very abundant and was the most valued species in the early Lake Ontario fishery before it was extirpated in the late 1800s (Christie 1973). Although adult lake trout exhibit diel vertical migration into offshore pelagic waters, the pelagic piscivore niche occupied by Atlantic salmon and lake trout (a facultative pelagic piscivore) in Lake Ontario was not as efficiently exploited by lake trout alone in the other Great Lakes. Six species of deepwater ciscoes *Coregonus* spp., endemic forms in the Great Lakes (Smith and Todd 1984), made up the bulk of the fish biomass and the forage base (Smith 1995). This species complex of large-bodied planktivores was ecologically and morphologically diverse, but it was specialized to prey upon both *Diporeia* spp. and the vertically migrating *Mysis relicta* (Beeton 1960; Eshenroder et al. 1995a). These specializations were important for energy transfer in offshore waters; these two macroinvertebrates made up the greatest biomass of large, high-energy prey for planktivores (Sprules et al. 1991). Lake trout in the Great Lakes evolved similarly to prey successfully on the deepwater ciscoes (Eshenroder and Burnam-Curtis 1999).

The Great Lakes fish community and major food webs changed markedly as a result of overharvest, habitat destruction, eutrophication in Lake Erie, and invasions by exotic marine species, most notably alewife *Alosa pseudoharengus* and sea lamprey *Petromyzon marinus* (Christie 1972; Smith and Tibbles 1980; Smith 1995; Kitchell and Crowder1986). Alewife entered Lake Ontario through the Erie Canal during the mid-1800s (Ihssen et al. 1992), at a time when the piscivores were in decline as a result of overharvest and habitat destruction. The alewife became abundant in Lake Ontario by 1870 (Smith 1970). The proliferation of alewife throughout the Great Lakes was made possible by the construction of the Welland Canal, but their establishment in the upper Great Lakes was delayed initially by the abundance of large piscivores (Miller 1957; Brown 1972). The alewife never became well established in Lake Superior, most likely as a result of coldwater temperatures and possibly because of predation by lake trout (Lawrie and Rahrer 1973; Bronte et al. 1991). Nevertheless, in Lakes Huron, Michigan, and Ontario, the alewife ultimately became the dominant species, having reduced endemic fish populations by feeding on the pelagic eggs and larvae of native fish (Pritchard 1929; Crowder 1980; Eck and Wells 1987). Alewife were also shown to be direct predators on lake trout fry as they emerged from spawning shoals (Krueger et al. 1995).

Similarly, the sea lamprey was first observed in Lake Ontario in 1835, having entered via the Erie Canal, and Lake Erie following the completion of the Welland Canal in 1921 (Smith 1995; Eshenroder and Burnham-Curtis 1999). Sea lamprey subsequently migrated into the other Great Lakes and, in the absence of predators, were well established throughout the Great Lakes by mid-1940. Sea lamprey were incredibly destructive and caused severe damage to lake trout, whitefish, and burbot populations. Uncontrolled sea lamprey predation and fear that lamprey would ultimately destroy all stocks also served as a catalyst for increased overharvest of lake trout and coregonids.

By the mid-1950s, the large piscivores had been decimated by the combined effects of overharvest, habitat destruction, and predation by alewife and sea lamprey. Introductions of Atlantic and chinook salmon in the late 1800s and 1950s had been failures (Parsons 1973). Atlantic salmon had been extirpated in Lake Ontario, and lake trout had been extirpated in Lakes Ontario, Erie, and Michigan and significantly reduced in most of Lake Huron and the nearshore waters of Lake Superior. Three species of deepwater ciscoes had become extinct (Webb and Todd 1995; Eshenroder and Burnham-Curtis 1999). The bulk of the fish biomass in the Great Lakes shifted from the deepwater ciscoes to the exotic alewife (Smith 1970). Having little commercial value and subject to massive die-offs, the alewife was considered to be a nuisance that periodically polluted beaches and harbors and clogged municipal water intakes.

Response of Fisheries Management

In 1955, the Great Lakes Fishery Commission (GLFC) began a program of sea lamprey control that effectively reduced the populations of sea lamprey to at least 10% of historic abundance (Smith and Tibbles 1980; Heinrich et al., in press; Lavis et al. 2002). This significant reduction in sea lamprey caused an increase in the survival of stocked salmonids that fed on alewives, grew to a large size, and lived to reach adulthood (Goddard 1998).

Initially, management agencies tried to control alewives by direct commercial harvest and the development of markets (O'Gorman and Stewart 1999). Subsequently, the stocking of salmonines was seen as an effective method to suppress alewife populations and, over time, the management of alewives became linked to the management of chinook salmon. Nevertheless, the purpose of the reintroductions of offshore pelagic chinook salmon and the introduction of coho salmon into the Great Lakes in the mid-1960s was twofold:

- To elevate the fisheries resources of the Great Lakes to their maximum potential for recreational fishing.
- To convert the annual production of alewives into an abundance of sport fishes for the recreational fisherman (Tody and Tanner 1966).

The reduction in alewife populations would also benefit the on-going binational effort to rehabilitate the depressed stocks of native fish species, such as lake trout and deepwater ciscoes (O'Gorman and Stewart 1999).

The stocking of Pacific salmon in the Great Lakes was an unprecedented success—recreational fisheries flourished (Talhelm 1988) and alewife populations declined substantially. In many localities, Pacific salmon have become the most popular recreational fish in the Great Lakes and led to the widespread development of the charter fishing industry (see Tanner and Tody, Chapter 7). As a result of the Pacific salmon introductions, the social, economic, and political weight of the Great Lakes fishing industry shifted from the commercial fishery to the recreational fishery, with Lake Erie being the possible exception, although it, too, is undergoing a shift in focus from commercial to recreational fishing by state and provincial management agencies (Nepszy 1999). This popularity is well illustrated by the fact that all 800,000 chinook annually stocked into the Canadian waters of Lake Huron are raised not in government hatcheries but by angling groups in partnership with the Ontario Ministry of Natural Resources. It should be noted that coho salmon cost much more to raise in hatcheries than chinook salmon. Tremendous widespread support exists for the continuation of Pacific salmon in the Great Lakes and, interestingly, there is also strong support for the maintenance of large alewife populations as the forage base for the Pacific salmon.

Management Goals and Objectives

Fisheries management in the Great Lakes is coordinated through A Joint Strategic Plan for the Management of Great Lakes Fisheries (GLFC 1997). The eight Great Lakes states, the province of Ontario, and the fisheries agencies of the governments of Canada and the United States signed this agreement in 1981. The Chippewa-Ottawa Treaty Fisheries Management Authority and the Great Lakes Indian Fish and Wildlife Commission were added as signatories in 1989. These agencies reaffirmed their commitment to the joint strategic plan and their strong commitment to making fisheries management decisions by consensus when they revised it in 1997. Embracing the ecosystem concept, the agencies agreed that consensus must be achieved when management will significantly influence the interests of more than one jurisdiction. Through the joint strategic plan, the management agencies on the Great Lakes implemented a framework for achieving consensus and coordinated fisheries management via the lake committees of the GLFC. Each Great Lake has a lake committee that is composed of representatives of each fisheries management agency with jurisdiction on the lake. Decisions about stocking and harvest levels as they affect the interests of the other jurisdictions are made by the lake committees. (See Dochoda and Jones, Chapter 11, for additional information about the joint strategic plan and the lake committee process.)

The joint strategic plan elaborates a philosophy that each agency has a stake in the whole ecosystem and that, to achieve the common good, management authority should be exercised cooperatively rather than unilaterally. The goal statement of the joint strategic plan is

> To secure fish communities, based on foundations of stable self-sustaining stocks, supplemented by judicious plantings of hatchery-reared fish, and provide from these communities an optimum contribution of fish, fishing opportunities, and associated benefits to meet needs identified by society for wholesome food, recreation, cultural heritage, employment and income, and a healthy aquatic ecosystem (p. 2).

To achieve this goal, lake committees have defined objectives for the structure and function of desired fish communities for each lake and have identified the environmental and socioeconomic pressures that have potential to prevent their achievement. These fish community objectives rest on a set of ecological principles that are tempered with social values, provide direction to the fisheries management agencies, and serve to express management intent. As such, fish community objectives

- Recognize that there are limits to aquatic productivity.
- Stress the preservation and restoration of native species and habitat.
- Call for enhanced natural reproduction of native and desirable introduced species.
- Support the role of stocked fish.
- Protect and enhance threatened and endangered species.

The fish community objectives vary from lake to lake, but, with the exception of Lake Erie, all specifically call for an on-going role for Pacific salmon. In Lake Huron, for example, fish community objectives are established for the entire lake, and the overall objective is

> Over the next two decades, restore an ecologically balanced fish community dominated by top predators and consisting largely of self-sustaining indigenous and naturalized species capable of sustaining annual harvests of 8.9 million kg (DesJardine et al. 1995, p. 15).

The specific salmon and trout objective for Lake Huron is

> Establish a diverse salmonine community that can sustain an annual harvest of 2.4 million kg with lake trout the dominant species

> and anadromous (stream-spawning) species also having a prominent place (DesJardine et al. 1995, p. 15).

The supporting text further states

> Chinook salmon are now widely distributed in Lake Huron, because of planting and natural reproduction. Consistent with demand, chinook are expected to have an important future in the salmonine community. Coho salmon, however, currently present in low numbers, may persist because of limited natural reproduction or immigration from the other lakes, but will not be stocked because of potential conflicts with other riverine species, its relatively short period of availability for fishing, and poor returns from past stocking programs (DesJardine et al. 1995, p. 16).

Similarly, Lake Michigan fish community objectives call for the enhanced natural reproduction of native and desirable introduced fishes and acknowledge the role of planted fish. Specifically, for Pacific salmon, the objective is the establishment of a diverse salmonine community capable of sustaining an annual harvest of 2.7 million to 6.8 million kg, with chinook contributing 3.1 million kg and coho 0.7 million kg (Eshenroder et al. 1995a). Lake Superior is the deepest, coldest, and most oligotrophic of the Great Lakes. The original Lake Superior fish community objectives logically focus on the rehabilitation of native lake trout and whitefish yet call for an unspecified yield of "other salmonid predators" (Busiahn 1990). The working-draft revised 2000 objective for Lake Superior Pacific salmon is populations that are predominantly self-sustaining but that may be supplemented by stocking that is compatible with the restoration of native species.

In contrast, Lake Erie is the shallowest of the Great Lakes and predominantly mesotrophic, and its fish community objectives focus on percid fishes: walleye *Stizostedion vitreum vitreum* and yellow perch *Perca flavescens*. Lake Erie fish community objectives for the deeper eastern basin, however, do call for sustainable harvests of fish species including rainbow trout (steelhead; *Oncorhynchus mykiss*) and other valued salmonids.

Pacific salmon have a much greater role in the specific Fish Community Objectives for Lake Ontario, which were developed for three major overlapping and interacting habitat zones: nearshore, offshore pelagic, and offshore benthic zones (Stewart et al. 1999). The offshore pelagic fish community in Lake Ontario will be characterized by

- A diversity of salmon and trout.
- Chinook salmon as the top predator.
- Abundant populations of rainbow trout.

- Fishable populations of coho salmon and brown trout *Salmo trutta*.
- Populations of stocked Atlantic salmon at levels consistent with investigating the feasibility of restoring self-sustaining populations.
- Amounts of naturally produced (wild) salmon and trout, especially rainbow trout, that are consistent with fishery and watershed plans.
- A diverse prey fish community with the alewife as an important species (Stewart et al. 1999).

The fisheries management agencies, therefore, have agreed that there is a varying and on-going role for Pacific salmon in the Great Lakes.

Ecological Perspective

The population dynamics of Pacific salmon and alewife are tightly coupled through the predator–prey relationship (Kitchell and Crowder 1986); thus, stocked Pacific salmon are largely responsible for the reduction of the overabundant alewife populations (Stewart et al. 1981). Alewife are preyed upon by all salmonids, but chinook salmon consume more alewives than the other salmonids (Brandt 1986; Jude et al. 1987). The alewife populations in most of the lakes are now suppressed to levels that, as a result of decreased alewife predation, could permit an increased level of rehabilitation of native species, notably lake trout and deepwater ciscoes. The reduction in alewife abundance should also tend to increase fish community stability and integrity (Eshenroder 1985).

Making the population dynamics between Pacific salmon and alewife more complex, however, is early mortality syndrome (EMS), which is affecting Pacific salmon and other salmonids in the Great Lakes (Hornung et al. 1998). It appears that alewife, the preferred prey of Pacific salmon, is the primary link with EMS, which results in a lethal thiamine deficiency in the early life stages and causes a loss of equilibrium, inability or lack of feeding, and general lethargy (Honeyfield et al. 1998). To offset losses from this disease, hatcheries take extra eggs and subject them to thiamine washes. Quantitatively, the effect of EMS on naturally reproducing populations is unknown. Hornung et al. (1998, p. 132), however, state: "Of potentially greater concern than the ability of hatcheries to produce healthy progeny and maintain stocking quotas are the physical, biological, or ecological conditions in the Great Lakes that may be affecting the adult broodstock and causing this reproductive failure." It is probable, therefore, that EMS is playing a role in structuring the pelagic fish community by reducing reproduction and recruitment and limiting the opportunity for natural reproduction.

Natural Reproduction of Salmon

Natural reproduction of chinook and coho salmon is increasing, and both species have now become naturalized in essentially all of the Great Lakes (Busiahn 1990; DesJardine et al. 1995; Eshenroder et al. 1995a; Stewart et al. 1999; Ryan et al. 1999; Kocik and Jones 1999). During the next decade, an estimated 2 million chinook salmon will be naturally produced in Lake Ontario, and it has been speculated that this may well sustain the fishery (T. Stewart, personal communication, 1998). Natural reproduction, particularly of coho salmon, is occurring even in Lake Erie, in both Ontario and New York tributaries (Ryan et al. 1999). Hansen (1994) indicates that most chinook salmon in Lake Superior are produced in the wild, although abundance and returns from stocked chinook are declining. Peck et al. (1999) reported that stocked chinook made up only 25% of the total chinook population in Lake Superior and questioned whether this contribution was sufficient to justify continued stocking. Natural reproduction will be insufficient to maintain the chinook population in Lake Michigan, and on-going stocking will be required (Eshenroder et al. 1995b).

Natural reproduction of Pacific salmon is likely to continue to increase substantially in the Great Lakes, as a result of the on-going improvements in water quality and stream habitat (International Joint Commission 1998). Cooperative projects by fisheries agencies and angling groups to stabilize streambanks, create riffles, and establish streamside vegetation to provide shade and instream cover have significantly improved the spawning and nursery habitat for Pacific salmon. The installation of fish passage devices has also significantly increased access to spawning and nursery habitat. In addition, new legislation, such as that produced by the U.S. Federal Energy Regulatory Commission, contains provisions to provide run-of-river flow regimes that should increase survival of juvenile life stages. Nevertheless, the productive capacity of Great Lakes tributaries is finite, and these rivers also provide spawning and nursery habitat for other naturally reproducing populations of salmonids, notably rainbow trout, pink salmon *O. gorbuscha*, brown trout, native brook trout *Salvelinus fontinalis*, and Atlantic salmon (Kocik and Jones 1999). Although the competition for spawning and nursery habitat can be predicted somewhat, the long-term outcome of interactions between these species and Pacific salmon is difficult to predict accurately. Again, it must be pointed out that EMS could offset potential gains in natural reproduction of Pacific salmon.

Role of Hatcheries

The success of natural reproduction will be critical to the long-term success of Pacific salmon because the hatchery process itself faces challenges.

A major challenge facing the hatcheries is the maintenance of the genetic integrity of the stocks. The genetic variability in the original stocks imported to the Great Lakes was limited, and there have essentially been no new infusions of genetic variability from the west coast. The hatchery process relies on high survival from a relatively small parental population, whereas natural production has a relatively low survival from a large parental population. Inbreeding documented in Lake Michigan (Weeder 1997) reveals that the limited genetic variability is declining, reducing species resilience. The maintenance of genetic variability is essential to the ability of Pacific salmon to adapt to ecological and environmental changes, such as changes in water quality and the invasions of exotic species. A decrease in returns to weirs in relation to numbers stocked suggests a loss of fitness in chinook salmon in Lake Michigan (Benjamin 1998). Hatcheries are also facing pressure from environmental (water quality) groups and cottage associations to reduce discharges of nutrients. The required installation of technologies to essentially eliminate nutrient discharge is on-going but expensive. At the same time, despite tremendous support from the public and the reality that hatchery closures require great feats of political courage, the on-going cost of operation of hatcheries is being reviewed critically by many fisheries management agencies as they are being "rationalized" and downsized. Natural reproduction of Pacific salmon, where possible, is clearly a more cost-effective and preferred option in the long run (Carl 1982).

The stocking of Pacific salmon also requires maintaining a delicate balance between predator and prey abundance (Jones et al. 1993). In the 1980s, a large population of stocked chinook salmon in Lake Michigan resulted in fewer alewives, a switch from feeding primarily on large alewives to eating smaller alewives and other prey species, and reductions in salmon growth by the end of the decade (Stewart and Ibarra 1991). These changes appear to have led to a die-off of chinook salmon in the late 1980s, which in turn led to a 90% reduction in recreational harvest. It is hypothesized that inadequate nutrition resulting from the reduced alewife forage base triggered an outbreak of bacterial kidney disease (BKD), which accounted for the chinook die-off (Rybicki and Clapp 1996; Eshenroder et al. 1995a; Hansen and Holey, Chapter 8). This decline in salmon harvest ultimately led to a reduction in the number of chinook salmon stocked in some states in subsequent years. Stocking was reduced to relieve pressure on alewives and thereby diminish the risk of another collapse of the chinook salmon fishery. Similarly, in 1999, despite an adequate forage base, chinook salmon stocking in Lake Michigan was reduced by 20% because of continuing concerns about the predator–prey relationship.

Predator–Prey Balance

It is an interesting management dilemma: angler demand for Pacific salmon requires maintaining large alewife populations to support the salmon fishery, even though Pacific salmon were originally stocked to reduce alewife populations to allow the rehabilitation of native species, in addition to providing a recreational fishery. The maintenance of Pacific salmon populations, therefore, requires fisheries management agencies to stock salmon at levels that will not collapse the alewife forage base yet continue to suppress alewife populations to levels that will allow rehabilitation of native species.

Ideally, to realize these objectives, stocking levels should change annually to reflect predicted changes in the forage base. The majority of prey consumed by a chinook salmon occurs three years poststocking, so the annual adjustment of the stocking numbers requires the ability to accurately predict prey abundance three years hence as well as the likely contribution of wild fish to the cohort. Neither can be done with sufficient reliability. Although the salmon-alewife interaction has been modeled as an aid to fisheries management, a high level of uncertainty exists associated with model predictions (Jones et al. 1993). Populations maintained by stocking lack the critical natural population regulatory feedback mechanisms that are inherent in self-sustaining populations (Christie et al. 1987). It is not certain that these stocking objectives can be realized.

This predator–prey balancing act is further complicated by rapid establishment of nonnative species such as the zebra mussel *Dreissena polymorpha*, quagga mussel *D. bugensis*, ruffe *Gymnocephalus cernuus*, and round goby *Neogobius melanostomus* (Mills et al. 1993). The roles of these exotic species in the Great Lakes food web and their long-term impacts on the forage base are not known, but collectively these species are consuming a great deal of primary productivity, rerouting energy flow, and altering food webs with unpredictable consequences (Dermott and Munawar 1993; Hoyle et al.1999). It is also possible that gobies, by eating zebra/quagga mussels, may be recycling energy shunted from the pelagic food web to the benthos by the mussels. This constraint on the salmon food web is occurring at a time when the overall productivity of Lakes Erie and Ontario is also decreasing, as a result of success in reducing nutrient loading (primarily phosphorus) to improve water quality. Therefore, the forage base will most likely continue to decline in future years as a result of decreasing productivity and the diversion of energy away from forage fishes to nonnative species.

It is probable that, in addition to a declining pelagic planktivore biomass, there will be a concomitant shift in species composition as the species suppressed by alewife expand, which will also affect Pacific salmon energetics. In Lake Michigan, for example, the decline in alewife resulted in increased bloater *Coregonus hoyi* populations, which were less accessible to salmon as prey (Eshenroder and Burnham-Curtis 1999). The reduced forage base and altered species composition, therefore, will likely result in a reduced Pacific salmon community.

The decreasing forage base is providing a difficult problem for fisheries management agencies to address, as angler expectation for the Pacific salmon fishery may be greater than the Great Lakes can realistically sustain. This may result in pressure for the management agencies to reduce or abandon rehabilitative efforts for lake trout and deepwater ciscoes in favor of continuing Pacific salmon programs (Gale 1987; Negus 1995).

Summary

It is not clear how the Great Lakes ecosystem will evolve because of a number of natural and human-induced fundamental changes in the system. The Great Lakes ecosystem changed significantly in the past two centuries as a result of habitat destruction, overharvest, and the arrival of nonnative species. Following successful sea lamprey control, chinook and coho salmon were introduced to provide a recreational fishery and to control the exotic alewife. A tremendously popular recreational fishery developed for Pacific salmon, and alewife populations were reduced. Although Pacific salmon have become naturalized in all of the Great Lakes, the fishery continues to be supported by stocking programs.

The fishery management agencies have clearly stated that Pacific salmon will continue to have a major role as a pelagic piscivore in the Great Lakes and are managing and stocking to achieve that objective. Nevertheless, a complex predator–prey relationship exists between stocked Pacific salmon and alewife, and the alewife appears to be linked to two diseases, bacterial kidney disease and early mortality syndrome, which are affecting salmon survival and reproduction. Moreover, it is a management dilemma that the large populations of alewife required to maintain a large Pacific salmon fishery would likely preclude the rehabilitation of native lake trout, another management objective. In addition, the Great Lakes are experiencing a reduced capacity to support Pacific salmon because of the recent arrival and expansion of nonnative species such as the zebra mussel and round goby, which are diverting a great deal of the primary productivity away from the pelagic forage base. It is likely, however, that the current stocking levels, at least in Lakes Michigan and Ontario, exceed the levels that will be sustainable in the future (Kocik and Jones 1999). Furthermore, the overall productivity of Lakes Erie and Ontario is declining because of effective controls on nutrient loading.

Pacific salmon, however, will continue to provide a recreational fishery throughout the Great Lakes and control alewife. Natural reproduction will increasingly contribute to Pacific salmon populations, except perhaps in Lake Superior. The combined effects of aquatic nuisance species and reduced primary productivity will reduce and alter the community structure of the forage base, which will likely result in smaller Pacific salmon populations in the future.

References

Beeton, A. M. 1960. The vertical migration of *Mysis relicta* in Lakes Huron and Michigan. Journal of the Fisheries Research Board of Canada 17:517–539.

Benjamin, D. A. 1998. Chinook salmon (*Oncorhynchus tshawytscha*) population dynamics in Lake Michigan, 1985 to 1996. Master's thesis, Michigan State University.

Brandt, S. B. 1986. Food of trout and salmon in Lake Ontario. Journal of Great Lakes Research 12(3):200–205.

Bronte, C. R., J. H. Selgeby, and G. L. Curtis. 1991. Distribution, abundance, and biology of the alewife in U.S. waters of Lake Superior. Journal of Great Lakes Research 17(3):304–313.

Brown, E. H. 1972. Population biology of alewives, *Alosa pseudoharengus,* in Lake Michigan, 1949–1970. Journal of the Fisheries Research Board of Canada 29:477–500.

Busiahn, T. R., editor. 1990. Fish Community Objectives for Lake Superior. Great Lakes Fishery Commission, Special Publication 90-1.

Carl, L. M. 1982. Natural reproduction of coho salmon and chinook salmon in some Michigan streams. North American Journal of Fisheries Management 4:375–380.

Christie, W. J. 1972. Lake Ontario: effects of exploitation, introductions, and eutrophication on the salmonid community. Journal of the Fisheries Research Board of Canada 29:913–929.

Christie, W. J. 1973. A review of the changes in the fish species composition of Lake Ontario. Great Lakes Fishery Commission, Technical Report 23. Ann Arbor, Michigan

Christie, W. J., G. R. Spangler, K. H. Loftus, W. L. Hartman, P. J. Colby, M. A. Ross, and D. R. Talhelm. 1987. A perspective on Great Lakes fish community rehabilitation. Canadian Journal of Fisheries and Aquatic Sciences 44:486–499.

Crowder, L. B. 1980. Alewife, rainbow smelt, and native fishes in Lake Michigan: competition or predation. Environmental Biology of Fishes 5:225–233.

Dermott, R., and M. Munawar. 1993. Invasion of Lake Erie offshore sediments by *Dressina* and its ecological consequences. Canadian Journal of Fisheries and Aquatic Sciences 50:2298–2304.

DesJardine, R. L., T. K. Gorenflo, R. N. Payne, and J. D. Schrouder. 1995. Fish-community objectives for Lake Huron. Great Lakes Fishery Commission, Special Publication 95-1. Ann Arbor, Michigan

Eck, G. W., and L. Wells. 1987. Recent changes in Lake Michigan's fish community and their probable causes, with emphasis on the role of the alewife (*Alosa pseudoharangus*). Canadian Journal of Fisheries and Aquatic Science 44:53–60.

Edwards, C. J., R. A. Ryder, and T. R. Marshall. 1990. Using Lake trout as a surrogate of ecosystem health for oligotrophic waters of the Great Lakes. Journal of Great Lakes Research 16(4):591–608.

Eshenroder, R. L., editor. 1985. Presented papers from the Council of Lake Committees plenary session on Great Lakes predator prey issues, March 20, 1985. Great Lakes Fishery Commission, Special Publication 85-3. Ann Arbor, Michigan.

Eshenroder, R. L., E. J. Crossman, G. K. Meffe, C. H. Olver, and E. P. Pister. 1995a. Lake trout rehabilitation in the Great Lakes: an evolutionary, ecological, and ethical perspective. Journal of Great Lakes Research 21(Supplement 1):518–529.

Eshenroder, R. L., M. E. Holey, T. K. Gorenflo, and R. D. Clark. 1995b. Fish-community objectives for Lake Michigan. Great Lakes Fishery Commission, Special Publication 95-3. Ann Arbor, Michigan.

Eshenroder, R. L., and M. K. Burnham-Curtis. 1999. Species succession and sustainability of the Great Lakes fish community. Pages 145-184 *in* W. W. Taylor and C. P. Ferreri, editors. Great Lakes fishery policy and management: a binational perspective. Michigan State University Press, East Lansing, Michigan.

Gale, R. P. 1987. Resource miracles and rising expectations: a challenge to fishery managers. Fisheries 12(5):8–13.

Goddard, C. I. 1998. Knowing our limits: realistic expectations - the Great Lakes experience. DFO Special Publication. Casting into the future: sustaining Canada's recreational fisheries. Proceedings of the 1998 National Recreational Fisheries Conference, 19–21 November 1998, Toronto, Canada.

GLFC (Great Lakes Fishery Commission). 1997. A joint strategic plan for management of Great Lakes fisheries (supersedes 1994 version). Ann Arbor, Michigan.

Hansen, M. J., editor. 1994. The state of Lake Superior in 1992. Great Lakes Fishery Commission, Special Publication 94-1. Ann Arbor, Michigan.

Heinrich, J. W., K. M. Mullett, M. J. Hansen, G. C. Christie, G. T. Klar, D. A. Johnson, R. J. Young, and J. V. Adams. In press. Sea lamprey control and abundance in Lake Superior 1957-1999. Journal of Great Lakes Research.

Honeyfield, D. C., J. D. Fitzsimons, S. B. Brown, S. V. Marcquenski, and G. McDonald. 1998. Introduction and overview of early life stage mortality. Pages 1–7 *in* G. McDonald, J. D. Fitzsimons, and D. C. Honeyfield, editors. Early life stage mortality syndrome in fishes of the Great Lakes and Baltic Sea. American Fisheries Society, Symposium 21, Bethesda, Maryland.

Hornung, M. W., L. Miller, R. E. Peterson, S. Marcquenski, and S. B. Brown. 1998. Efficacy of thiamine, Astaxanthin, B-carotene, and thyroxine treatments in reducing early mortality syndrome in Lake Michigan salmonid embryos. Pages 124–134 *in* G. McDonald, J. D. Fitzsimons, and D. C. Honeyfield, editors. Early life stage mortality syndrome in fishes of the Great Lakes and Baltic Sea. American Fisheries Society, Symposium 21, Bethesda, Maryland.

Hoyle, J. A., T. Schaner, J. M.Casselman, and R. Dermott. 1999. Changes in lake whitefish (*Coregonus clupeaformis*) stocks in eastern Lake Ontario following *Dreissena* mussel invasion. Great Lakes Research Review 4(2):5–10.

Ihssen, P. E., G. W. Martin, and D. W. Rogers. 1992. Allozyme variation of Great Lakes alewife, *Alosa pseudoharengus:* genetic differentiation and affinities of a recent invader. Canadian Journal of Fisheries and Aquatic Sciences 49:1770–1777.

International Joint Commission (IJC). 1998. Ninth biennial report on Great Lakes water quality. IJC, Washington, D.C. and Ottawa, Canada.

Jones, M. L., J. F. Koonce, and R. O'Gorman. 1993. Sustainability of hatchery-dependent salmonine fisheries in Lake Ontario: the conflict between predator demand and prey supply. Transactions of the American Fisheries Society 122:1002–1018.

Jude, D. J., F. J. Tesar, S. F. Deboe, and T. J. Miller. 1987. Diet and selection of major prey species by Lake Michigan salmonines 1973–1982. Transactions of the American Fisheries Society 116:677–691.

Kitchell, J. F., and L. Crowder. 1986. Predator-prey interactions in Lake Michigan: model predictions and recent dynamics. Environmental Biology of Fishes 16:205–211.

Kocik, J. F., and M. L. Jones. 1999. Pacific salmonines in the Great Lakes basin. Pages 455-488 *in* W. W. Taylor and C. P. Ferreri, editors. Great Lakes fishery policy and management: a binational perspective. Michigan State University Press, East Lansing, Michigan.

Krueger, C. C., D. L. Perkins, E. L. Mills, and J. E. Marsden. 1995. Predation by alewives on lake trout fry in Lake Ontario: role of an exotic species in preventing the restoration of a native species. Journal of Great Lakes Research 21(Supplement 1):348–363.

Lavis, D. S., M. P. Hanson, D. A. Johnson, E. M. Coon, and D. J. Ollila. 2002. A case history of sea lamprey control in Lake Michigan: 1979-1999. Journal of Great Lakes Research.

Lawrie, A. H., and J. F. Rahrer. 1973. Lake Superior: a case history of the lake and its fisheries. Great Lakes Fishery Commission, Technical Report 60. Ann Arbor, Michigan.

Miller, R. B. 1957. Origin and dispersal of the alewife *Alosa pseudoharangus* and the gizzard shad *Dorosoma cepedianum* in the Great Lakes. Transactions of the American Fisheries Society 81:97–111.

Mills, E. L., J. H. Leach, J. T. Carlton, and C. L. Secor. 1993. Exotic species in the Great Lakes: a history of biotic crises and anthropogenic introductions. Journal of Great Lakes Research 19:1–54.

Negus, M. T. 1995. Bioenergetics modeling as a salmonine management tool applied to Minnesota waters of Lake Superior. North American Journal of Fisheries Management 15:60–78.

Nepszy, S. J. 1999. The changing fishery regime in Lake Erie. Pages 233-239 *in* M. Munawar, editor. State of Lake Erie –past, present, and future. Backhuys Publishers, Leiden, The Netherlands.

O'Gorman, R., and T. J. Stewart. 1999. Ascent, dominance, and the decline of the alewife in the Great Lakes: food web interactions and management strategies. Pages 489–513 *in* W. W. Taylor and C. P. Ferreri, editors. Great Lakes fishery policy and management: a binational perspective. Michigan State University Press, East Lansing, Michigan.

Parsons, J. A. 1973. History of salmon in the Great Lakes, 1850–1970. U.S. Department of the Interior, Fish and Wildlife Service, Technical Paper 68.

Peck, J. W., T. S. Jones, W. R. MacCallum, and S. T. Schram. 1999. Contribution of hatchery-reared fish to chinook salmon populations and sport fisheries in Lake Superior. North American Journal of Fisheries Management 19:155–164.

Pritchard, A. L. 1929. The alewife *Pomolbus pseudoharangus.* Publication of the Ontario Fisheries Research Laboratory 38.

Ryan, P. A., L. D. Witzel, J. Paine, M. Freeman, M. Hardy, S. Scholten, L. Sztramko, and R. MacGregor. 1999. Recent trends in fish populations in eastern Lake Erie in relation to changing lake trophic state and food web. Pages 241-289 *in* M. Munawar, editor. State of Lake Erie –past, present, and future. Backhuys Publishers, Leiden, Netherlands.

Rybicki, R. W., and D. F. Clapp. 1996. Diet of chinook salmon in eastern Lake Michigan, 1991-1993. Michigan Department of Natural Resources Fisheries Research Report 2027. Lansing, Michigan.

Smith, B. R., and J. J. Tibbles. 1980. Sea lamprey (*Petromyzon marinus*) in Lakes Huron, Michigan, and Superior: history of invasion and control, 1936-1978. Canadian Journal of Fisheries and Aquatic Science 37:1780–1801.

Smith, G. R., and T. N. Todd. 1984. Evolution of species flocks of fishes in north temperate lakes. Pages 45–64 *in* A. A. Echelle and I. Kornfield, editors. Evolution of fish species flocks. University of Maine at Orono Press, Orono, Maine.

Smith, S. H. 1968. Species succession and fishery exploitation in the Great Lakes. Journal of the Fisheries Research Board of Canada 25:667–693.

Smith, S. H. 1970. Species interactions of the alewife in the Great Lakes. Transactions of the American Fisheries Society 99:754–765.

Smith, S. H. 1972. Factors of ecological succession in oligotrophic fish communities of the Laurentian Great Lakes. Journal of the Fisheries Research Board of Canada 29:717–730.

Smith, S. H. 1995. Early changes in the fish community of Lake Ontario. Great Lakes Fishery Commission, Technical Report 60. Ann Arbor, Michigan.

Sprules, W. G., S. B. Brandt, D. J. Stewart, M. Munawar, E. H. Jin, and J. Love. 1991. Biomass size spectrum of the Lake Michigan pelagic food web. Canadian Journal of Fisheries and Aquatic Sciences 48:105–115.

Steedman, R. J., and H. A. Regier. 1987. Ecosystem science for the Great Lakes: perspectives on degradative and rehabilitative transformations. Canadian Journal of Fisheries and Aquatic Science 44(Supplement 2):95–103.

Stewart, D. J., J. F. Kitchell, and L. B. Crowder. 1981. Forage fishes and their salmonid predators in Lake Michigan. Transactions of the American Fisheries Society 110:751–763.

Stewart, D. J., and M. Ibarra. 1991. Predation and production by salmonines fishes in Lake Michigan: 1978–88. Canadian Journal of Fisheries and Aquatic Science 48:909–922.

Stewart, T. J., R. E. Lange, S. D. Orsatti, C. P. Schneider, A. Mathers, and M. E. Daniels. 1999. Fish-community objectives for Lake Ontario. Great Lakes Fishery Commission, Special Publication 99–1. Ann Arbor, Michigan.

Talhelm, D. R. 1988. Economics of Great Lakes fisheries: a 1985 assessment. Great Lakes Fishery Commission, Technical Report 54. Ann Arbor, Michigan.

Tody, W. H., and H. A. Tanner. 1966. Coho salmon for the Great Lakes. State of Michigan, Department of Conservation. Fish Management Report 1. Lansing, Michigan.

Webb, S. A., and T. N. Todd. 1995. Biology and status of the shortnose cisco *Coregonus reighardi* Koelz in the Laurentian Great Lakes. Archiv fuer Hydrobiologie. Special Issues in Advanced Limnolology 46:71–77. July 1995 International Symposium on Biology and Management of Coregonid Fishes.

Weeder, J. A. 1997. A genetic comparison of Lake Michigan chinook salmon (*Oncorhynchus tshawytscha*) to their source population. Michigan Department of Natural Resources Fisheries Research Report 2032.

Part III Pacific Salmon

Sustaining North American Salmon: Perspectives Across Regions and Disciplines, 261–276

Chapter 13

History of the Salmon Fisheries of the Pacific Northwest Coast

Terry Glavin

Introduction

People and salmon have maintained an elaborate and complex relationship throughout North America's northwest coast for several thousand years. For most of the 20th century, a prevailing assumption among North Americans has been that salmon were not intensively exploited until the industrial period. In fact, long before the canneries came, salmon were harvested in highly productive fisheries, and salmon were a crucial component of the diet of aboriginal peoples well into the continental interior.

The center of fishing effort in the aboriginal fisheries was not along the Pacific coast but well into the continent's river systems. Aboriginal fisheries were generally stock-specific, governed according to customary laws. They were generally species-selective and run-selective in nature, which tended to conserve biological diversity and guard against losses in spatial diversity and long-term abundance. These characteristics stand in stark contrast to the industrial fisheries, in which the center of fishing effort was at the mouths of rivers and well into mixed-stock areas in the saltwater environment. Such fisheries management features have long been associated with overfishing, the extirpation of co-migrating stocks, losses in biological diversity, and losses in spatial diversity and abundance.

Cultures change, and at the outset of the 21st century, the industrial values that informed the early settler cultures of North America's west coast have given way to new values. The societies of the U.S. northwestern states, British Columbia, and southeastern Alaska have come to share a common understanding that the salmon is important, not just for its commercial or recreational value and genetic information, but also for the role that salmon play in ecosystems, for salmon's heritage value, and for the intrinsic value of salmon as a defining feature of the regional landscape.

Canadian and American institutions have been slow to reflect these changed values, as the 1999 agreement under the Pacific Salmon Treaty shows. However, these changed values may ultimately prove the most effective defense for salmon against the threats salmon face in these changing and uncertain times.

People, Salmon, and the Antiquity of Highly Productive North Pacific Salmon Fisheries

Seven North American species of the genus *Oncorhynchus* fall under the definition of Pacific salmon and spend part of their lives in the Pacific Ocean. They are sockeye *O. nerka*, chinook *O. tshawytscha*, coho *O. kisutch*, pink *O. gorbuscha*, chum *O. keta*, steelhead *O. mykiss,* and sea-run cutthroat trout *O. clarki* (Groot and Margolis 1991). Salmon descend to the sea from rivers that arise sometimes deep within the continent, as far inland as Wyoming. In the final decades of the 20th century, harvests of salmon throughout the North Pacific—primarily harvests of sockeye, pink, and chum—reached 800,000 metric tons annually. Alaskan production accounts for more than all other North American jurisdictions combined (CAFÉ 1998; Beamish and Bouillon 1995; National Research Council 1996).

Salmon have contributed significantly to the terrestrial/marine ecotone of northwestern North America as it has evolved during the past 10,000 years—that period of history since the close of the Pleistocene epoch. Seals, sea lions, killer whales, and a variety of coastal fish and bird species directly and indirectly rely on salmon. Salmon also provide a key food source to 22 forest mammals. Analyses of the bones of grizzly bears from the Upper Columbia watershed, several hundred miles from the sea, show that 90% of the carbon and nitrogen in the bears' diets came from salmon (Hebda and Whitlock 1997; Levy 1997).

If the early Holocene epoch is taken as the beginning of time for the purposes of this chapter, then people, like salmon, have been here from the beginning. From the time that ice-age glaciers began to retreat from the coast, aboriginal cultures settled down and set about the business of co-evolving with the salmon populations that reestablished themselves within the post-Pleistocene landscape. Archeologists tell us that on the Pacific coast, people have been catching salmon for almost 10,000 years (R. Carlson 1992, unpublished data).

It has been conventionally held that native communities presented little threat to salmon abundance because of low levels of human population and because of technological restraints. The conventional wisdom has been: "There weren't many people here, and they didn't have the capacity to do much damage," and that is why there was salmon in such abundance when Europeans arrived in the late 1700s. By the late 20th century, however, that conventional wisdom was failing to withstand scrutiny. Much research was finding that aboriginal salmon harvests in parts of the Northwest appear to have been comparable to the harvests associated with the industrial fisheries of the 19th and 20th centuries (Chisholm et al. 1983). The question of how many fish did aboriginal people catch is related to the question of how many people there were. These are difficult questions to answer with any precision.

There are a variety of estimates of the precontact aboriginal population of the Northwest coast culture area, which ranges, very roughly, from the Eel River in California to Alaska's Yakutat Bay (Suttles and Ames 1997). Precontact aboriginal population estimates have been consistently revised upwards (Mooney 1910), with a general consensus among anthropologists that the Northwest coast culture area may have supported populations in the hundreds of thousands—certainly the highest human population levels in North America north of the Valley of Mexico. The role of salmon in supporting such dense human populations cannot be understated. It is also important to remember that salmon were not simply a coastal resource. In fact, salmon probably played as important a role, if not a more important role, in the diets of upriver, interior peoples (Taylor 1999). Analysis of nitrogen isotopes in human remains from archeological sites several hundred miles up the Fraser River show that salmon was an even more important source of protein for people than land mammals (Chisholm 1986).

Probably the most exhaustive coast-wide examination of the how-many-fish question was undertaken by Gordon Hewes in 1947, a question to which he returned in the early 1970s. Hewes estimated that precontact harvests of salmon by aboriginal peoples from California to Alaska, on the basis of conservative population estimates and caloric-requirement analyses, would have amounted to as much as 130 million pounds annually. This is the equivalent of 25 million sockeye per year (Hewes 1973), which is slightly more than half the volume of salmon associated with the peak years of the industrial fisheries, from the 1860s to the 1920s, from California to Alaska. Anthropologist Randall Shalk, meanwhile, estimated that the precontact aboriginal salmon harvest in the Columbia basin may have amounted to an annual catch in excess of 40 million pounds, which is roughly the same as the average industrial salmon production associated with the Columbia River during the peak years of the Columbia's commercial fisheries, between 1883 and 1919 (Taylor 1999).

The Fraser River was by far the largest salmon-producing river on the whole coast. Other salmon-fishing areas are more productive than the

Fraser—such as Alaska's Bristol Bay—but the salmon catches of those areas are attributable to several river systems. The Columbia River was primarily a chinook river, with preindustrial stock sizes estimated at 16 million fish (Cone and Ridlington 1996). The Fraser always supported more but generally smaller salmon: the Fraser is primarily a sockeye and pink river. Preindustrial stock sizes for Fraser sockeye have been estimated as high as 50 million fish in the big years of the Fraser sockeye's four-year cycle (Ricker 1987), but these stocks' sizes declined fivefold as a consequence of early industrial overfishing and the Hell's Gate slide of 1913 (Northcote and Burwash 1991). Conservative estimates of the aboriginal population of the Fraser basin put the number of people in the precontact era at about 60,000. At that population level, using an estimate of 160 dried fish per person per year—which is about what Hewes' calculations suggest as a Fraser basin average—the aboriginal catch would have produced a harvest equivalent to about 6 million sockeye salmon per year. To put this in context, the average annual total abundance of Fraser River sockeye, between 1894 and 1986, was about 6 million fish.

Whatever the final numbers, they all add up to a lot of people and a lot of salmon. How this ended up happening is a long story, but it is at a point about 5,000 years ago that a confluence of events produced the conditions that resulted in the populous, salmon-fishing societies explorers encountered on the coast in the early days of contact (Suttles and Ames 1997). It was about 5,000 years ago that gradually sloping beaches were beginning to form, all up and down the west coast, exposing tidal flats and productive clam beds. It was also about this time that a being arrived on the coast, known to the Salish people of British Columbia's southwest coast as Long-Life-Maker. Cedar, the great warehouse of Northwest coast culture, was establishing itself as the dominant tree species after a long migration north, eventually finding a foothold as far north as Alaska's Baranof Islands.

Cedar became the raw material for clothing, ocean-going canoes, great works of monumental art, and a lumber supply for cedar-planked houses of a type that required as much as 70,000 board feet of timber just to build and keep maintained over the years. These weren't simply houses—they were fish-processing and storage plants, which also served, in the words of anthropologists Wayne Suttles and Kenneth Ames, as "banquet halls, theaters and temples."

The advent of cedar meant people could hang significant quantities of fish to dry in the rafters, archeologist Phil Hobler observes. That fact alone might be as important as anything else behind the revolutionary changes that were taking place in coastal societies all those years ago (P. Hobler 1998, personal communication), but the advent of large communal houses also signals the development of elaborate social organization, stratification and division of labor.

It was also about this time, 5,000 years ago (about 1,000 years before ancient Britons began building Stonehenge), that people all over the west coast started building stone fish traps, tidal traps, fish weirs, and other fixed-gear fishing methods. This is important because related to the question "How many fish did aboriginal people catch?" is the crucial question "How did they catch so many fish?"

To my knowledge, a study by University of British Columbia student Patricia Ann Berringer is the only comprehensive analysis of the salmon-fishing technologies that were deployed throughout the Northwest coast culture area. What is noticeable is the preponderance of fixed or "terminal" gear. There were hook-and-line trolling gear, trawls, gaffs, spears, harpoons, dip nets, and reef nets, but mainly, fishing technology was about traps and weirs: stone tidal traps, grid traps, cylindrical traps, tumbleback traps, box traps, basin traps, trough traps, log dams, and stone dams. From the Yurok people in the south to the Tlingit and Eyak fisheries in the north, most of these gear types were used (Berringer 1982).

I am not aware of any complete, coastal archeological inventory of fixed-gear technologies, but for the British Columbia coast, Roy Carlson of Simon Fraser University put one together in 1992. Carlson came across 455 reports of sites with weirs, fish traps, or probable fish traps in the files of the government's archeology branch (Carlson, 1992 unpublished report).

The archival record, meanwhile, also presents a picture of fisheries that were, in the main, terminal and selective fisheries.

The first written description of west coast fishery methods comes from the explorer Alexander MacKenzie, eight years before the Americans Lewis and Clark reached the mouth of the Columbia River. He settled on the term "machines" to describe what fishing devices he encountered in the final days of his trek across the continent to the Pacific in 1793 (Lamb 1970). At Friendly Village, on the Bella Coola River, about 50 kilometers above salt water, MacKenzie encountered one of several dam-and-trap complexes operating on the river at the time. The Friendly Village dam was a permanent or semipermanent structure of logs and gravel built by Nuxalk people at a point in the river about 50 m across. The dam constricted about two-thirds of the river from both banks, forcing the current through a narrow opening in the middle.

"Beneath it the machines are placed, into which the salmon fall when they attempt to leap over," Mackenzie wrote. "On either side there is a large frame of timber-work six feet above the level of the upper water, in which passages are left for the salmon leading directly into the machines which are taken up at pleasure. At the foot of the fall, dipping nets are also employed."

One of the last written descriptions of this type of technology comes from 1905, in the reports of Canadian fisheries officer Hans Helgeson and what he encountered in the Upper Skeena watershed in northern British

Columbia (Helgeson et al. 1906). His descriptions of the fish traps on the Babine River are particularly noteworthy: "The barricades were constructed of an immense quantity of materials, on scientific principles," Helgeson reported. "There were posts driven into the bed of the river, which is 200 ft wide, and from two to four feet deep, and running swiftly at the intervals of 6 or 8 ft. . . .Then sloping braces well bedded in the bottom and fastened to the top of posts, then strong stringers all the way on top and bottom. . . . Altogether the barricades presented a most formidable and imposing appearance."

The most elaborate of the weir fisheries Helgeson observed was the twinned barricade on the main stem of the Babine River, about 10 kilometers downstream from Babine Lake. He wrote: "The banks of the Babine River have a lovely appearance at this place and a most wonderful sight met our eyes when we beheld the immense array of dried salmon. On either side, there were no less than 16 houses, 30 × 27 × 8 ft, filled with salmon from the top down so low that one had to stoop to get into them and also an immense quantity of racks, filled up outside. If the latter had stood close together they would have covered acres and acres of ground, and though it was impossible to form an estimate, we judged it to be nearly three quarters of a million fish at those two barricades. . . though the whole tribe had been working together for six weeks and a half it was a wonder that so much salmon could be massed together in that time."

The purpose of Officer Helgeson's journey through the Upper Skeena was to culminate a campaign by the salmon canneries to have the upriver traps and weirs dismantled to force the Skeena Indians to come down to the coast, where they could be gainfully employed under Her Majesty's guidance and authority in the canneries of Port Edward, Port Essington and other such cannery towns around the mouth of the Skeena River. Which brings us to the next chapter in this 10,000-year epic.

The Rise and Fall of the Industrial Salmon Fisheries

The industrial salmon fisheries of North America's west coast began with the first salmon cannery, established in 1864 by the firm of Hapgood, Hume, and Company near what is now the foot of K Street in Sacramento, California. Andrew Hapgood and the Hume brothers managed to pack and sell 2,000 cases of salmon that year. Within three years, the Humes had moved to the Columbia River, and in 1867, at Eagle Cliff, Washington, the company packed 4,000 cases of 48 one-pound cans. On the Fraser River, salmon canning began in 1870 with a primitive business begun by Alexander Loggie at Annieville and a small operation established by Edward Stamp, at nearby Sapperton, the same year. On the Skeena River, salmon canning began in the early 1870s, and by 1876, the North Western Commercial Company was running a bustling enterprise at Woodcock's Landing, near the mouth of the Skeena. Before the end of the 19th century, scores of canneries were in operation between the Columbia River and Alaska (Newell 1989).

It is difficult to describe just how rapidly and how significantly the canneries transformed the coast. The industrial revolution transformed coastal societies, which had been decimated by smallpox and other epidemic diseases. On the Canadian coast, residential schools set about the work of disassembling aboriginal society. On the American coast, there were disease, smallpox, and war. Although in remnants, aboriginal communities nevertheless played crucial roles in the industrialization of the salmon fisheries, and the culture born in those canneries was one of the more fascinating in the history of the so-called New World. Aboriginal people from different tribal groups, speaking different languages, were all thrown in together, working alongside Japanese immigrants, Chinese immigrants, Americans, Hawaiians, Orkney Islanders, Englishmen, Finns, and Canadians. The cannery culture even had its own Creole, known as the "Chinook" jargon, which was born in the maritime fur trade period.

White, Native, and Japanese men dominated the harvest sector. Typically, salmon canneries were supplied by fleets of gill-net boats, first powered by oar and sail, and then by gas and diesel engines. Trollers and big-boat seine vessels were common gear types from California to Alaska by the 1940s, but in the early part of this century, the industrial salmon fisheries also relied upon industrial-scale traps, weirs, beach seines, and fish wheels. Fish were usually brought to the cannery directly by fishermen, although sometimes by a packer vessel that collected salmon from the fishing grounds. Cannery labor was segregated along lines of race and gender. Crews of Native women gutted the salmon and removed the heads and tails. Chinese crews cut the fish into can-sized pieces in assembly-line fashion until that work was replaced by a mechanical gang knife known as the "Iron Chink." Native women filled the cans and were paid piece-work rates. Native women then weighed the cans, which were then moved along to more Chinese workers who soldered, cooked, sealed, and tested the cans (Muszynski 1996).

What happened during the comparatively brief period of the cannery culture is a story of consolidation, centralized canning and cold-storage capacity, rapid technological advances in the transportation of raw fish, railroads, and a shift in world markets toward fresh and frozen fish products.

It is also a story of salmon declines as a result of overfishing; dam construction; habitat loss, arising from poor logging and agricultural practices; and urban development. By the 1980s, the cannery culture had all but vanished. Salmon canning persisted as a seasonal mainstay of the Alaskan coastal economy, relying on local labor and students from the "Lower 48" earning summer wages. A handful of salmon canneries remained on the British Columbia coast. By the 1990s, salmon canneries south of the 49th parallel were mainly a distant memory.

Through the 1990s, about 2,000 salmon trollers were licensed to fish the California coast, but it was difficult to tell in any given year how many of those boats, dependent mainly upon the depleted chinook runs of the Sacramento River, actually went fishing. In Oregon, the number of troll licenses rose and fell dramatically through the last two decades of the 20th century. Washington's fleet was reduced to a few hundred vessels, mainly gill-net boats and small seiners, most of which earned the lion's share of their income by traveling to the Alaskan fishing grounds every summer. British Columbia's fleet declined from about 7,000 vessels in the early 1960s to fewer than 2,600 vessels by the close of the 1990s.

By the beginning of the 21st century, Pacific salmon had disappeared from almost half their spawning range on the North American continent (National Marine Fisheries Service 1998). Endangered Species Act maps made the western states look like a color-coded guide to biological trauma wards. In B.C., hundreds of salmon populations were either extinct or at high or moderate risk of extinction (Slaney et al. 1996).

The Tribal Fisheries and the Commercial Fisheries: What's The Difference?

Given the fact that aboriginal fisheries were of such significant magnitude and persisted over such a long period of time, it is noteworthy that these fisheries produced no known impact upon the genetic and spatial diversity of salmon populations—impacts that are easily attributable, in many instances, to industrial fisheries.

An analysis of the distinguishing features of the salmon fisheries conducted in the precontact period shows that, though opportunistic saltwater fishing was common—mainly in the form of trolling for passing coho and chinook salmon—most aboriginal salmon fisheries occurred at the mouths of small rivers and well into the river systems. The center of fishing effort occurred predominantly in "terminal" fishing areas and was governed by elaborate customary laws that tended to take the needs of spawning escapement into account (Berringer 1982).

To the extent that mixed-stock fishing occurred, the very nature of the fisheries undertaken downstream of fixed-gear systems tended to produce low harvest rates. Fishing effort was spread out, but the fisheries that produced high harvest rates were undertaken far up the major rivers and in the tributaries.

These defining characteristics allowed for stock-specific and selective fisheries undertaken according to rules that might be described as adaptive, in-season management (Glavin 1996). These characteristics provided for significant harvests over a long period of time that did not produce losses in biological diversity, spatial diversity, or long-term abundance.

Generally speaking, the industrial fisheries took that same fisheries-management model and turned it upside down. The center of fishing effort moved down the rivers, out of the estuaries, and into the ocean. Technology progressed through the late 19th century and early 20th century in accordance with a "race for the fish" pattern (Meggs 1991; Taylor 1999). While industrial fishing effort was often aimed at harvestable surpluses of relatively healthy salmon populations, a consequence of this practice was often the unavoidable depletion, and probably, in some cases, the extinction of smaller, co-migrating salmon populations. The most serious consequence of the loss of small salmon runs is a crippling decline in the biological diversity in salmon, a requirement for long-term abundance. As the 20th century progressed, salmon came to be harvested increasingly in mixed-stock fishing areas, far from the salmon runs' river of origin, in non-selective fisheries. Snake River sockeye, for instance, once harvested well into the interior of Washington and Idaho, came to be harvested far from the spawning grounds, as far away as southeastern Alaska.

It is hard to say what role mixed-stock imprecision and its consequent overfishing may have played in historic declines. It would be unfair to suggest that declines in abundance, in spatial distribution, and in genetic diversity in salmon can be attributed solely to industrial fishing practices. But because of these declines, mixed-stock fishing may prove a luxury we will find it increasingly difficult to afford. In light of the present state of salmon, the necessary rebuilding strategies that Canadian and American fisheries managers are obliged to pursue, and the threats that salmon are expected to face in the coming years, mixed-stock fishing is becoming a difficult practice to defend.

The Prospect of New Management Methodologies, Informed By Old Traditions, in Changing and Uncertain Times

In its historic 1996 report, *Upstream: Salmon and Society in the Pacific Northwest*, the U.S. National Research Council included this observation: "To achieve long term sustainability, which requires sufficient genetic diversity, fishing should occur only where the identity (i.e., the originating population) of the salmon is known, when total fishing mortality is consistent with the productivity of the fish, and when the catching technology ensures minimal mortality in depleted *demes* (local breeding populations). This will require fishing methods that allow different degrees of fishing effort on various salmon populations and that allow identification of fish taken from depleted *demes* so that they can be avoided or released alive" (Committee on Protection and Management of Pacific Northwest Anadromous Salmonids, National Research Council 1996, p. 13).

That statement made eminent sense, and by the 1990s, Canadian and American salmon fisheries managers were attempting to retool many of the coast's salmon fisheries to make them selective and stock-specific. Reducing the impacts of mixed-stock fishing can be achieved by eliminating mixed-stock fishing entirely, significantly lowering the harvest rates produced by fisheries conducted in mixed-stock areas, closing certain fishing areas and moving fisheries to more terminal areas, changing gear types, and amending fishing times. In the coming years, moving back to more stock-specific and selective fisheries will likely be made more necessary by the sheer weight of scientific uncertainty that follows upon broad-scale disruptions underway in Pacific salmon populations. These disruptions are occurring not only in spawning and rearing habitat in the terrestrial environment but also within the broad-scale ecosystems of the Pacific Ocean.

Recent changes in the marine environment began to take their toll on the marine survival rates of salmon off the California coast. These changes are apparently becoming more frequent and more dramatic. And they are moving north. These disruptions have wreaked havoc with coho, steelhead, and chinook populations from California to British Columbia. In 1997 and 1998, Alaskan fisheries scientists were confounded by a completely unanticipated collapse of Alaska's most productive sockeye fishery, the Bristol Bay sockeye runs. In September 1997 and in July 1998, anomalous plankton blooms, never before observed in such intensity so far north, covered such vast areas of the Bering Sea that they were visible from space (Schumacher et al. 1998). In 1998, severe shortfalls occurred in returns of chinook salmon to the Yukon River and chum salmon bound for the Kuskowim River. In 1998 and 1999, meanwhile, B.C.'s salmon fisheries suffered the worst years of the 20th century, mainly because of the collapse of the Fraser sockeye runs.

Some scientists expect that climate change—whether human-induced, part of a poorly understand long-term trend, or as a consequence of factors associated with the Pacific decadal oscillation—will result in continuing salmon declines in the Pacific for at least the first quarter of the 21st century (Klyashtorin 1998). Others foresee the Pacific Ocean becoming increasingly and fatally inhospitable to salmon to the extent that that salmon's ocean pasture will shrink and all but disappear south of the Bering Sea within the first 50 years of the 21st century (Welch et al. 1999). Whatever the case, at the very least, climate change is a factor, as the North Pacific Anadromous Fish Commission concluded in 1998, that must be recognized as deserving of consideration equal to that given every other factor that determines salmon abundance throughout the Pacific.

Fisheries management agencies have taken many necessary steps to curb the harmful effects of mixed-stock fisheries that have predominated in the industrial salmon fisheries. The trend away from mixed-stock fishing has been underway in California and Oregon for some time. Only a few healthy runs of salmon remain that far south, and management efforts are aimed at ensuring that harvests are conducted only upon salmon runs with known harvestable surpluses. In Washington, the Boldt decision of the 1970s, which guaranteed the various tribes a 50% share of the commercial fishery, has obliged fisheries managers to move to more selective fisheries to comply with the law. In May of 1998, in British Columbia, the federal government vowed to make selective fisheries the "cornerstone of salmon management" throughout the coast. But there is still much resistance to this policy in the commercial fishing industry. Mixed-stock fishing is still a feature of Alaskan salmon fisheries. In truth, Alaska has managed its own mixed-stock fisheries quite well. After years of trial and error, Alaska's in-season management regimes appear to work fairly well, owing mainly to a fairly effective system of monitoring a series of indicator populations. But Alaska's success involves Alaska's salmon runs, not the many salmon runs harvested in Alaska that are bound for B.C., Washington. and Oregon rivers. And even Alaska's domestic management may be sorely tested in the years to come.

The last great bastion of mixed-stock fishing occurs in the interception fisheries undertaken by both Canada and the United States that have been more or less entrenched by the Pacific Salmon Treaty. The Pacific Salmon Treaty of 1985 had its origins in a binational convention negotiated in the 1930s, and in many ways it was a product of notions about salmon that were fashionable back then but are now thoroughly discredited. The treaty failed to meet even its own objectives, which were to avoid harvest-share disputes between Canada and the United States and to control overfishing by the fleets of both countries. It failed to contain two very dangerous "salmon wars" that erupted between Canada and the United States, and it caused severely strained relations between Canada and the province of British Columbia. The Pacific Salmon Treaty also codified the continuation of management regimes that produced disputes between Alaska and the Pacific Northwest states about salmon runs that Washington, Oregon, and Idaho were attempting to conserve and restore.

The "equity" dilemma, which caused fiery debates between Canadian and American negotiators since the treaty was signed, resulted in endless and arcane arguments. For instance, even in the event of an agreement that an equity imbalance had been produced by, say, a relatively recent Alaskan fishery, Canadian negotiators were often left with little choice but to attempt to negotiate restraints on Washington state fishermen, 1,000 mi away, whose traditions of harvesting Fraser sockeye went back into the 19th century, and who often knew little and cared less about Alaskan fishing.

Equity disputes proved impossible to resolve because of disagreements between Canada and the United States over how many fish each country's fleet was catching. These were not minor quibbles. In 1987, for instance, Canadian fisheries managers estimated that Canadian fishermen caught 904,000 Alaska-bound pink salmon. That same year, Alaskan fisheries managers reckoned that Canadians had actually caught 3.3 million Alaskan pink salmon. In 1988, Canadian fisheries managers said Canadian fishermen caught 725,000 Washington state coho. U.S. fisheries managers said the Canadians actually caught 1.34 million Washington state coho that year. In 1990, Canadians said Washington state fishermen caught 229,000 B.C. coho. Americans said the Washington fleet caught only 63,000 B.C. coho that year (Glavin 1998).

Through the 1990s, the two sets of figures for B.C. and Washington chinook, sockeye, and pink salmon started to look more or less the same. But then disputes flared over how the estimates of each country's biologists should be characterized and whether the figures should be made public. In the years before a new set of salmon treaty provisions was negotiated by Canada and the United States in 1998, there was no complete set of fish-catch records, either in the form of agreed-upon figures or conflicting estimates, available for public release by fisheries managers in either country.

In 1999, however, owing largely to the diminished influence of commercial salmon-fishing interests in both Canada and the United States and heightened public concern about salmon on both sides of the border, it became possible for Canada and the United States to conclude a series of new arrangements under the Pacific Salmon Treaty in what amounted to a completely new treaty.

The main features of the June 1999 agreement included: caps on Alaskan harvests of Skeena-bound salmon; a fixed U.S. share of Fraser-bound sockeye and pink catches; new controls on the Canadian catch of U.S.-bound chinook and coho salmon; strengthened institutional arrangements, such as an improved dispute-resolution mechanism and processes allowing for greater cooperation between the fisheries scientists of both countries; and joint programs for the protection and restoration of salmon habitat (Agreement under the Pacific Salmon Treaty between Canada and the U.S., June 1999).

The agreement was clearly an improvement over the situation that had previously prevailed, but the 1999 deal also entrenched mixed-stock fisheries, particularly fisheries for chinook and coho in southeastern Alaska, in northern B.C. waters, and on the west coast of Vancouver Island. To control the adverse impacts of mixed-stock fisheries, the 1999 deal provided for the introduction of an aggregate, abundance-based management (AABM) system, based on the Alaskan model. But an assessment of the new system prepared for the Pacific Fisheries Resource Conservation Council, a federal-provincial "watchdog" body, found several shortcomings (Peterman and Pyper 2000). Among the agreement's observed shortcomings: the new fishing rules would not be adequate to maintain certain Canadian chinook populations; mechanisms to restrain fishing when stocks were low would not be triggered frequently enough or when needed; and the agreement did not take the precautionary principle into account and did not sufficiently account for scientific uncertainty.

It is an irony of the June 1999 agreement that it was made possible because of heightened public concern about poor fishing practices—concern that was broadly shared by both Canadians and Americans—but the main thrust of the agreement was to authorize the persistence of dangerous fishing practices and add some safeguards to contain the damage done by those practices. In hindsight, it is worth considering whether a Canada-U.S. treaty setting out the rules for mixed-stock fishing is a good idea at all. It may be unrealistic to argue that Canadians should catch only Canadian fish and Americans should only catch American fish. But it is not unreasonable to expect that a salmon treaty between Canada and the United States should reflect not only the legitimate interests of fishermen but also the public interest in salmon, which is a classic common-property resource in the United States and a resource owned by the Crown in Canada. It is not unreasonable to expect that such a treaty should be fully informed by international standards in the protection of biological diversity and well recognized, risk-averse management practices rather than simply short-term industrial and commercial purposes. Neither is it unreasonable to expect that a salmon treaty should reflect the changed and perilous circumstances surrounding wild salmon at the advent of the 21st century.

Indeed, just such a treaty has become politically and culturally possible. From San Francisco to Sitka, a far more mature public attitude has developed with respect to salmon. People are beginning to understand that salmon, in ways that are often not very easy to articulate, are important, and that they are important in their own right, not just as a marketable commodity, as food for people, or for the genetic information they carry, but also for the role they play in ecosystems. Aboriginal, commercial, and recreational fishermen may be entitled to salmon, but eagles need salmon, too, as do seals, sea lions, killer whales, wolves, and bears. This is not just about the new environmental consciousness that is supposed to be so entrenched in the Pacific Northwest. It is all that, but it is more than that.

Over time, the cultural associations that aboriginal peoples maintained with salmon have become manifest in a nonaboriginal affection for salmon, a sort of a loyalty to salmon that goes beyond mere sentimentality, and it is reasonable to be optimistic because of it.

Some bold moves toward the establishment of necessary habitat protection regimes appear to be underway in the Pacific Northwest under the auspices of the U.S. Endangered Species Act. It has become possible for politicians in the northwestern states to publicly support the decommissioning of hydroelectric dams, even the closure of salmon hatcheries where hatchery fish are reasonably believed to be restraining the recovery of wild salmon populations. In British Columbia, thousands of schoolchildren raise chum salmon fry in their classrooms, release them into neighborhood creeks, and talk to one another about how much they hope the fish will come home some day. At least 10,000 volunteers can be found all across the province, spending their weekends in the rain clearing trash out of salmon creeks, monitoring water quality, and counting salmon fry. Similar programs exist in Washington state and Oregon. Many if not most of these people have never fished but simply want salmon to return and flourish.

Throughout the Pacific Northwest, wild salmon have become emblematic of our sense of ourselves, our pride of place. Salmon have come to serve as a symbol of our common wish, as Canadians and Americans, that our shared destiny is not unalterable. And that, in the end, may be the single greatest factor in ensuring a place for wild salmon in our common future.

References

Beamish, R. J., and D. R. Bouillon. 1995. Marine fish production trends off the Pacific coast of Canada and the United States. Climate change and northern fish populations, Canadian Special Publications in Fisheries and Aquatic Sciences 121.

Berringer, P. A. 1982. Northwest coast traditional salmon fisheries: systems of resource utilization. Master's thesis, Faculty of Graduate Studies, University of British Columbia.

CAFE Global Outlook. 1998. World salmon production. Miscellaneous data sources, World Seafood Report, Program Planning and Economics Branch, Department of Fisheries and Oceans.

Chisholm, B. 1986. Reconstruction of prehistoric diet in British Columbia using stable-carbon isotope analyses. Doctoral thesis in archeology, Simon Fraser University.

Chisholm B. S., D. E. Nelson, and H. P. Schwarz. 1983. Marine and terrestrial protein in prehistoric diets in the British Columbia coast. Current Anthropology 24(3).

Cone, J., and S. Ridlington, editors. 1996. The northwest salmon crisis: a documentary history. Oregon State University Press, Corvallis, Oregon.

Glavin, T. 1996. Dead reckoning: confronting the crisis in Pacific fisheries. David Suzuki Foundation/Greystone Books, Vancouver.
Glavin, T. 1998. Last call: the will to save pacific salmon. A report of the Pacific Salmon forests project, David Suzuki Foundation, Vancouver.
Groot, C., and L. Margolis, editors. 1991. Pacific salmon life histories. Department of Fisheries and Oceans and University of British Columbia Press, Vancouver.
Hebda, R., and C. Whitlock. 1997. Environmental history. The rainforests of home: profile of a North American bioregion. Island Press, Washington, D.C.
Helgeson, H., et al. 1906. Thirty-eighth annual report of the Department of Marine and Fisheries, 1905. Pages 206–211, John T. Williams, Inspector of Fisheries, Ottawa.
Hewes, G. W. 1973. Indian fisheries productivity in pre-contact times in the Pacific salmon area. Northwest Anthropological Research Notes 7(2).
Klyashtorin, L. B. 1998. Cyclic climate changes and Pacific salmon stock fluctuations: a possibility for long-term forecasting. Russian Federal Institute of Fisheries and Oceanography, Moscow. Technical report. Workshop on Climate Change and Salmon Production, North Pacific Anadromous Fisheries Commission, Vancouver.
Lamb, W. K. 1970. The journals and letters of Sir Alexander MacKenzie. MacMillan and Company, Toronto.
Levy, S. 1997. Ultimate sacrifice. New Scientist, September 6.
Meggs, Geoff. 1991. Salmon: the decline of the British Columbia fishery. Douglas and McIntyre, Vancouver.
Mooney, J. 1910. Population. Bureau of American Ethnology Bulletin No. 30.
Muszynski, A. 1996. Cheap wage labour: race and gender in the fisheries of British Columbia. McGill-Queen's University Press.
National Marine Fisheries Service. 1998. Geographic extent of Pacific salmon ESUs. May 4.
National Research Council. Committee on Protection and Management of Pacific Northwest Anadromous Salmonids. 1996. Upstream: salmon and society in the Pacific Northwest. National Academy of Sciences, Washington, D.C.
Newell, D., editor. 1989. The development of the Pacific salmon canning industry: a grown man's game. McGill-Queen's University Press.
Northcote, T. G., and M. D. Burwash. 1991. Fish and fish habitats of the Fraser basin. In A. H. J. Dorcey and J. R. Griggs, editors. Water and sustainable development: exploring our common future in the Fraser river basin. Westwater Research Centre, University of British Columbia.
Peterman, R. M., and B. J. Pyper. 2000. Review of the coho and chinook salmon sections of the agreement under the Pacific Salmon Treaty between Canada and the United States, dated June 30, 1999: a report to the Pacific Fisheries Resource Conservation Council. PFRCC, Vancouver.
Ricker, W. E. 1987. Effects of the fishery and of obstacles to migration on the abundance of Fraser river sockeye salmon. Canadian technical report, Fisheries and Aquatic Sciences 1522.
Schumacher, J. D., P. J. Stabeno, N. A. Bond, and J. M. Napp. 1998. Ecosystem anomalies in the Eastern Bering sea during 1997. Workshop on climate change

and salmon production technical report. North Pacific Anadromous Fisheries Commission, Vancouver.

Slaney, T. L., K. D. Hyatt, T. G. Northcote, and R. J. Fielden. 1996. Status of anadromous salmon and trout in British Columbia and the Yukon. Fisheries 20(10).

Suttles, W., and K. Ames. 1997. Pre-european history. In P. K. Schoonmaker, B. Von Hagen, and E. C. Wolf, editors. The rain forests of home. Ecotrust/Interrain, Island Press, Washington, D.C.

Taylor, J. E. 1999. Making salmon: an environmental history of the northwest fisheries crisis. University of Washington Press, Seattle.

Welch, D. W., Y. Ishida, and K. Nagasawa. 1999. Thermal limits and ocean migrations of sockeye salmon (*Oncorhynchus nerka*): long-term consequences of global warming. Canadian Journal of Fisheries and Aquatic Sciences 55.

Chapter 14

Ecological Perspectives on Pacific Salmon: Can We Sustain Biodiversity and Fisheries?

E. Eric Knudsen

Introduction

Pacific salmon *Oncorhynchus* spp. may arguably be the most studied group of exploited fishes. Yet, with many populations already extinct, a number listed as threatened or endangered, and many others depressed (Nehlsen et al. 1991; Slaney et al. 1996; NRC 1996a), biologists remain unable to answer critical questions for the public regarding sustainability of these salmon populations. Natural salmon production is significantly down in many of the remaining U.S. Pacific Northwest and some British Columbia populations, leading to significant reductions of important subsistence, commercial, and sport fisheries. Further, although many Alaskan and some British Columbia populations have been at all-time highs over the past several decades (e.g., Holmes and Burkett 1996; Wertheimer 1997), there have been serious declines during the past several years in some important Alaskan populations relative to record high previous years (e.g., Kruse 1998). The obvious question in light of these pervasive declines is: how can we sustain biodiversity and fisheries?

Sustaining Pacific salmon includes both recovery for listed populations and protecting and reinvigorating healthy populations (Huntington et al. 1996). A thorough understanding of the various aspects of salmon biodiversity is a basic requirement for sustainability. To find answers, we must explore the interplay between Pacific salmon ecology and human activities. We need to incorporate what we currently know into improved management, both at the domestic and international levels. We also need to identify changes in research and management that are necessary to do a better job.

The purpose of this chapter is to describe how integrated awareness of key ecological concepts helps to explain the dramatic declines in salmon populations in some portions of their range, and why healthy runs are also susceptible to decline. The concepts also help create a framework that reveals the physical, ecological, political, social, and economic requirements for restoration and sustainability. These ideas are not necessarily new, but recent scientific and technological developments are rapidly providing an expanded basis for synthesized thinking about salmon management. Integration of these ideas into habitat and harvest management will be critical to the long-term survival of salmon populations and the fisheries they support.

Important Definitions for Ecological Perspectives of Pacific Salmon

To avoid ambiguity in the discussion of ecological concepts in this chapter, I define some core terms specific to this discussion of Pacific salmon:

Salmon ecosystem—The salmon ecosystem for each population includes every location where a given population occurs throughout its life cycle: the spawning location, freshwater rearing areas, all downstream migration waters, the estuary, nearshore marine habitats, oceanic regions where that population migrates, and all estuarine and freshwater upstream migration waters (Lichatowich et al. 1995; Hartman et al. 2000). This entire complex of habitats must be included when considering the health and management of each population. This implies that separate populations' ecosystems overlap in downriver, estuarine, and marine environments.

Sustainability—The long-term viability and productivity of aquatic ecosystems, natural population biodiversity, and biomass that support healthy aboriginal, sport, and commercial fisheries and vital communities throughout the range of salmon for generations to come (Kendall 1999; Knudsen et al. 2000a). Sustainability in this context is not to be confused with reliability or consistency. Significant fluctuations in abundance are part of the natural state of salmon populations (discussed further below).

Biodiversity—The naturally evolved (i.e., genetically based) array of hierarchical population structures (Allendorf and Waples 1995), life history strategies (Mangel 1994), phenotypes (Healey and Prince 1995), and unique population production capacities (Pearcy 1992) exhibited by salmon. Most of the biodiversity in anadromous salmonids is based on local adaptation mediated largely by the habitats in their ecosystem as defined by their migratory and homing tendency.

Populations—The smallest freely interbreeding spawning aggregations not mixing with other populations, other than by natural background straying. These units are analogous to the demes described by Riddell (1993) and NRC (1996a).

Management unit—One or more populations, grouped together for management purposes such as setting escapement goals, managing for harvest rates, and implementing regulations (Knudsen 2000).

Key Ecological Concepts for Salmon Sustainability

Salmon populations, like all organisms, tend to expand their range and fully utilize available habitat through the evolution of survival strategies that allow them to be successful. This expansion is limited by the interactions of natural physical, hydrographic, and biological features, gradual climatic change, and short-term weather-induced events. Superimposing human-induced harvest, habitat alteration, and artificial salmon production on natural salmon systems reduces, constrains, or alters the populations' natural expansion capacities (NRC 1996a). I use four essential ecological concepts, relevant to natural expansion and constraint, to set the stage for understanding how human activities limit salmon production capacity and, more importantly, what steps can be taken to restore depleted populations and sustain healthy runs. The concepts are habitat suitability; population biodiversity; population biomass; and migrations, straying, and colonization.

Habitat Suitability

Salmon abundance and distribution are strongly influenced by basic habitat availability and by natural variation in the physical environment. Salmon populations gradually adapt to a variety of subtle habitat differences (e.g., Allendorf and Waples 1995). In the shorter term, favorable conditions allow expansion and population growth; unfavorable conditions constrain populations and production. Much has been written about the habitat requirements for successful salmon production (e.g., Groot and Margolis 1991). Several recent scientific developments, however, deserve emphasis here.

Riparian zones

Riparian zones provide salmon streams a connection to and a buffer from upland areas. These zones contribute to salmonid health by providing shade, streambank stabilization, sediment control, litter input, large woody debris, and nutrients (Spence et al.1996 and references therein). The priority for riparian zones is to protect all existing healthy streamside areas (Dominguez and Cederholm 2000). For restoration, streamside vegetation should be replanted where necessary and protected from livestock grazing, agriculture, logging, urbanization, and heavy recreational use. Vegetation provides relatively short-term restoration and forms the basis for long-term riparian ecosystem function by anchoring streamside soils, providing overhanging and undercut streambanks, increasing habitat complexity, forming pools and braiding, enhancing flows between instream and hyporheic zones (Beschta 1997), and eventually supplying large woody debris to the stream system.

Riparian zones have been dramatically altered, often eliminated, in urban and suburban areas (Boule and Bierly 1987; Langer et al. 2000). Logging and agriculture have also substantially and extensively affected riparian areas (Gregory et al. 1987; NRC 1996a). Streambank destruction resulting from livestock grazing, channelization, logging in riparian areas, urbanization, and streamside recreation is extensive and profoundly reduces important physical stream habitat features, especially for juvenile salmonid rearing (NRC 1996a).

Instream habitat complexity

Despite inherent differences among streams and stream reaches, it is clear that habitat complexity and dynamics are important features of aquatic systems (see Spence et al. 1996 and Burger 2000 for reviews). Juvenile salmonids depend on a variety of depths, substrates, habitat structures, flows, and off-channel habitats to supply food, feeding stations, and cover. These features are usually present in streams having a variety of woody debris and pool/riffle configurations (e.g., Gregory and Bisson 1997). Adults depend on adequate depth, substrate, flows, and instream structure for upstream migration, cover while waiting to spawn, and successful spawning (e.g., Burger 2000). Complex stream structure also helps to retain organic material, essential to the nutritional basis of the aquatic ecosystem (Gregory et al. 1991).

Large wood is an essential construction vector that influences stream erosive forces, helping to create additional habitat complexity and spawning and rearing habitat (Cederholm et al. 1997; Beechie and Sibley 1997). Woody debris has been demonstrated as crucial for juvenile salmonids in contributing to cover habitat, flood refuge, food production, and nutrient retention (e.g., Bryant 1983; Cederholm et al. 1997). Large wood has also been implicated in providing habitat diversity in estuarine and marine habitats (Sedell et al. 1988).

Natural habitat suitability has been greatly compromised through activities that reduce stream habitat complexity. Gregory and Bisson (1997) summarized studies that document broad-scale stream habitat simplification through reduction of large instream wood and complexity of pool habitats. Early logging depended on log driving, the practice of flushing logs downstream, as the preferred mode of transportation (Wendler and Deschamps 1955). Loggers removed as many obstacles from the streams as possible. Furthermore, large trees, the primary source of large instream wood, were logged and cleared from the streambanks and the rest of the watershed, resulting in a loss of future wood for natural stream restoration (Bisson et al. 1987). The logging industry later went through a shift to land transportation, during which much debris was allowed to accumulate in streams. Ironically, the recognition of this as a problem for fish passage and water quality (Bisson et al. 1987) eventually led to a period of stream cleaning when nearly all woody debris was again removed from some streams. The problem of too little or too much wood debris in streams has now been clarified, and most current forestry practices include consideration for maintaining large woody debris in streams. In many areas, however, the long-term needs for large wood remain unmet because of the lack of large trees in the riparian zone.

Habitat restoration has included instream wood replacement, with mixed success. Although many instream wood placement projects have either failed or have not shown increases in fish production (e.g., Frissell and Nawa 1992), a number of projects have demonstrated short-term benefits to both instream habitat complexity and density of stream-rearing salmonids (e.g., Peters et al. 1993; Cederholm et al. 1997). Additional research is needed to determine whether those local, short-term improvements translate into increased adult salmon production. As Dominguez and Cederholm (2000) point out, prior knowledge about salmonid requirements and fluvial geomorphic processes is required for successful habitat restoration. The limiting aspects of properly functioning stream and riparian ecosystems need to be considered to create habitats that provide the interim structural framework until riparian and upland forests recover from historical disturbances (Dominguez and Cederholm 2000).

Other activities that diminish habitat complexity—such as gravel mining, channelization, diking, riprap, and stream clearing for flood control—reduce the carrying capacity for salmonids (see Gregory and Bisson 1997 for review). Channelized small streams are generally less hospitable for larger salmonids, although negative effects for all rearing salmonids increase with severity of channelization (Chapman and Knudsen 1980; Knudsen and

Dilley 1987). Because of sampling difficulties, there has been no successful research on how channelization affects salmonid use in larger streams. However, studies demonstrating the importance of habitat complexity (e.g., Peters et al. 1993) imply that simplification of larger stream habitat during channelization likely reduces the carrying capacity for salmonids. Diking is especially harmful in lower stream reaches because it prevents access of juveniles to important off-channel rearing habitats (Beechie et al. 1994).

Channelization and diking aggravate flooding effects on fish survival because the flow volume and velocity are restricted to the channel areas rather than being dissipated into off-channel areas. The higher flows tend to scour substrate and wash woody debris out of the system. Diking and channelization also reduce flows between the river and the hyporheic zone (Stanford and Ward 1993).

Spawning substrate

Clean gravel with adequate interstitial water flows is essential for successful egg and fry incubation and survival (Groot and Margolis 1991 and citations therein). Natural spawning habitat is degraded by both gradual, nonpoint deliveries of fine sediments from agricultural, urban, and logging landscapes, as well as accelerated landslides and excessive streambank erosion caused by increased flooding from altered landscapes (NRC 1996a; Gregory and Bisson 1997). Selection of spawning locations is likely a heritable characteristic and, considering the subtleties of criteria for optimal spawning habitat, relatively minor changes in the nature of the streambed can make the habitat unsuitable. Road building, especially associated with logging activities, has accelerated the natural rate of streamside landslides that deliver excessive silt, sand, and gravel to streams (e.g., Harr and Nichols 1993). Spawning substrate can be too unstable or fine sediments can compact the interstices, making the substrate unusable (e.g., Hartman and Scrivener 1990; Montgomery et al. 1996). Incubating eggs and fry in the gravel are more easily dislodged if the substrate is too unstable.

The hyporheic zone

The hyporheic zone of rivers, the area where water flows under the sand and gravel substrate, is extremely critical (Edwards 1998). This zone is often the connection between groundwater and alluvial flows. The extent of this zone, together with the relative volume of river flow, can dictate the relationship between rainfall, snowmelt, and groundwater over the seasons. Hyporheic zones are biological hotspots that contain intensive physical and chemical gradients (Edwards 1998; Dahm et al. 1998). Although the true extent is yet to be documented throughout the range of salmon, there are some important direct connections between upwelling and salmon success. For example, interior Alaskan chum salmon populations find incubation refuge from extreme winter weather in groundwater upwelling areas (Finn et al. 1998), and Hanford Reach chinook salmon *O. tshawytscha* spawn predominantly in areas of hyporheic discharge (Geist 2000). Furthermore, the movement of water, nutrients, and microorganisms into (downwelling) and out of (upwelling) the hyporheic zone is a critical ecological process in streams and lakes (Stanford 1998; Dahm et al. 1998) that likely has strong influence on the production processes upon which juvenile salmonids depend.

The biological importance of the hyporheic zone has been largely overlooked until recently. Physical changes—through gravel mining, channelization, diking, riprap, and flow alterations—have altered the groundwater/surface water interface (Stanford and Ward 1993). Reestablishing the natural channel configuration, which allows for natural amounts of woody debris for channel complexity, will be necessary for restoration of this important ecological interface (e.g., Wissmar and Beschta 1998).

Water quality and quantity

Salmon in every life history stage require water within acceptable temperature and oxygen levels and free of contaminants (Groot and Margolis 1991 and references therein). Thermal alterations potentially affect the survival and growth of every stage during the freshwater life cycle (NRC 1996a). Removal of riparian vegetation and changes in flow regimes have contributed to increased temperatures. Many urban and suburban watersheds are being overloaded with nutrients, resulting in abnormal plankton blooms and oxygen deficiency (e.g., Fresh and Lucchetti 2000). Toxic chemicals, as well as nutrients, are washed into streams from urban and suburban landscapes, particularly under altered flood runoff patterns. Industrial activities continue to add toxic chemicals and oxygen-depleting nutrients to water bodies (e.g., Seiler 1989).

Runoff patterns critically influence the degree to which a given stream reach is conducive to salmon production. Timing and intensity of stream runoff greatly influence success or failure of instream production. Floods have been recognized as a beneficial and important vector in the long-term natural disturbance regime to which salmon have adapted (Bisson et al. 1997). Problems occur, however, when natural flow patterns are exaggerated by human-induced landscape changes. Watershed landscapes have been dramatically altered through logging, agriculture, and urban and suburban development. Removal of vegetation, increased impervious surfaces in urban and suburban areas, and flood control projects have led to major changes in runoff patterns. In the short term, flooding can limit survival by disrupting spawning activity or dislodging incubating eggs or fry (see Gregory and Bisson 1997 for review). In the longer term, excessive flooding also alters stream morphology through erosion and increases the frequency of streamside landslides, thereby increasing bedload movement and sedimentation to the detriment of salmon (Gregory and Bisson 1997 and references therein).

Urban areas particularly suffer from increased frequency and intensity of flooding (e.g., Fresh and Lucchetti 2000). Runoff patterns from agricultural and clear-cut landscapes have also changed. Increased likelihood of streambed scouring and resultant egg or alevin mortality are contributing to salmon population losses (Montgomery et al. 1996; Gregory and Bisson 1997). Increased flood flows also tend to wash important woody debris from rivers and streams and cause increased erosion of streambank habitats (Gregory and Bisson 1997).

Conversely, low flows can dewater redds, limit juvenile habitat, or preclude successful migrations (Hicks et al. 1991; NRC 1996a). Water that flows off altered landscapes more rapidly than unaltered watersheds not only causes increased frequency and intensity of flooding but is then unavailable to recharge groundwater, exacerbating summer low flows (Wissmar and Beschta 1998). Aquifers are the primary source for summer flows, an important determinant of the physical rearing habitat for juvenile salmon (Gregory and Bisson 1997).

Water is often removed from streams and rivers for municipal water supplies, agricultural irrigation, industry, and power generation, compounding the negative effects of natural low flows. Dewatered stream and river reaches have reduced spawning and rearing habitat as well as increased predation on juveniles, barriers to migration, and stream temperatures (NRC 1996a). For example, up to 50% mortality of juveniles was attributed to flows diverted from the Sacramento River (Kjelson and Brandes 1989). Furthermore, most of the removed water is not returned to the stream, and, of the water that is returned, most is substantially degraded (NRC 1996a). Varying discharges downstream of power production facilities result in stranding of juvenile salmon (Bradford 1997).

Systemic watershed health

Awareness of the importance of systemic watershed health is increasing. Functional watersheds provide a relatively stable mix of conditions conducive to salmon production and survival. For example, forested landscapes dampen the extremes of spawning substrate-scouring rain and snow runoff events (Wissmar and Beschta 1998), and forested riparian lands provide gradual input of habitat-creating large wood (Domiguez and Cederholm 2000). Disturbance at various scales is also part of the natural watershed process (Bisson et al. 1997), but the cumulative effects of human changes to the landscape have in many cases dramatically exaggerate watershed disturbance to the detriment of salmon. Therefore, protection and restoration of watershed function depends on catchment-wide understanding of where problems occur and how to remediate them. Most now agree that specific instream restoration does not help unless the actions are integrated into a systemic watershed restoration program (e.g., Beschta 1997).

Systemic watershed health can be maintained by keeping intact the full range of aquatic and riparian conditions generated by natural disturbance events at landscape scales (Bisson et al. 1997), including actions that restore natural run-off patterns, recruitment of large wood to streams, and connections to off-channel habitats. Several authors have described important new approaches for watershed-oriented restoration (e.g., Bisson et al. 1997; Williams et al. 1999; Fresh and Lucchetti 2000). Fresh and Lucchetti (2000) present a strategy for combining protection and restoration while considering the scale at which watershed functions occur. They emphasize the importance of considering how restorative or protective actions at one scale can either enhance or negate actions at another scale. Success depends largely on understanding the physical and biological dynamics among the scales as well as coordination among political jurisdictions. The new paradigm for watershed health appears to be an ecosystem approach (MacDonald et al., Chapter 15) guided by watershed analysis together with adaptive learning (Naiman et al. 1992).

Arguably the single most effective habitat-oriented action for salmon sustainability is to protect existing good habitat (see Frissell 1993; Lichatowich et al. 2000). Salmon refuges should be designed to create connectivity and coincide with areas of genetic resources (e.g., Thurow et al. 2000). Federal forests and wilderness areas already serve as a core of protected area (Sedell et al. 1997). The concept of preserving the few remaining key watersheds is crucial to maintenance and recovery (NRC 1996a; Bisson et al. 1997; Williams and Williams 1997). U.S. federal plans for recovery include identifying key watersheds for analysis and then restoring those watersheds located on federal lands that meet the following criteria: those having one or more anadromous fish populations at risk of extinction, those consisting of high quality remaining habitat, those that include degraded watersheds with a high potential for restoration, or any combination of the three (Williams and Williams 1997). Local governments must also implement protective measures for small watersheds threatened by encroaching urbanization. Counties and municipalities have tremendous influence over salmon habitat because their actions directly affect land use patterns. Watershed function must be protected and improved where necessary through conservation easements and other protective measures that maintain riparian and instream integrity (Fresh and Lucchetti 2000). Another shift is necessary, from simple prescribed-width buffers to identification and protection of critical and/or sensitive habitats, such as unstable slopes and off-channel habitats (NRC 1996a). These critical locations can be identified through processes such as watershed analysis (e.g., Armantrout 2000).

Changes in estuarine and marine habitat

Naturally complex estuarine habitats are important to juveniles on their way to the ocean (Simenstad et al. 1982). Estuaries are particularly important to chinook, chum *O. keta*, pink *O. gorbscha*, and coho *O. kisutch* salmon for rearing as they make the transition from freshwater to salt water (Simenstad et al. 1982). In many places, access to intertidal marshes has been eliminated, particularly through diking for agriculture and flood control and for urban port development (NRC 1996a). For example, Simenstad et al. (1982) determined that about 90% of estuarine habitat has been eliminated in many developed Puget Sound river deltas.

Extensive nonsalmonid fishing is physically altering marine benthic habitat, and marine food webs are being changed through removals of significant biomass in both directed and incidental fishing (e.g., NRC 1996b). Chemical pollution by persistent organics, heavy metals, and radioisotopes has been identified as cause for concern, at least in the Bering Sea (NRC 1996b). Though there are no definitive studies on the effects of these activities on salmonid production, concerns have been raised about reverberations through the marine ecosystem (e.g., NRC 1996b).

Marine productivity

The importance of marine productivity limitations for salmon is increasingly apparent. Marine survival and production of salmon have been correlated with oceanographic and marine productivity indices (Holtby et al. 1990; Beamish and Bouillon 1993; Pearcy 1997). There appears to be a marine carrying capacity that, in some cases, can be exceeded and that varies annually (variation discussed below). Density-dependent marine growth has been implied by Ishida et al. (1993) and Rogers and Ruggerone (1993), and there are indications that salmon survival is density-dependent (see Thomas and Mathisen 1993; Perry 1995; Hilborn and Eggers 2000). Further evidence for limits to marine carrying capacity includes recent progress indicating that seemingly small changes in the ocean environment can strongly affect marine survival (e.g., Welch et al. 1998). Also, studies of data from Japanese chum hatchery production indicate poor survival and reduced size of returning salmon at sea in years of large hatchery releases (e.g., Ishida et al. 1993). Hilborn and Eggers (2000) also listed early marine competition as one possible explanation for reduced wild pink salmon production in the face of large expansions in the Prince William Sound, Alaska, hatchery program.

Environmental variation

Interactions between habitat suitability and environmental fluctuations strongly influence survival at all life stages: spawning, incubation, rearing, marine foraging, and migration (e.g., Bisson et al. 1997). A critical concept here is scale. Salmon survival and production can be influenced locally by a single storm event, but also on much larger scales by gradual, long-term shifts in regional climate (e.g., Kruse 1998). Montgomery et al. (1996) concluded that chum salmon egg burial depths, just below the bedload scour depth at mean annual high flows, were probably a finely tuned adaptation. This implies that larger floods from a single event could wipe out a brood class by dislodging eggs from redds. In the long run, as climate fluctuates, the frequency and intensity of devastating floods will ebb and flow.

In the marine environment, salmon production appears to be related to fluctuations in climate on several important scales. There is evidence that upwelling, which exhibits tremendous annual variation, significantly influences survival of Oregon coastal coho (e.g., Pearcy 1997). The Pacific decadal oscillation has been identified as a cycle in which ocean conditions dramatically change approximately every 20 to 30 years, relative to salmon survival needs, and has strong influence on salmon production patterns (Hare et al. 1999). Recent work by Welch et al. (1998) has indicated that ocean growth and survival may be limited by warm temperatures and that long-term climate warming could push the southern boundary of temperature tolerance farther north, reducing the total range of suitable sockeye salmon *O. nerka* grazing habitat (Figure 1).

Uncertainty is an important characteristic of environmental fluctuations and their influence on salmon production (NRC 1996a; Francis 1997). We will continue to learn about the relationships between environment and salmon production, but we will never be able to predict with certainty what will happen over the next several salmon generations. Still, our predictive powers are improving; in 1997, the most recent El Niño event was predicted months in advance (O'Brien 1998). That is primarily because meteorologists had studied the links sufficiently to understand that when the equatorial low becomes established, the El Niño is likely to follow. I believe science and technology can gradually reduce ignorance and uncertainty about salmon production drivers. Until technical knowledge improves, though, we need to manage without full knowledge and consider the effects of uncertainty on our decisions (Francis 1997). Caution is also necessary when interpreting the causes of good or poor run returns, so that they are not attributed to the wrong factors. For example, we might falsely interpret habitat restoration as futile in years of poor marine survival (Hare et al. 1999).

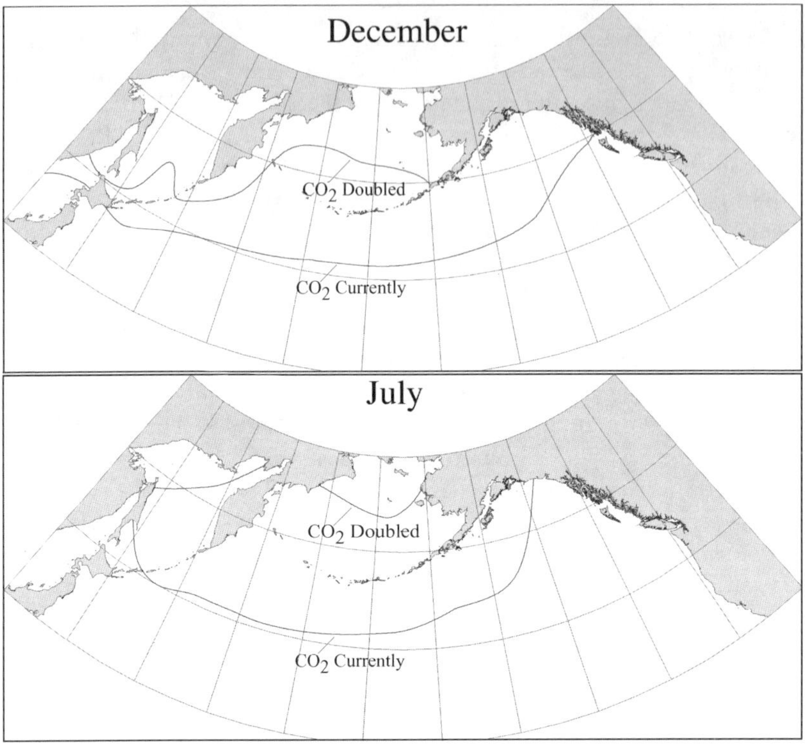

Figure 1. An illustration of how a doubling of atmospheric CO_2 might push the southern boundary of acceptable winter (top) and summer (bottom) sockeye salmon distribution northward, thereby seriously reducing the amount of marine habitat available for rearing (from Welch et al. 1998).

Population Biodiversity

Local adaptation

Because of the tendency to home to their natal streams, Pacific salmon have evolved a diversity of genetic and phenotypic population characteristics (Waples 1991a). Every spawning population is potentially a unique genotype (Healey and Prince 1995); there is even evidence of genetically based divergence within a single, relatively small spawning area (Woody et al. 2000). Examples of apparently heritable ecological strategies for success include variations in body size correlated with differences in stream flows (e.g., Beacham and Murray 1987), run timing for spawning and incubation survival (e.g., Smoker et al. 1998), duration of egg incubation (Woody

1998), and a variety of freshwater rearing strategies (e.g., Wood et al. 1987; Bisson et al. 1997).

That anadromous behavior is adapted to local and specific conditions is further evidenced by numerous failed attempts to establish self-sustaining runs of Pacific salmon through transplanting, at least within the native range (Withler 1982). Yet transplants of the nonanadromous form of sockeye salmon, kokanee, have often been successful (Wood 1995). Since numerous kokanee transplants have been successful while anadromous sockeye salmon transplants have generally not (Wood 1995), critical local adaptation apparently also includes the marine component of the salmon ecosystem. Successful transplants are apparently more likely if the habitat and geographic orientation of the watershed, relative to marine migrations, are similar to the habitat from which the donor population originated. For example, chinook salmon introductions into New Zealand rivers have been successful only in locations where the marine migration direction matches that of the donor population (T. Quinn, University of Washington, personal communication). The importance of local adaptation was further demonstrated in an unusually successful introduction of sockeye populations transplanted into previously unused habitat in Frazer Lake, Alaska. Genetic analysis revealed that populations established themselves in spawning habitats most similar to their source habitat types (Burger et al. 2000).

Because salmon populations have gradually adapted to maximize survival in relation to subtleties in their habitat, changes in the habitat, such as those experienced in many Pacific Northwest locations, reduce the likelihood of long-term survival. Although populations could conceivably evolve to further adapt to the changes within several decades (Healey and Prince 1995), this will not occur at a pace in accordance with society's urgency to restore salmon sustainability. Because of the relationship between local adaptation and habitat, the appropriate conservation unit is the population within its habitat (Healey and Prince 1995). To maximize biodiversity and productivity, every population should be managed conservatively.

Hierarchical population structure

Numerous researchers (e.g., Reisenbichler et al. 1992; Riddell 1993; Wood 1995) have described the hierarchical nature of broad-scale genetic population structuring in salmon. Intraspecific genetic distance generally parallels geographical distance (Waples 1991a; Allendorf and Waples 1995). The locally adapted populations are genetically more similar to their neighboring populations, resulting in a branched, hierarchical organization with populations grouped into metapopulations and metapopulations grouped into biological races (Riddell 1993). The races have been demonstrated, at least on a relatively broad scale, to exhibit and be defined by significant intraspecific run timing differences within large watersheds (Allendorf and

Waples 1995). Races tend to be more genetically different from one another than the differences among metapopulations within races, implying that the different run timings were based on unique historical colonization and evolutionary events (e.g., Utter et al. 1995; Burger et al. 1997).

The importance of individual, locally adapted populations to the long-term success of metapopulations should not be overlooked (Riddell 1993). As climate gradually changes or natural disturbances occur, they may be the basis for recolonization of habitat. Marginal populations may have a high adaptive significance to the metapopulation because they can contribute variation back into the central populations at times of contraction (Scudder 1989). The implications of local population differentiation and population structuring are important in fisheries management decisions regarding the scale at which populations are protected from habitat and harvest threats (Riddell 1993; NRC 1996a).

The diversity of Pacific salmon population structuring, as well as demographics, is clearly in peril in some areas. As of June 2000, there were 26 evolutionarily significant units (ESUs) listed under the U.S. Endangered Species Act (ESA) as threatened or endangered and another five candidate ESUs. Decisions about which populations to include in listings are based on the concept of the ESU (Waples 1991b, 1995), which is often applied at the metapopulation, race, or multiple-race levels. There is, however, an important distinction between actions to prevent extinctions, as defined by ESA, and actions required for achieving sustainability. Whether the ESUs are listed or not, the ESA process will not solely prevent continued losses of local populations or demographic declines. Because ESUs are broadly defined, some constituent populations may effectively drop out, regardless of the overall performance of the ESU by ESA criteria. Sustainability requires activities that maintain and enhance diverse population substructuring through protecting all local populations, allowing continued local adaptation and maintaining adequate abundance.

Harvest impacts on biodiversity

Salmon management should be based on the premise that local reproductive populations are genetically different from one another and valuable to long-term salmon production (NRC 1996a). When harvesting reduces the number of adults reaching the spawning grounds, several aspects of biodiversity can be affected. First, extinctions of local spawning populations (e.g., Walters and Cahoon 1985; Nehlsen et al. 1991) decrease the overall genetic and phenotypic diversity of metapopulations, as well as the total biomass produced from conglomerate populations. Though the definition of reproductive isolation and the degree of protection afforded to recognizably different spawning populations remain debatable (Waples 1995), it is clear that some amount of diversity is lost with every lost population (Reisenbichler 1997). Genetic variation within and between the hierarchical levels of salmon population structure are the resource base for Pacific salmon, both for long-term sustainability and continuing evolutionary process, and it should be vigorously preserved (Riddell 1993).

Second, when harvest reduces a population to some minimum number of spawners, genetic diversity is threatened by inbreeding and/or genetic drift (Reisenbichler 1997). Resultant reduced fitness contributes to a downward spiral of survival, with the population increasingly susceptible to random downturns in environmental variables that cause poor survival (e.g., Lawson 1993). Minimum population sizes are impossible to estimate with certainty; the actual value likely varies among populations. Depending on the circumstances, however, even populations larger than 500 individuals may be at risk of extinction from these factors (Reisenbichler 1997). Genetic diversity is also threatened, even in abundant populations, by the selective forces of fishing—for example, by potentially changing run timing, spawn timing, size at age, or age of maturity (Reisenbichler 1997).

Third, the mixed-population fishery is a particularly challenging threat to population biodiversity. Many smaller, less productive populations can be chronically overharvested when harvest rates are set for entire management units based on spawner-recruit information for the more productive constituent populations (NRC 1996a; Knudsen 2000). Hatchery-reared populations are particularly troublesome when they are mixed with wild populations because they can sustain very high harvest rates. The problem is worse the farther the fishery occurs from terminal spawning areas. As populations migrate toward natal streams, they separate from one another in time and space, potentially increasing the ability to harvest from various populations. Still, decisions are made to incidentally harvest commingled weak runs while targeting healthy runs, even in terminal areas (Copes 2000). The ideal situation is either to harvest each population separately or to moderate harvest rates so as to not overharvest the least productive population in a mixture (e.g., Van Alen 2000). The latter is problematic because, in many cases, the weakest population can withstand only low or no harvest. This would result in foregoing harvest on other, possibly many, populations in the mixture. In practice, there are numerous combinations of management alternatives, ranging from a single harvest rate for a large number of populations, as in Alaska (e.g., Van Alen 2000), to extensive closures to protect weak populations, as in Canadian coho conservation measures (Fisheries and Oceans Canada 1999). With notably declining run sizes and ESA listings, management scenarios are changing rapidly and extensively (e.g., WDFW 1999).

Hatchery impacts on biodiversity

Many authors have addressed the issues of how artificial enhancement can negatively affect natural populations (e.g., Krueger and May 1991; Hilborn

1992; NRC 1996a). Artificial enhancement can take a variety of forms along a continuum, from small-scale captive brood programs for maintaining and expanding the last vestiges of a dwindling population to augmentation projects designed to improve survival and production in a natural run to large production hatcheries. While most would agree on the need for rescuing dangerously depleted populations through artificial measures, debate has ensued over the social and biological costs and benefits of artificial production in projects designed primarily to produce more fish. Most of the following discussion pertains to artificial enhancement programs at the larger production end of the spectrum.

Artificial enhancement can result in genetic effects for both the population managed in the facility and neighboring populations (Campton 1995), although observed differences between wild and hatchery fish may be complicated by management decisions such as population transfers (Nielsen et al. 1994). Pacific salmon reared in hatcheries quickly become domesticated—they adapt to the unnatural conditions of the hatchery environment at the expense of adaptation for living in natural streams (Reisenbichler 1997). Genetic changes in hatchery populations result from: random drift, which can increase the proportion of individuals homozygous for deleterious recessive alleles, reducing fitness through inbreeding depression; selection, caused by both intentional selection of mating crosses, as well as natural selection for individuals better suited to hatchery survival; or the practice of crossing various, sometimes geographically distant, stocks within a single hatchery (Waples 1991a). Although these genetic changes may result in a population more amenable to hatchery rearing, hatchery fish often do not survive the marine environment as well as wild fish (NRC 1996a), and marine survival of hatchery fish may deteriorate with increasing generations in the hatchery (Reisenbichler 1997).

Deleterious genetic effects also occur when hatchery fish spawn with natural fish. This results not only when hatchery fish stray into natural populations, but especially when managers try to augment natural production with a hatchery population (Reisenbichler 1997). Waples (1991a) describes the deleterious effects of hybridization between two gene pools (hatchery and wild) through reduction in genetic variation between populations, which tends to genetically homogenize populations away from their local adaptations, and outbreeding depression, which causes reduction in fitness, hence survival, when parental populations of increasing genetic distance are crossed.

Both inter- and intraspecific competition occur naturally for salmon in freshwater, estuarine, and marine environments (Fresh 1997). The release of hatchery fish complicates the natural limitations imposed by competition. Several important studies have demonstrated that juvenile hatchery fish displace wild fish in streams. Hatchery coho juveniles rearing in streams were found to outcompete wild fish for food (Nielsen 1994). A number of authors have attributed reductions in juvenile coho instream survival and densities to competition with hatchery fish (e.g., Nickelson et al. 1986; Flagg et al. 1995). Both Nickelson et al. (1986) and Flagg et al. (1995) speculated that observed decreases of wild juvenile coho in streams stocked with hatchery juveniles were likely due to the larger hatchery fish forcing wild fish from feeding locations. McMichael et al. (1997) found that residual hatchery steelhead *O. mykiss* had a negative effect on wild steelhead growth in semicontrolled conditions. Competition may be especially critical in estuarine and nearshore marine areas, where juveniles can encounter food limitations that limit growth and survival while they are spatially concentrated and have similar diets (Fresh 1997). If there are limits to Pacific Ocean carrying capacity for salmon, as discussed in the section on marine productivity, excessive hatchery releases may aggravate poor survival, at least in years when conditions limit production.

Wild strays of exotic Atlantic salmon *Salmo salar* into Pacific salmon habitats are of particular emerging concern. The number of escapees from aquacultural net pens, particularly in British Columbia, has been increasing recently. Natural spawning and rearing have recently been confirmed, raising concerns about negative competitive impacts on native Pacific salmon (Volpe et al. 2000).

The effects of genetic dilution and competition are often exacerbated when natural populations in the area around the hatchery are struggling to survive in the face of high hatchery harvest rates. The case of Lower Columbia River coho salmon serves as a good example. That region has been managed for hatchery production, despite extensive quality coho habitat in the area. When it was recognized that very few truly wild coho remained in the streams and a petition for ESA listing was considered, genetic analysis could not discern wild from hatchery fish. The fish found in local streams likely were mostly hatchery strays (Wright 1993). For these and other reasons, primarily related to various hatchery practices throughout the region, Lower Columbia coho salmon were not listed under ESA (Flagg et al. 1995), but only because any remaining wild fish could not be differentiated from hatchery fish. Furthermore, the extensive habitat remains mostly devoid of natural production.

Population Biomass

The ecological importance of Pacific salmon spawning biomass to the freshwater environment may have often been overlooked (e.g., Cederholm et al. 1999; Gresh et al. 2000). Spawner-recruit models, upon which salmon management has depended for decades, tell us that limits to natural reproductive capacity can not only be affected by too few spawners, but also that too many spawners can actually reduce production through density-dependent mortality (Ricker 1975). Returning adults that exceed the perceived capacity of the spawning and/or rearing habitat have traditionally been considered surplus available for harvest. With very little data or research on natural, unexploited salmon populations, however, it is possible that the so-called surplus may actually serve purposes previously misunderstood.

Carcasses as nutrients

Recent scientific evidence has shown that salmon carcasses provide a significant source of nutrients to freshwater ecosystems (Cederholm et al. 1989; Kline et al. 1993; Bilby et al. 1996; Cederholm et al. 1999). Not only do the carcasses support the basis of a nutrient pathway via primary production, but flesh and eggs are directly consumed by aquatic insects (Wipfli et al. 1999) and by rearing fish (Bilby et al. 1996). This biological feedback loop thereby ensures future salmon production.

Carcasses also supply nutrients to streamside vegetation (Ben-David et al. 1998) and wildlife such as bears, martens, eagles, and even songbirds (Willson and Halupka 1995; Ben-David et al. 1997; Hildebrand et al. in press). Retention of carcasses within the freshwater system is also critical to maintaining their effectiveness as a basis for production (Cederholm et al. 1989). When large woody debris is reduced in stream habitats, the network of carcass-trapping physical structure is eliminated.

Probably the most important implication for Pacific salmon is that the production relationship (returning adults per spawner) is influenced not only by the number of eggs deposited in the gravel, but also by the amount of biomass delivered and retained in the watershed (e.g., Cederholm et al. 1999). The carrying capacity for freshwater production depends on both the physical space available and the amount of nutrients provided to the system. This varies, depending on the freshwater life history of the species and the nutrient interdependence among species, but in any case, there is a feedback mechanism relating the number of adults allowed to escape harvest directly to the productivity of the system. This biological control factor must be considered in the spawner-recruit relationship.

Freshwater carrying capacity

One seemingly inexorable problem in Pacific salmon management is the difficulty of truly understanding the capability of the freshwater habitat to support salmon production. Current observations are limited to populations that are subject to exploitation, have had their habitats altered, or some of both. Inferential estimates have been made of much greater historic salmon production in areas now altered by habitat degradation (Chapman 1986; Michael 1998; Gresh et al. 2000). But even where habitat is pristine, without observations of unexploited spawners and smolts, there is no sure way of knowing production potential. Theory behind spawner-recruit models tells us that the model expresses the population's expansion capacity, the spawners required for one-to-one replacement, and the spawning escapement at which production is maximized *if the model includes a wide array of escapement and production values and the data are accurate* (Hilborn and Walters 1992). Unfortunately, data for the vast majority of populations do not include values prior to exploitation that define the true equilibrium replacement point on the curve.

Recent technological advancements support reevaluating the efficacy of the spawner-recruit relationship for salmon. For example, sediment core samples of marine-derived nitrogen in Karluk Lake, Alaska, presumably from sockeye salmon carcasses, were well correlated with sockeye salmon escapement records during the recent 100 years and therefore could be extended as a salmon production surrogate back 300 years (Finney et al. 2000). These data indicate that salmon run sizes declined significantly and stayed lower during the most recent 100 years of exploitation (Schmidt et al. 1998; Finney et al. 2000). The nutrient data also supported recommendations for increased escapement goals relative to those estimated from spawner-recruit modeling with exploited population data alone (Schmidt et al. 1998).

I used the basic concept of these important findings to test a hypothesis: that the spawner-recruit model equilibrium point may be underestimated when the model is based on data collected after exploitation has begun. I generated hypothetical but realistic data for a salmon population before and after exploitation (Figure 2, top panel). Spawner-recruit relationships were then plotted for the unexploited population (Figure 2, middle panel) to compare with the spawner-recruit relationship after exploitation (Figure 2, bottom panel). One can easily see that the production of recruits under exploitation is much lower than when the population is pristine, *from exactly the same habitat.*

This is thought to result from the fact that fewer nutrients are entering the system to support freshwater growth and survival, thereby essentially reducing the carrying capacity, even in a physically unaltered watershed (Cederholm et al. 1999 and references therein). From a management perspective, the situation has been exacerbated because a Ricker model of the exploited population, without knowledge of the true equilibrium point (i.e., carrying capacity), may provide erroneously low estimates of escapement at maximum sustainable yield (MSY) but that appear to make sense. When the reduced escapement goal is implemented, it aggravates the nutrient deficiencies, thereby further reducing production and setting the stage for sequential reductions in the escapement goals. I believe this management error is pervasive throughout the geographic range of Pacific salmon, depending variously on species and local management scenarios. This dynamic is preventing attainment of full sustainability and helps explain why

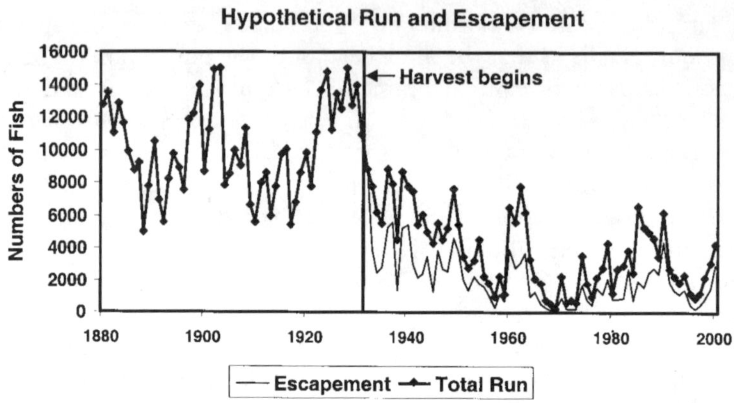

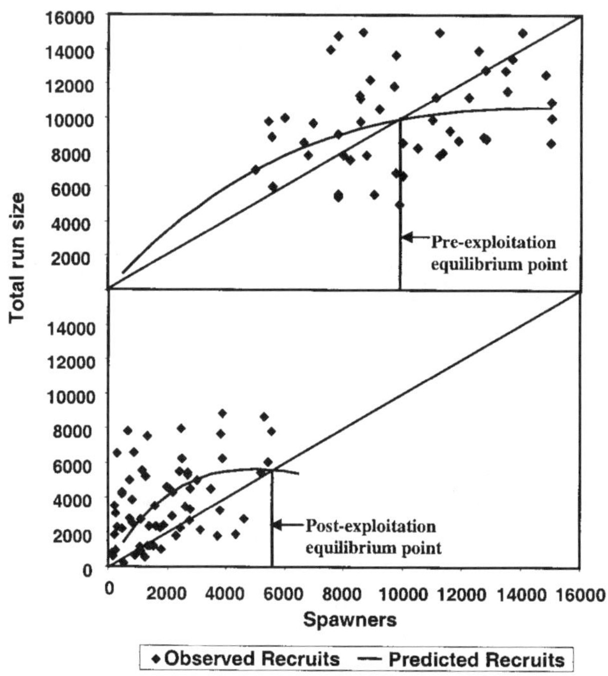

Figure 2. Top panel: Hypothetical escapement and total run size data of a salmon population before and after fishing ensues. Middle panel: Spawner-recruit relationship before fishing begins. The data are spread around the equilibrium population replacement point, which defines the freshwater system's capacity to produce salmon when it is fully seeded and fertilized. Bottom panel: Spawner-recruit relationship for the period of fishing. Notice the estimated MSY spawners are much fewer than before fishing, and the calculated equilibrium point is dramatically less, even though the physical habitat is still the same.

populations are declining, even in locations with healthy habitat. When the biological downward spiral is combined with habitat alteration, the effect is even more dramatic.

Natural production buffer

The number of recruits produced per spawner can vary by more than an order of magnitude (Cramer 2000). Spawner-recruit models tend to average that variation and incorporate it into a simplistic view of the relationship. The most common outcome of the simple spawner-recruit relationship is a tendency to overestimate the exploitation rate and underestimate the spawning escapement needed for perpetuation (Hilborn and Walters 1992; NRC 1996a). Some of the perceived "excess" in run sizes may in actuality be a hedge against those years when natural, constraining disturbances occur (Hare et al. 1999). Although spawner-recruit models indicate that too many spawners may somewhat diminish production of the next cohort, there are often years when floods, drought, or poor marine food production will reduce survival, making extra spawners critical to long-term production. Straightforward use of spawner-recruit models for predicting harvestable surplus also does not account for measurement error (Hilborn and Walters 1992; NRC 1996a; Cramer 2000).

Harvest management implications

We have become accustomed to assuming that every salmon population has a quantifiable and reliable surplus production that can be harvested each year, that the only factors determining production capacity are the number of parents and the physical space required for spawning and rearing, and that humans are the only harvesters of returning adult salmon. Sustainable salmon populations will require a new escapement management paradigm that includes allowances for seeding of the freshwater incubation and rearing habitat, annual run size variability, carcass biomass to support optimal freshwater carrying capacity, and food for wildlife (e.g., Michael 1998).

Salmon harvest management is complex and difficult, fraught with challenges caused by inaccurate or insufficient data (Knudsen 2000), poor predictive models (Knudsen 2000), and a complex state, tribal, federal, and international management structure (Rutter 1997; MacDonald et al., Chapter 15). Today's salmon management challenges range from the dilemma of too many salmon, such as in some southeastern Alaska fisheries, to populations in dire straits with only a few remaining individuals. In all cases, managers share the common problems of poor data, weak models, poor understanding of production capacity, uncertainty about factors controlling future production, and difficulties in managing fishing effort to optimize harvest while ensuring adequate escapement. Furthermore, decisions to harvest salmon can sometimes be too strongly influenced by the desire to fish as opposed to the need to conserve (Fraidenburg and Lincoln 1985).

Ocean variability has strong influences on salmon escapement (NRC 1996a). It is critically important that harvest rates be adjusted accordingly, so that adequate escapements can be met even in years of low ocean survival (Lawson 1993). This may mean low or no harvest in years of poor ocean survival, but this is a key to sustainability (NRC 1996a). Maintaining optimal freshwater habitat seeding in all years will ensure the best smolt production, hence the greatest possible marine production, given variable marine survival.

Migrations, Straying, and Colonization

A final natural ecological attribute of Pacific salmon is their tendency to migrate to and throughout the ocean, return home to natal spawning areas, variably stray and mix with other populations, and colonize new habitats. These extensive migrations are a hallmark of salmon, thought to enhance reproduction and survival by allowing a fish to take advantage of the greater marine food supply, resulting in larger body size, which is useful for extensive upstream migrations and in digging deeper redds, also benefiting survival of eggs and fry.

Upstream migrations

Pacific salmon may spawn anywhere from the intertidal zone of stream mouths, as in some pink and chum populations, to the upper reaches of huge watersheds sometimes more than 1,500 km from the ocean. Successful reproduction obviously depends on unimpeded upstream access. Blocks to migration prevent adults from reaching spawning grounds, preclude offspring from using rearing areas, and cause mortality of juveniles moving downstream. The most notable losses are caused by main-stem dams on large rivers (e.g., Hassemer et al. 1997), although dams erected for power generation, water control, irrigation, or shipping enhancement on any size stream have resulted in significant losses (NRC 1996a). Some large dams have completely blocked access to extensive upstream areas. At other large dams, upstream passage facilities are less than fully effective and/or migrants suffer from "fallback" (NRC 1996a). Small dams and other barriers have been blamed for the extirpation of some runs (NRC 1996a). Culverts and other types of road construction have also often created migration barriers.

A unique and relatively unstudied source of upstream migration mortality may be attributable to various forms of catch-and-release angling. Limited studies on sport catch and release of chinook salmon indicate 10% or less direct hooking mortality (Bendock and Alexandersdottir 1993).

However, because salmon have evolved marine feeding behaviors to acquire a finite energy resource for their nonfeeding upstream migration and spawning, it is conceivable that females subjected to one or more capture-and-release events may prematurely expire before depositing their full complement of eggs into the gravel. This topic should be further researched. Although not specifically attributable to upstream mortality, additional research is also required for the effects of capture and subsequent release of salmon in nonretention commercial fisheries. Wertheimer (1988) estimated 25% mortality of chinook caught and released in troll fisheries. Candy et al. (1996) also estimated about 23% short-term mortality of chinook caught and released from purse seiners targeting other species. They also noted that viable released salmon remained in the area for the subsequent 24 h and consequently were susceptible to multiple captures. Future population models should account for these fisheries-induced mortalities.

Downstream migrations

Smolts moving downstream suffer extensive natural losses due to predation, competition for food, stranding by variable flow levels, and inaccessibility to offstream rearing habitats. These natural mortalities are exaggerated by human activities of dams, diversions, and diking. Mortality of downstream migrants passing through turbines, from gas supersaturation, or from increased predation can be extreme at large, main-stem dams, and the effects are compounded if they must pass through multiple dams (e.g., Raymond 1988). Juvenile salmonids are subjected to stranding due to flow reductions downstream of water diversions and hydroelectric installations (e.g., Bradford et al. 1995). Downstream migrating juveniles are also lost into all sizes of unscreened or poorly screened diversions (e.g., Kjelson and Brandes 1989). Numerous studies have established the importance of off-channel freshwater and estuarine rearing habitats (e.g., Swales and Levings 1989). Culverts and road construction have been shown to block upstream access of juveniles to important off-channel rearing habitat (NRC 1996a). Dikes and levees built to create agricultural land from former wetlands or control floods also block access to juvenile rearing areas (e.g., Beechie et al. 1994).

Natural straying and colonization

There are abundant data indicating that, although salmon possess a strong natural tendency to home to the natal stream, most salmon populations exhibit some degree of natural straying to nonnatal spawning locations (see Quinn 1993 for review). Stray rates tend to vary substantially among and within species and among locations (Quinn 1993). The ecological significance of straying is thought to help perpetuate genetic diversity, but it may also be important for establishing new populations or revitalizing populations faltering after a sudden or long-term ecological setback. Populations geographically closer to each other may tend to intermingle somewhat more than those farther apart (e.g., Klamath River chinook, as described by NRC 1996a).

Anadromous salmonids appear to be well adapted for reinvading areas of suitable habitat (Bisson et al. 1997). Adults are strong swimmers with high fecundity, and juveniles often disperse widely in search of productive rearing habitats (Groot and Margolis 1991). Research in Glacier Bay, Alaska, has revealed the rapid anadromous salmonid colonization of habitats recently exposed by retreating glaciers (Milner and Bailey 1989). Pink salmon and coho salmon juveniles have been found rearing in streams as young as 40 years (author's data). These findings imply that salmon will colonize suitable habitat as soon as their straying nature allows a spawning population to become established. Streams naturally exhibit ecological succession (Milner et al. 2000). If streams are at all amenable to salmonid colonization (i.e., having appropriate gradient and substrate and no insurmountable barriers), they gradually become increasingly suitable for colonization. Emigrants from established populations can then occupy this seemingly marginal habitat in which productivity would normally be low but which feasibly can become better than in source locations (Scudder 1989). As anadromous salmonids become established, their carcasses supply nutrients important to further biological development of the system (Milner et al. 2000). Streams that have been altered by humans have the same potential to recover through natural processes, become more available for colonization, and have salmon gradually reestablish themselves.

Successful restoration may ultimately depend on colonization by strays. Relative abundance of returning runs, directly affected by havest rates, is important to successful colonization because, assuming the stray rate is constant, run size determines the absolute numbers of strays to 'neighboring streams (Riddell 1993). Seemingly good habitat in some locations remains unused, probably because insufficient strays have been available to colonize habitat. If escapements to nearby streams were sufficient to supply the strays, we would expect to see fish straying into and colonizing unused habitats.

Achieving Sustainability

Sustainability of Pacific salmon populations and the socially and economically important fisheries that depend on them require a significant paradigm shift that emphasizes naturally functioning watershed ecosystems, with allowance for natural variation and diversity (Frissell et al. 1997; MacDonald et al., Chapter 15). Recovery and/or perpetuation of healthy salmon systems also depends on whether fisheries managers, decision makers, and the public are willing to support certain protective and remedial actions. Several key, recent publications have provided both the background and and many of the specifics required for sustainability (see NRC 1996a; Stouder et al.1997; Williams et al. 1999; Knudsen et al. 2000b).

Numerous fishery, habitat, and artificial production practices can and should be implemented to restore and maintain sustainable populations. Because the deleterious forces are multifaceted, the recovery and maintenance process requires action on all fronts simultaneously. Table 1 indicates, in broad categories, the kinds of management activities that should be considered for every salmon population. This table includes only general recommendations on each topic; expanded details will be required for population-specific remedies or protection measures. References cited are known examples that contain more specific information on each strategic management topic. Managers should carefully evaluate the requirements for recovery or maintenance of each population. It is essential that all relevant actions for each population be invoked simultaneously—leaving even one limiting factor unaddressed will prevent success.

Can We Sustain Salmon Biodiversity and Fisheries?

Just as salmon populations exhibit incredible diversity in the extent of their range, life history strategies, and productive capacities, there is also much diversity in the types and extent of ecological insults that salmon populations face. Almost no populations remain unexploited. Most are subjected to some degree of harvest, habitat, or hatchery threat. Without widespread restorative action, some populations will thrive, but many will falter or even disappear. It will be a constant battle to keep struggling populations from declining further or completely disappearing and to ensure that healthier populations achieve full production and sustainability.

So, how would we answer the question "can we sustain salmon biodiversity and fisheries?" I believe it is possible, but an affirmative outcome will depend on society choosing a combination of actions that allow salmon to become and remain a thriving component of the ecosystem. With the four ecological concepts and related topics described above as a framework, society can make choices that minimize negative effects on salmon and, more often than not, actually benefit salmon. I believe the review presented in this paper and in recent volumes on the subject (especially NRC 1996a; Stouder et al. 1997; Knudsen et al. 2000b) demonstrate that scientists and managers have sufficient understanding of what salmon need, what diminishes them, and what could be done to protect and restore them. This does not reduce the need for increased scientific research to improve our abilities—numerous technological and information deficiencies have been described in this chapter. It does mean, however, that we can and should do everything possible to support salmon recovery and sustainability

Table 1. Generalized management activities that, when implemented simultaneously as appropriate for each population, should lead to recovery and sustainability of healthy salmon populations and fisheries.

	Management activity		
Ecosystem and life history requirements	Habitat management	Harvest and population management	Artificial production and stocking
Habitat suitability			
Systemic watershed health	Protect watersheds through salmon refuges and other methods[1,2]. Prevent further destruction of freshwater, estuarine, and marine habitats[3]. Use watershed analysis to prescribe protection and restoration needs[4]. Restore degraded watershed functions[4,5,6]. Ameliorate extremes in runoff patterns[6]. Protect and restore critical habitats, such as off-channel rearing areas[7,8].	Ensure sufficient escapement to fully seed all availble rearing habitat[3].	Use the most local populations for supplementation[9,10]. Supplement, when necessary, with populations with phenotypically similar life history[11].
Riparian zones	Maintain substantial streamside buffers[4]. Repair habitats where appropriate[12].		
Instream habitat complexity	Restore and/or maintain healthy watersheds and buffer zones[4]. Add wood where appropriate[12,13].	Include salmon carcasses in considerations of stream complexity[14].	
Spawning substrate	Prevent gravel removal. Manage watersheds to reduce human-induced intensity and frequency of flooding and associated erosion[15].	Ensure sufficient spawners to use all available substrate. Ensure sufficient spawners to maintain spawning substrate quality[16].	

Table 1. continued.

	Management activity		
Ecosystem and life history requirements	Habitat management	Harvest and population management	Artificial production and stocking
Hyporheic zone	Maintain natural flows and groundwater recharge patterns[4]. Repair and prevent alterations of streambanks and riparian zones to maintain uninterrupted hyporheic zones[4].	Ensure sufficient carcasses to fertilize the hyporheic zone[17].	
Water quality and quantity	Manage watersheds for natural hydrologic regime to prevent extreme floods and maintain summer flows[6]. Ensure adequate flows for instream spawning, incubation, and rearing[18]. Ensure sufficient water is supplied to off-channel rearing habitats[19].	Allow sufficient spawners to replace nutrients required for aquatic productivity[20].	Ensure that fish cultural operations do not reduce water quality.
Estuarine and marine habitats	Prevent any further losses of estuarine and intertidal habitats[21]. Maintain and restore access to habitats[22].		
Marine productivity		Ensure maximum natural smolt production[23].	Minimize hatchery smolt releases where necessary[24].
Environmental variation	Repair and protect all local spawning and rearing habitats.	Maximize smolts, especially in periods of poor marine survival[25]. Increase research to account for variability in predictive production models, setting escapement goals, fishery economics[26].	

Table 1. continued.

Ecosystem and life history requirements	Management activity: Habitat management	Harvest and population management	Artificial production and stocking
Production bottlenecks	Identify whether production bottlenecks are habitat-limited[23]. Take corrective action where possible.	Be sure production is not limited by too few spawners for maximizing recruits or supplying marine-derived nutrients to the system[27].	Ensure that natural production is not limited by competition from hatchery populations[28,29]. Prevent overfishing of natural populations in mixed fisheries with hatchery populations[19].
Population biodiversity			
Local adaptation	Protect local spawning and rearing habitats[30].	Ensure sufficient spawners to all possible spawning habitats[19]. Consider genetic refuges as possible recolonization sources[31, 10]. Align harvest management units more closely with specific populations[32].	Use only the most local populations for supplementation[33]. Use hatchery supplementation of wild populations only as a temporary fix[9]. If supplementation is used, consider harvest implications for other local populations in the same fishery[28].
Hierarchical population structure	Protect all local spawning and rearing habitats[30].	Shift harvest emphasis toward maximizing all populations of spawners in their habitats[34]. Reduce the number of populations in each management unit[26]. Account for differences in popula-	Do not supplement one local population without consideration for genetic and abundance-based impacts on other local populations[19]. Minimize differences between hatchery and wild fish[35]. Minimize interbreeding

Table 1. continued.

	Management activity		
Ecosystem and life history requirements	Habitat management	Harvest and population management	Artificial production and stocking
		tion productivities in harvest manage-management[26].	by hatchery and wild fish[35]. Use conservation hatcheries to propagate populations in jeopardy[36]. Ensure that all hatchery operations have carefully defined objectives, including monitoring and evaluation[28].
Biomass and productivity			
Maximizing escapements	Protect and repair all local spawning and rearing habitats to ensure production opportunities[23, 25].	Harvest a fixed proportion, rather than a fixed number, and never fish below the minimum biological escapement threshold[19]. Improve escapement goal methods[26]. Improve catch and escapement data quality[26]. Increase target specificity of fisheries, including selective fisheries[19]. Continue the use of quotas and limited entry[26]. Improve accuracy of in-season management[26].	Locate hatcheries only in areas where harvest of hatchery fish will not affect wild populations[19,28].
Carcasses as nutrients	Consider fertilizer supplied by carcasses as a part of the freshwater habitat[37].	Maximize spawners to all possible spawning habitats. Implement research to determine target	Distribute excess carcasses to local streams where appropriate.

Table 1. continued.

Ecosystem and life history requirements	Management activity: Habitat management	Harvest and population management	Artificial production and stocking
		levels required for full nutrient levels[37]. Temporarily fertilize where necessary to restore production[20].	
Freshwater carrying capacity	Maintain quality of and access to freshwater habitat[23, 25]. Conduct additional research on carrying capacity, including considerations for freshwater nutrients[26].	Ensure sufficient spawners to seed all available spawning and rearing habitat[27]. Conduct additional research on predictive models to optimize escapements and, ultimately, harvests[26].	Minimize competition between hatchery and wild fish[38].
Natural production buffer	Protect all local spawning and rearing habitats[23, 25].	Account for poor production conditions in escapement goals[39]. Ensure sufficient escapement to meet wildlife needs[37].	Minimize hatchery smolt releases where necessary to avoid competition in years of low productivity[40].
Migrations, straying, and colonization			
Upstream migration	Ensure access to all possible spawning areas[19]. Improve passage at existing facilities[19]. Remove dams where necessary[41, 42].	Allow sufficient escapement to colonize newly opened areas.	Preferably use only locally adapted populations for restoration[43,11].
Downstream migration	Reduce mortalities at mainstem dams[19, 5]. Ensure access to all possible rearing areas and create habitat where appropriate[7, 22].		

Table 1. continued.

Ecosystem and life history requirements	Management activity: Habitat management	Harvest and population management	Artificial production and stocking
	Reduce losses in diversions[19]. Open access to off-channel freshwater and estuarine rearing areas[22].		
Straying	Protect all local spawning and rearing habitats[23, 25].	Maximize spawners to increase the absolute numbers of strays[19].	Prevent straying of non-native populations into native populations[35].
Colonization	Repair and protect all local spawning and rearing habitats[23, 25].	Maximize spawners to increase the absolute number of strays into new, restored, and vacant habitats[19].	Consider the trade-offs between natural re-recolonization and introduction of non-native populations[9,11]. If non-native introductions are made, use geographically closest populations having phenotypically similar life histories[11].

1. Frissell 1993. 2. Lichatowich et al. 2000. 3. Williams and Williams 1997. 4. Bisson et al. 1997. 5. Williams et al. 1999. 6. Fresh and Luchetti 2000. 7. Cederholm et al. 1988. 8. Richards et al. 1992. 9. Allendorf and Waples 1995. 10. Thurow et al. 2000. 11. Burger et al. 2000. 12. Beschta 1997. 13. Dominguez and Cederholm 2000. 14. Cederholm et al. 2000. 15. Gregory and Bisson 1997. 16. Montgomery et al. 1996. 17. Ben-David et al. 1998. 18. Beecher 1995. 19. NRC 1996a. 20. Stockner et al. 2000. 21. Levings 2000. 22. Beechie et al. 1994. 23. Lawson 1993. 24. Hilborn and Eggers 2000. 25. Hare et al. 1999. 26. Knudsen 2000. 27. Michael 1998. 28. Hilborn 1999. 29. Hilborn and Eggers 2000. 30. Healey and Prince 1995. 31. Li et al. 1995. 32. Mundy et al. 1995. 33. Hard 1995. 34. Hyatt and Riddell 2000. 35. Resisenbichler 1997. 36. Stickney 1994. 37. Cederholm et al. 1999. 38. McMichael et al. 2000. 39. Cramer 2000. 40. Francis 1997. 41. Wunderlich et al. 1994. 42. Marmorek and Peters 1997. 43. Riddell 1993.

now, based on the best current information, while further research and monitoring build and enhance the information base.

Ideally, each population would be managed independently with a firm understanding of its production capacity, the spawning escapement required for perpetuation, and the harvestable surplus. The population would also be managed in the context of the watershed and ecosystem in which it evolved (Healey and Prince 1995; Fresh and Lucchetti 2000). The new paradigm appears to be a salmon life history ecosystem approach guided by watershed analysis together with adaptive learning (Naiman et al. 1992; Williams and Williams 1997; Williams et al. 1999; MacDonald et al., Chapter 15).

The challenges, then, lie in the realm of the public process. Probably the single most important issue is the matter of priority. Recent ESA listings in the Pacific Northwest, and the public response to those listings seem to indicate that most people place a relatively high priority on maintenance and recovery of salmon. Another positive development is the recent accord under the Pacific Salmon Treaty that led to reductions of fisheries to help rebuild runs in both Canada and the United States. Further progress will depend on positive outcomes of difficult decisions on such diverse and seemingly disconnected topics as local land use, growth management, alternative power generation, and societal changes in fishing communities, among many others. Success will be achieved when humans can bring their economic and social systems back into harmony with the natural system that supports salmon. That transition depends on a combination of community involvement and public intentions that provide the necessary support for effective management programs and new science (MacDonald et al., Chapter 15). If these things fall into place, salmon populations will thrive wherever their biological needs are met and their tolerance limits are not exceeded. The biodiversity of salmon populations and the continuation of healthy fisheries can be sustained if we value our ecological heritage sufficiently to protect and repair watersheds, ensure healthy escapements, and preserve the genetic integrity of salmon populations.

Acknowledgements

This contribution would not have been possible without frequent technical and philosophical discussions with my esteemed colleagues in the Pacific salmon science and management realm. Those friends are too numerous to mention here but are no less appreciated. I am grateful for the thoughtful and thorough reviews of Don MacDonald and three anonymous reviewers.

References

Allendorf, F. W., and R. S. Waples. 1995. Conservation and genetics of salmonid fishes. Pages 238–281 *in* J. C. Avise and J. L. Hamrick, editors. Conservation genetics: case histories from nature. Chapman and Hall, New York.

Armantrout, N. B. 2000. Using watershed analysis to plan, and evaluate habitat restoration. Pages 601–616 *in* E. E. Knudsen, C. S. Steward, D. D. MacDonald, J. E.

Williams, and D. W. Reiser, editors. Sustainable fisheries management: Pacific salmon. Lewis Publishers, Boca Raton, Florida.

Beacham, T. D., and C. B. Murray. 1987. Adaptive variation in body size, age, morphology, egg size, and developmental biology of chum salmon (*Oncorhynchus keta*) in British Columbia. Canadian Journal of Fisheries and Aquatic Sciences 44: 244–261.

Beamish, R. J., and D. R. Bouillon. 1993. Pacific salmon production trends in relation to climate. Canadian Journal of Fisheries and Aquatic Sciences 50:1002–1016.

Beecher, H. A. 1995. Elkhorn ruling boosts state authority to set instream flows on federal projects. Fisheries 20(3):18–19.

Beechie, T., E. Beamer, and L. Wasserman. 1994. Estimating coho salmon rearing habitat and smolt production losses in a large river basin, and implications for habitat restoration. North American Journal of Fisheries Management 14:797–811.

Beechie, T. J., and T. H. Sibley. 1997. Relationships between channel characteristics, woody debris, and fish habitat in northwestern Washington streams. Transactions of the American Fisheries Society 126:217–229.

Ben-David, M., R. W. Flynn, and D. M. Schell. 1997. Annual and seasonal changes in diets of martens: evidence from stable isotope analysis. Oecologia 111:280–291.

Ben-David, M., T. A. Hanley, and D. M. Schell. 1998. Fertilization of terrestrial vegetation by spawning Pacific salmon: the role of flooding and predator activity. Oikos 83:47–55.

Bendock, T., and M. Alexandersdottir. 1993. Hooking and mortality of chinook salmon released in the Kenai River, Alaska. North American Journal of Fisheries Management 13:540–549.

Beschta, R. L. 1997. Restoration of riparian and aquatic systems for improved aquatic habitats in the upper Columbia River basin. Pages 475–491 *in* D. J. Stouder, P. A. Bisson, and R. J. Naiman, editors. Pacific salmon and their ecosystems: status and future options. Chapman and Hall, New York.

Bilby, R. E, B. R. Fransen, and P. A. Bisson. 1996. Incorporation of nitrogen and carbon from spawning coho salmon into the trophic system of small streams. Canadian Journal of Fisheries and Aquatic Sciences 53(1):164–173.

Bisson, P. A., R. E. Bilby, M. D. Bryant, C. A. Dolloff, G. B. Grette, R. A. House, M. L. Murphy, K. V. Koski, and J. R. Sedell. 1987. Large woody debris in forested streams in the Pacific northwest: past, present, and future. Pages 143–190 *in* E. O. Salo and T. W. Cundy, editors. Streamside management: forestry and fisheries interactions. College of Forest Resources, University of Washington, Seattle.

Bisson, P. A., G. H. Reeves, R. E. Bilby, and R. J. Naiman. 1997. Watershed management and Pacific salmon: desired future conditions. Pages 447–474 *in* D. J. Stouder, P. A. Bisson, and R. J. Naiman, editors. Pacific salmon and their ecosystems: status and future options. Chapman and Hall, New York.

Boule, M. E., and K. F. Bierly. 1987. History of estuarine wetland development and alteration: what have we wrought? Northwest Environmental Journal 3(1):43–61.

Bradford, M. J. 1997. An experimental study of stranding of juvenile salmonids on gravel bars and in side channels during rapid flow decreases. Regulated Rivers: Research and Management 13:395–401.

Bradford, M. J., G. C. Taylor, J. A. Allan, and P. S. Higgins. 1995. An experimental study of the stranding of juvenile coho salmon and rainbow trout during rapid flow decreases under winter conditions. North American Journal of Fisheries Management 15:473–479.

Bryant, M. D. 1983. The role and management of woody debris in west coast salmonid nursery streams. North American Journal of Fisheries Management 3:322–330.

Burger, C. V. 2000. The needs of salmon, and steelhead in balancing their conservation and use. Pages 15–29 *in* E. E. Knudsen, C. S. Steward, D. D. MacDonald, J. E. Williams, and D. W. Reiser, editors. Sustainable fisheries management: Pacific salmon. Lewis Publishers, Boca Raton, Florida.

Burger, C. V., W. J. Spearman, and M. A. Cronin. 1997. Genetic differentiation of sockeye salmon subpopulations from a geologically young Alaskan lake system. Transactions of the American Fisheries Society 126:926–938.

Burger, C. V., K. T. Scribner, W. J. Spearman, C. O. Swanton, and D. E. Campton. 2000. Genetic contributions of three introduced life history forms of sockeye salmon to colonization of Frazer Lake, Alaska. Canadian Journal of Fisheries and Aquatic Sciences 57:2096–2111.

Campton, D. E. 1995. Genetic effects of hatchery fish on wild populations of Pacific salmon and steelhead: what do we really know? Pages 337–353 in H. L. Schramm, Jr. and R. G. Piper, editors. Uses and effects of cultured fishes in aquatic ecosystems. American Fisheries Society, Symposium 15, Bethesda, Maryland.

Candy, J. R., E. W. Carter, T. P. Quinn, and B. E. Riddell. 1996. Adult chinook salmon behavior and survival after catch and release from purse-seine vessels in Johnstone Strait, British Columbia. North American Journal of Fisheries Management 16:521–529.

Cederholm, C. J., R. E. Bilby, P. A. Bisson, T. W. Bumstead, B. R. Fransen, W. J. Scarlett, and J. W. Ward. 1997. Response of juvenile coho salmon and steelhead to placement of large woody debris in a coastal Washington stream. North American Journal of Fisheries Management 17:947–964.

Cederholm, C. J., D. B. Houston, D. L. Cole, and W. J. Scarlett. 1989. Fate of coho salmon (*Oncorhynchus kisutch*) carcasses in spawning streams. Canadian Journal of Fisheries and Aquatic Sciences 46:1347–1355.

Cederholm, C. J., D. H. Johnson, R. E. Bilby, L. G. Dominguez, A. M. Garrett, W. H. Graeber, E. L. Greda, M. D. Kunze, B. G. Marcot, J. F. Palmisano, R. W. Plonikoff, W. G. Pearcy, C. A. Simenstad, and P. C. Trotter. 2000. Pacific salmon and wildlife–ecological contexts, relationships, and implications for management. Washington Department of Fish and Wildlife, Olympia, Washington.

Cederholm, C. J., M. D. Kunze, T. Murota, and A. Sibatani. 1999. Pacific salmon carcasses: essential contributions of nutrients and energy for aquatic and terrestrial ecosystems. Fisheries 24(10):6–15.

Cederholm, C. J., W. J. Scarlett, and N. P. Peterson. 1988. Low-cost enhancement technique for winter habitat of juvenile salmon. North American Journal of Fisheries Management 8:438–441.

Chapman, D. W. 1986. Salmon and steelhead abundance in the Columbia River in the nineteenth century. Transactions of the American Fisheries Society 115:662–670.

Chapman, D. W., and E. E. Knudsen. 1980. Channelization and livestock impacts on salmonid habitat and biomass in western Washington. Transactions of the American Fisheries Society 109:357–363.

Copes, P. 2000. Aboriginal fishing rights, and salmon management in British Columbia; matching historical justice with the public interest. Pages 75–91 *in* E. E. Knudsen, C. S. Steward, D. D. MacDonald, J. E. Williams, and D. W. Reiser, editors. Sustainable fisheries management: Pacific salmon. Lewis Publishers, Boca Raton, Florida.

Cramer, S. P. 2000. The effect of environmentally-driven recruitment variation on sustainable yield from salmon populations. Pages 485–503 *in* E. E. Knudsen, C. S. Steward, D. D. MacDonald, J. E. Williams, and D. W. Reiser, editors. Sustainable fisheries management: Pacific salmon. Lewis Publishers, Boca Raton, Florida.

Dahm, C. N., N. B. Grimm, P. Marmonier, H. M. Valett, and P. Vervier. 1998. Nutrient dynamics at the interface between surface waters and groundwaters. Freshwater Biology 40:427–451.

Dominguez, L. G., and C. J. Cederholm. 2000. Rehabilitating stream channels using large woody debris with considerations for salmonid life history and fluvial geomorphic processes. Pages 545–563 *in* E. E. Knudsen, C. S. Steward, D. D. MacDonald, J. E. Williams, and D. W. Reiser, editors. Sustainable fisheries management: Pacific salmon. Lewis Publishers, Boca Raton, Florida.

Edwards, R. T. 1998. The hyporheic zone. Pages 399–429 *in* R. J. Naiman and R. E. Bilby, editors. River ecology and management: lessons from the Pacific coastal ecoregion. Springer, New York.

Finn, J. E., S. MacLean, H. Hander, and E. E. Knudsen. 1998. Yukon chum progress report. United States Geological Survey, Alaska Biological Science Center, Anchorage, Alaska

Finney, B. P., I. Gregory-Eaves, J. Sweetman, M. Douglas, and J. P. Smol. 2000. Impacts of climatic change and fishing on Pacific salmon abundance over the past three hundred years. Science 290:795–799.

Fisheries and Oceans Canada. 1999. Pacific region 1999 salmon troll management plan, area G, west coast Vancouver Island. Fisheries and Oceans Canada, Vancouver, British Columbia.

Flagg, T. A., F. W. Waknitz, D. J. Maynard, G. B. Milner, and C. V. W. Mahnken. 1995. The effect of hatcheries on native coho salmon populations in the lower Columbia River. Pages 366–375 *in* H. L. Schramm and R. G. Piper, editors. Uses and effects of cultured fishes in aquatic ecosystems. American Fisheries Society, Symposium 15, Bethesda, Maryland.

Fraidenburg, M. E., and R. H. Lincoln. 1985. Wild chinook salmon management: an international conservation challenge. North American Journal of Fisheries Management 5:311–329.

Francis, R. C. 1997. Managing resources with incomplete information: making the best of a bad situation. Pages 513–524 *in* D. J. Stouder, P. A. Bisson, and R. J. Naiman, editors. Pacific salmon and their ecosystems: status and future options. Chapman and Hall, New York.

Fresh, K. L. 1997. The role of competition and predation in the decline of anadromous salmonids in the Pacific northwest. Pages 245–276 *in* D. J. Stouder, P. A. Bisson, and R. J. Naiman, editors. Pacific salmon and their ecosystems: status and future options. Chapman and Hall, New York.

Fresh, K. L., and G. Lucchetti. 2000. Protecting and restoring the habitats of anadromous salmonids in the Lake Washington watershed, an urbanizing ecosystem. Pages 525–544 *in* E. E. Knudsen, C. S. Steward, D. D. MacDonald, J. E. Williams, and D. W. Reiser, editors. Sustainable fisheries management: Pacific salmon. Lewis Publishers, Boca Raton, Florida.

Frissell, C. A. 1993. Topology of extinction and endangerment of native fishes in the Pacific Northwest and California. Conservation Biology 7:342–354.

Frissell, C. A., W. J. Liss, R. E. Greswell, R. K. Nawa, and J. L. Ebersole. 1997. A resource in crisis: changing the measure of salmon management. Pages 411–444 *in* D. J. Stouder, P. A. Bisson, and R. J. Naiman, editors. Pacific salmon and their ecosystems: status and future options. Chapman and Hall, New York.

Frissell, C. A., and R. K. Nawa. 1992. Incidence and causes of physical failure of artificial habitat structures in streams of western Oregon and Washington. North American Journal of Fisheries Management 12:182–197.

Geist, D. R. 2000. Hyporheic discharge of river water into fall chinook salmon (*Oncorhynchus tshawytscha*) spawning areas in the Hanford Reach, Columbia River. Canadian Journal of Fisheries and Aquatic Sciences 57:1647–1656.

Gregory, S. V., and P. A. Bisson. 1997. Degradation and loss of anadromous salmonid habitat in the Pacific Northwest. Pages 277–314 *in* D. J. Stouder, P. A. Bisson, and R. J. Naiman, editors. Pacific salmon and their ecosystems: status and future options. Chapman and Hall, New York.

Gregory, S. V., G. A. Lamberti, D. C. Erman, K. V. Koski, M. L. Murphy, and J. R. Sedell. 1987. Influence of forest practices on aquatic production. Pages 234–255 *in* E. O. Salo and T. W. Cundy, editors. Streamside management: forestry and fisheries interactions. College of Forest Resources, University of Washington, Seattle.

Gregory, S. V., F. J. Swanson, and W. A. McKee. 1991. An ecosystem perspective of riparian zones. Bioscience 40:540–551.

Gresh, T., J. Lichatowich, and P. Schoonmaker. 2000. An estimation of historic and current levels of salmon production in the northwest Pacific ecosystem: evidence of a nutrient deficit in the freshwater systems of the Pacific Northwest. Fisheries 25(1):15–21.

Groot, C., and L. Margolis, editors. 1991. Pacific salmon life histories. University of British Columbia Press, Vancouver, British Columbia.

Hard, J. J. 1995. A quantitative genetic perspective on the conservation of intraspecific diversity. Pages 304–328 *in* J. L. Nielsen, editor. Evolution and the aquatic ecosystem: defining unique units in population conservation. American Fisheries Society, Symposium 17, Bethesda, Maryland.

Hare, S. R., N. J. Mantua, and R. C. Francis. 1999. Inverse production regimes: Alaska and west coast Pacific salmon. Fisheries 24(1):6–14.

Harr, R. D., and R. A. Nichols. 1993. Stabilizing forest roads to help restore fish habitats: a Northwest Washington example. Fisheries 18(4):18–22.

Hartman, G. F., C. Groot, and T. G. Northcote. 2000. Science and management in sustainable salmonid fisheries: the ball is not in our court. Pages 31–50 *in* E. E. Knudsen, C. S. Steward, D. D. MacDonald, J. E. Williams, and D. W. Reiser, editors. Sustainable fisheries management: Pacific salmon. Lewis Publishers, Boca Raton, Florida.

Hartman, G. F., and J. C. Scrivener. 1990. Impacts of forestry practices on a coastal stream ecosystem, Carnation Creek, British Columbia. Department of Fisheries and Oceans, Ottawa.

Hassemer, P. F., S. W. Kiefer, and C. E. Petrosky. 1997. Idaho's salmon: can we count every last one? Pages 113–125 *in* D. J. Stouder, P. A. Bisson, and R. J. Naiman, editors. Pacific salmon and their ecosystems: status and future options. Chapman and Hall, New York.

Healey, M. C., and A. Prince. 1995. Scales of variation in life history tactics of Pacific salmon and the conservation of phenotype and genotype. Pages 176–184 *in* J. L. Nielsen, editor. Evolution and the aquatic ecosystem: defining unique units in population conservation. American Fisheries Society, Symposium 17, Bethesda, Maryland.

Hicks, B. J., R. L. Beschta, and R. D. Harr. 1991. Long-term changes in streamflow following logging in western Oregon and associated fisheries implications. Water Research Bulletin 27(2):217–226.

Hilborn, R. 1992. Hatcheries and the future of salmon in the northwest. Fisheries 17(1):5–8.

Hilborn, R. 1999. Confessions of a reformed hatchery basher. Fisheries 24(5):30–31.

Hilborn, R., and D. Eggers. 2000. A review of the hatchery programs for pink salmon in Prince William Sound and Kodiak Island, Alaska. Transactions of the American Fisheries Society 129:333–350.

Hilborn, R. C., and C. J. Walters. 1992. Quantitative fisheries stock assessment. Chapman and Hall, New York.

Hildebrand, G. V., C. C. Schwatz, C. T. Robbins, et al. In press. The importance of meat, particularly salmon, to body size, population productivity, and conservation of North American brown bears. Canadian Journal of Zoology.

Holmes, R. A., and R. D. Burkett. 1996. Salmon stewardship: Alaska's perspective. Fisheries 21(10):30–32.

Holtby, L. B., B. C. Andersen, and R. K. Kadowaki. 1990. Importance of smolt size and early ocean growth to interannual variability in marine survival of coho salmon (*Oncorhynchus kisutch*). Canadian Journal of Fisheries and Aquatic Sciences 47:2181–2194.

Huntington, C., W. Nehlsen, and J. Bowers. 1996. A survey of healthy native stocks of anadromous salmonids in the Pacific Northwest and California. Fisheries 21(3):6–14.

Hyatt, K. D., and B. E. Riddell. 2000. The importance of "stock" conservation definitions to the concept of sustainable fisheries. Pages 51–61 *in* E. E. Knudsen, C. S. Steward, D. D. MacDonald, J. E. Williams, and D. W. Reiser, editors. Sustainable fisheries management: Pacific salmon. Lewis Publishers, Boca Raton, Florida.

Ishida, Y., S. Ito, M. Kaeriyama, S. McKinnell, and K. Nagasawa. 1993. Recent changes in age and size of chum salmon (*Oncorhynchus keta*) in the North Pacific ocean and possible causes. Canadian Journal of Fisheries and Aquatic Sciences 50:290–295.

Kendall, R. L. 1999. Sustainability. Fisheries 24(3):4.

Kjelson, M. A., and P. L. Brandes. 1989. The use of smolt survival estimates to quantify the effects of habitat changes on salmonid stocks in the Sacramento-San Joaquin Rivers, California. Pages 100–115 *in* C. D. Levings, L. B. Holtby, and M. A.

Henderson, editors. Proceedings of the National Workshop on Effects of Habitat Alteration on Salmonid Stocks. Department of Fisheries and Oceans, Ottawa, Ontario.

Kline, T. C., J. J. Goering, O. A. Mathisen, and P. H. Poe. 1993. Recycling of elements transported upstream by runs of Pacific salmon: II. 15N and 13C evidence in the Kivchak River watershed, Bristol Bay, southwestern Alaska. Canadian Journal of Fisheries and Aquatic Sciences 50:2350–2365.

Knudsen, E. E. 2000. Managing Pacific salmon escapements: the gaps between theory, and reality. Pages 237–272 *in* E. E. Knudsen, C. S. Steward, D. D. MacDonald, J. E. Williams, and D. W. Reiser, editors. Sustainable fisheries management: Pacific salmon. Lewis Publishers, Boca Raton, Florida.

Knudsen, E. E., D. D. MacDonald, and C. R. Steward. 2000a. Setting the stage for a sustainable Pacific salmon strategy. Pages 3–14 *in* E. E. Knudsen, C. S. Steward, D. D. MacDonald, J. E. Williams, and D. W. Reiser, editors. Sustainable fisheries management: Pacific salmon. Lewis Publishers, Boca Raton, Florida.

Knudsen, E. E., C. S. Steward, D. D. MacDonald, J. E. Williams, and D. W. Reiser. 2000b. Sustainable fisheries management: Pacific salmon. Lewis Publishers, Boca Raton, Florida.

Knudsen, E. E., and S. J. Dilley. 1987. Effects of riprap bank reinforcement on juvenile salmonids in four western Washington streams. North American Journal of Fisheries Management 7:351–356.

Krueger, C. C., and B. May. 1991. Ecological and genetic effects of salmonid introductions in North America. Canadian Journal of Fisheries and Aquatic Sciences 48(Supplement 1):66–77.

Kruse, G. H. 1998. Salmon run failures in 1997–1998: a link to anomalous ocean conditions? Alaska Fishery Research Bulletin 5(1):55–63.

Langer, O. E., F. Hietkamp, and M. Farrell. 2000. Human population growth and the sustainability of urban salmonid streams in the lower Fraser valley. Pages 349–364 *in* E. E. Knudsen, C. S. Steward, D. D. MacDonald, J. E. Williams, and D. W. Reiser, editors. Sustainable fisheries management: Pacific salmon. Lewis, New York.

Lawson, P. W. 1993. Cycles in ocean productivity, trends in habitat quality, and the restoration of salmon runs in Oregon. Fisheries 18(8):6–10.

Levings, C. D. 2000. An overview assessment of compensation, and mitigation techniques used to assist fish habitat management in British Columbia estuaries. Pages 341–348 *in* E. E. Knudsen, C. S. Steward, D. D. MacDonald, J. E. Williams, and D. W. Reiser, editors. Sustainable Fisheries Management: Pacific salmon. Lewis Publishers, Boca Raton, Florida.

Li, H. W., K. Currens, D. Bottom, S. Clarke, J. Dambacher, C. Frissell, P. Harris, R. M. Hughes, D. McCullough, A. McGie, K. Moore, R. Nawa, and S. Thiele. 1995. Safe havens: refuges and evolutionarily significant units. Pages 371–380 *in* J. L. Nielsen, editor. Evolution and the aquatic ecosystem: defining unique units in population conservation. American Fisheries Society, Symposium 17, Bethesda, Maryland.

Lichatowich, J., L. Mobrand, L. Lestelle, and T. Vogel. 1995. An approach to the diagnosis and treatment of depleted salmon populations in Pacific Northwest watersheds. Fisheries 20(1):10–18.

Lichatowich, J. A., G. R. Rahr, and S. M. Whidden. 2000. Sanctuaries for Pacific salmon. Pages 675–686 *in* E. E. Knudsen, C. S. Steward, D. D. MacDonald, J. E. Williams, and D. W. Reiser, editors. Sustainable fisheries management: Pacific salmon. Lewis Publishers, Boca Raton, Florida.

Mangel, M. 1994. Life history variation and salmonid conservation. Conservation Biology 8(3):879–880.

Marmorek, D. R., and C. Peters. 1997. Plan for analyzing and testing hypotheses (PATH): report of retrospective analysis for fiscal year 1997. ESSA Technologies Limited, Vancouver, British Columbia.

May, C. W. 1996. Assessment of cumulative effects of urbanization on small streams in the Puget Sound Lowland Ecoregion. Ph.D. Dissertation, University of Washington, Seattle.

McMichael, G. A., T. N. Pearsons, and S. A. Leider. 2000. Minimizing ecological impacts of hatchery-reared juvenile steelhead trout on wild salmonids in a Yakima Basin watershed. Pages 365–380 *in* E. E. Knudsen, C. S. Steward, D. D. MacDonald, J. E. Williams, and D. W. Reiser, editors. Sustainable fisheries management: Pacific salmon. Lewis Publishers, Boca Raton, Florida.

McMichael, G. A., C. S. Sharpe, and T. N. Pearsons. 1997. Effects of residual hatchery-reared steelhead on growth of wild rainbow trout and spring chinook salmon. Transactions of the American Fisheries Society 126:230–239.

Michael, Jr., J. H. 1998. Pacific salmon spawner escapement goals for the Skagit River watershed as determined by nutrient cycling considerations. Northwest Science 72(4):239–248.

Milner, A. M., and R. G. Bailey. 1989. Salmonid colonization of new streams in Glacier Bay, Alaska. Aquaculture and Fisheries Management 20:179–192.

Milner, A. M., E. E. Knudsen, C. Soiseth, A. L. Robertson, D. Schell, I. T. Phillips, and K. Magnusson. 2000. Colonization and development of stream communities across a two hundred year gradient in Glacier Bay National Park, Alaska, USA. Canadian Journal of Fisheries and Aquatic Sciences 57:2319–2335.

Montgomery, D. R., J. M. Buffington, N. P. Peterson, D. Schuett-Hames, and T. P. Quinn. 1996. Streambed scour, egg burial depths, and the influence of salmonid spawning on bed surface mobility and embryo survival. Canadian Journal of Fisheries and Aquatic Sciences 53:1061–1070.

Mundy, P. R., T. W. H. Backman, and J. M. Berlson. 1995. Selection of conservation units for Pacific salmon: lessons from the Columbia River. Pages 28–40 *in* J. L. Nielsen, editor. Evolution and the aquatic ecosystem: defining unique units in population conservation. American Fisheries Society, Symposium 17, Bethesda, Maryland.

Naiman, R. J., T. J. Beechie, L. E. Benda, D. R. Berg, P. A. Bisson, L. H. MacDonald, M. D. O'Connor, P. L. Olson, and E. A. Steel. 1992. Fundamental elements of ecologically healthy watersheds in the Pacific Northwest coastal ecoregion. Pages 127–188 *in* R. J. Naiman, editor. Watershed management: balancing sustainability and environmental change.

Nehlsen, W., J. E. Williams, and J. A. Lichatowich. 1991. Pacific salmon at the crossroads: stocks at risk from California, Oregon, Idaho, and Washington. Fisheries 16(2):4–21.

Nickelson, T. E., M. F. Solazzi, and S. L. Johnson. 1986. Use of hatchery coho salmon (*Oncorhynchus kisutch*) presmolts to rebuild wild populations in Oregon coastal streams. Canadian Journal of Fisheries and Aquatic Sciences 43:2443–2449.

Nielsen, J. L. 1994. Invasive cohorts: impacts of hatchery-reared coho salmon on the trophic, developmental, and genetic ecology of wild stocks. Pages 361–385 *in* D. J. Stouder, K. L. Fresh, and R. J. Feller, editors. Theory and application in fish feeding ecology. University of South Carolina Press, Columbia, South Carolina.

Nielsen, J. L., C. Gan, and W. K. Thomas. 1994. Differences in genetic diversity for mitochondrial DNA between hatchery and wild populations of *Oncorhynchus*. Canadian Journal of Fisheries and Aquatic Sciences 5(Supplement 1):290–297.

NRC (National Research Council). 1996a. Upstream: salmon and society in the Pacific Northwest. National Academy Press, Washington, D.C.

NRC (National Research Council). 1996b. The Bering Sea ecosystem. National Academy Press, Washington, D.C.

O'Brien, J. J. 1998. The big picture - the ENSO of 1997–98. *<http://pices.ios.bc.ca/picespub/reports/report99.htm>*.

Pearcy, W. 1992. Ocean ecology of North Pacific salmonids. University of Washington Press, Seattle.

Pearcy, W. G. 1997. Salmon production in changing ocean domains. Pages 331–352 *in* D. J. Stouder, P. A. Bisson, and R. J. Naiman, editors. Pacific salmon and their ecosystems: status and future options. Chapman and Hall, New York.

Perry, E. A. 1995. Salmon stock restoration and enhancement: strategies and experiences in British Columbia. Pages 152–160 *in* H. L. Schramm and R. G. Piper, editors. Uses and effects of cultured fishes in aquatic ecosystems. American Fisheries Society, Symposium 15, Bethesda, Maryland.

Peters, R. J., E. E. Knudsen, J. C. Cederholm, W. J. Scarlett, and G. B. Pauley. 1993. Preliminary results of woody debris use by summer-rearing juvenile coho salmon (*Oncorhynchus kisutch*) in the Clearwater River, Washington. Pages 323–339 *in* L. Berg and P. W. Delaney, editors. Proceedings of the coho workshop, Nanimo, British Columbia.

Quinn, T. P. 1993. A review of homing and straying of wild and hatchery-produced salmon. Fisheries Research 18:29–44.

Raymond, H. L. 1988. Effects of hydroelectric development and fisheries enhancement on spring and summer chinook salmon and steelhead in the Columbia River basin. North American Journal of Fisheries Management 8:1–24.

Reisenbichler, R. R. 1997. Genetic factors contributing to the declines of anadromous salmonids in the Pacific Northwest. Pages 223–244 *in* D. J. Stouder, P. A. Bisson, and R. J. Naiman, editors. Pacific salmon and their ecosystems: status and future options. Chapman and Hall, New York.

Reisenbichler, R. R., J. D. McIntyre, M. F. Solazzi, and S. W. Landino. 1992. Genetic variation in steelhead of Oregon and northern California. Transactions of the American Fisheries Society 121:158–169.

Richards, C., P. J. Cernera, M. P. Ramey, and D. W. Reiser. 1992. Development of off-channel habitats for use by juvenile chinook salmon. North American Journal of Fisheries Management 12(4):721–727.

Ricker, W. E. 1975. Computation and interpretation of biological statistics of fish populations. Fisheries Research Board of Canada Bulletin 191, Ottawa, Ontario.

Riddell, B. E. 1993. Spatial organization of Pacific salmon: what to conserve? Pages 23–41 *in* J. G. Cloud and G. H. Thorgaard, editors. Genetic conservation of salmonid fishes. Plenum Press, New York.

Rogers, D. E., and G. T. Ruggerone. 1993. Factors affecting marine growth of Bristol Bay sockeye salmon. Fisheries Research 18:89–103.

Rutter, L. G. 1997. Salmon fisheries in the Pacific Northwest: how are harvest management decisions made? Pages 355–374 *in* D. J. Stouder, P. A. Bisson, and R. J. Naiman, editors. Pacific salmon and their ecosystems: status and future options. Chapman and Hall, New York.

Schmidt, D. C., S. R. Carlson, G. B. Kyle, and B. P. Finney. 1998. Influence of carcass-derived nutrients on sockeye salmon productivity of Karluk Lake, Alaska: importance in the assessment of an escapement goal. North American Journal of Fisheries Management 18:743–763.

Scudder, G. G. E. 1989. The adaptive significance of marginal populations: a general perspective. Pages 180–185 *in* C. D. Levings, L. B. Holtby, and M. A. Henderson, editors. Proceedings of the National Workshop on Effects of Habitat Alteration on Salmonid Stocks. Department of Fisheries and Oceans, Ottawa, Ontario.

Sedell, J. R., P. A. Bisson, F. J. Swanson, and S. V. Gregory. 1988. What we know about large trees that fall into streams and rivers. Pages 47–81 *in* C. Maser and J. R. Sedell, editors. From the forest to the sea: a story of fallen trees. U.S. Department of Agriculture, Forest Service, Pacific Northwest Research Station, Portland, Oregon.

Sedell, J. R., G. H. Reeves, and P. A. Bisson. 1997. Habitat policy for salmon in the Pacific Northwest. Pages 375–387 *in* D. J. Stouder, P. A. Bisson, and R. J. Naiman, editors. Pacific salmon and their ecosystems: status and future options. Chapman and Hall, New York.

Seiler, D. 1989. Differential survival of Grays Harbor Basin anadromous salmonids: water quality implications. Pages 123–135 *in* C. D. Levings, L. B. Holtby and M. A. Henderson, editors. Proceedings of the national workshop on effects of habitat alteration on salmonid stocks. Department of Fisheries and Oceans, Ottawa, Ontário.

Simenstad, C. A., K. L. Fresh, and E. O. Salo. 1982. The role of Puget Sound and Washington coastal estuaries in the life history of Pacific salmon: an unappreciated function. Pages 343–364 *in* V. S. Kennedy, editor. Esturaine comparisons. Academic Press, New York.

Slaney, T. K., K. D. Hyatt, T. G. Northcote, and R. J. Fielden. 1996. Status of anadromous salmon and trout in British Columbia and Yukon. Fisheries 21(10):20–35.

Smoker, W. W., A. J. Gharrett, and M. S. Stekoll. 1998. Genetic variation of return date in a population of pink salmon: a consequence of fluctuating environment and dispersive selection? Alaska Fishery Research Bulletin 5(1):46–54.

Spence, B. G., G. A. Lomnicky, R. M. Hughes, and R. P. Novitski. 1996. An ecosystem approach to salmonid conservation. Mantech Environmental, Inc., Corvallis, Oregon.

Stanford, J. A. 1998. Rivers in the landscape: introduction to the special issue on riparian and groundwater ecology. Freshwater Biology 40:402–406.

Stanford, J. A., and J. V. Ward. 1993. An ecosystem perspective of alluvial rivers: connectivity and the hyporheic corridor. Journal of the North American Benthological Society 12(1):48–60.

Stickney, R. R. 1994. Use of hatchery fish in enhancement programs. Fisheries 19(5):6–13.

Stockner, J. G., E. Rydin, and P. Hyenstrand. 2000. Cultural oligotrophication: causes and consequences for fisheries resources. Fisheries 25(5):7–14.

Stouder, D. J., P. A. Bisson, and R. J. Naiman, editors. 1997. Pacific salmon and their ecosystems: status and future options. Chapman and Hall, New York.

Swales, S., and C. D. Levings. 1989. Role of off-channel ponds in the life cycle of coho salmon (*Oncorhynchus kisutch*) and other juvenile salmonids in the Coldwater River, British Columbia. Canadian Journal of Fisheries and Aquatic Sciences 46:232–242.

Thomas, G. L., and O. A. Mathisen. 1993. Biological interactions of natural and enhanced stocks of salmon in Alaska. Fisheries Research 18:1–18.

Thurow, R. F., D. C. Lee, and B. E. Rieman. 2000. Status and distribution of chinook salmon and steelhead in the interior Columbia River basin and portions of the Klamath River basin. Pages 133–160 *in* E. E. Knudsen, C. S. Steward, D. D. MacDonald, J. E. Williams, and D. W. Reiser, editors. Sustainable fisheries management: Pacific salmon. Lewis Publishers, Boca Raton, Florida.

Utter, F. M., D. W. Chapman, and A. R. Marshall. 1995. Genetic population structure and history of chinook salmon of the upper Columbia River. Pages 149–168 *in* J. L. Nielsen, editor. Evolution and the aquatic ecosystem: defining unique units in population conservation. American Fisheries Society, Symposium 17, Bethesda, Maryland.

Van Alen, B. W. 2000. Status and stewardship of salmon stocks in southeast Alaska. Pages 161–194 *in* E. E. Knudsen, C. S. Steward, D. D. MacDonald, J. E. Williams, and D. W. Reiser, editors. Sustainable fisheries management: Pacific salmon. Lewis Publishers, Boca Raton, Florida.

Volpe, J. P., E. B. Taylor, D. W. Rimmer, and B. W. Glickman. 2000. Evidence of natural reproduction of aquaculture-escaped Atlantic salmon in a coastal British Columbia River. Conservation Biology 14(3):899–903.

Walters, C. J., and P. Cahoon. 1985. Evidence of decreasing spatial diversity in British Columbia salmon stocks. Canadian Journal of Fisheries and Aquatic Sciences 42:1033–1037.

Waples, R. S. 1991a. Definition of "species" under the endangered species act: application to Pacific salmon. National Marine Fisheries Service, Seattle.

Waples, R. S. 1991b. Genetic interactions between hatchery and wild salmonids: lessons from the Pacific Northwest. Canadian Journal of Fisheries and Aquatic Sciences 48(Supplement 1):124–133.

Waples, R. S. 1995. Evolutionarily significant units for Pacific salmon: lessons from the Columbia River. Pages 8–27 *in* J. L. Nielsen, editor. Evolution and the aquatic ecosystem: defining unique units in population conservation. American Fisheries Society, Symposium 17, Bethesda, Maryland.

WDFW (Washington Department of Fish and Wildlife). 1999. Final joint WDFW/tribal wild salmonid policy. *<http://www.wa.gov/wdfw/fish/wsp/joint/final/fwsptoc.htm>*.

Welch, D. W., Y. Ishada, and K. Nagasawa. 1998. Thermal limits and ocean migration of sockeye salmon (*Oncorhynchus nerka*): long-term consequences of global warming. Canadian Journal of Fisheries and Aquatic Sciences 55:937–948.

Wendler, H. O., and G. Deschamps. 1955. Logging dams on coastal Washington streams. Washington Department of Fisheries Research Papers 1, Seattle.

Wertheimer, A. 1988. Hooking mortality of chinook salmon released by commercial trollers. North American Journal of Fisheries Management 8:346–355.

Wertheimer, A. C. 1997. Status of Alaska salmon. Pages 179–197 *in* D. J. Stouder, P. A. Bisson, and R. J. Naiman, editors. Pacific salmon and their ecosystems: status and future options. Chapman and Hall, New York.

Williams, J. E., and C. D. Williams. 1997. An ecosystem-based approach to management of salmon and steelhead habitat. Pages 541–558 *in* D. J. Stouder, P. A. Bisson, and R. J. Naiman, editors. Pacific salmon and their ecosystems: status and future options. Chapman and Hall, New York.

Williams, R. N., P. A. Bisson, D. L. Bottom, L. D. Calvin, C. C. Coutant, M. W. Erho, Jr., C. A. Frissell, J. A. Lichatowich, W. J. Liss, W. E. McConnaha, P. R. Mundy, J. A. Stanford, and R. R. Whitney. 1999. Scientific issues in the restoration of salmonid fishes in the Columbia River. Fisheries 24(3):10–19.

Willson, M. F., and K. C. Halupka. 1995. Anadromous fish as keystone species in vertebrate communities. Conservation Biology 9(3):489–497.

Wipfli, M. S., J. P. Hudson, D. T. Chaloner, and J. P. Caouette. 1999. Influence of salmon spawner densities on stream productivity in Southeast Alaska. Canadian Journal of Fisheries and Aquatic Sciences 56:1600–1611.

Wissmar, R. C., and R. L. Beschta. 1998. Restoration and management of riparian ecosystems: a catchment perspective. Freshwater Biology 40:571–585.

Withler, F. C. 1982. Transplanting Pacific salmon. Canadian Technical Reports of Fisheries and Aquatic Sciences 1079.

Wood, C. C. 1995. Life history variation and population structure in sockeye salmon. Pages 195–216 *in* J. L. Nielsen, editor. Evolution and the aquatic ecosystem: defining unique units in population conservation. American Fisheries Society, Symposium 17, Bethesda, Maryland.

Wood, C. C., B. E. Riddell, and D. T. Rutherford. 1987. Alternative juvenile life histories of sockeye salmon (*Oncorhynchus nerka*) and their contribution to production in the Stikine River, northern British Columbia. Pages 12–24 *in* H. D. Smith, L. Margolis, and C. C. Wood, editors. Sockeye salmon (*Oncorhynchus nerka*) population biology and future management. Department of Fisheries and Oceans, Ottawa, Ontario.

Woody, C. A. 1998. Ecological, morphological, genetic, and life history comparison of two sockeye salmon populations, Tustumena Lake, Alaska. Doctoral thesis. University of Washington, Seattle.

Woody, C. A., J. Olsen, J. Reynolds, and P. Bentzen. 2000. Temporal variation in phenotypic and genotypic traits in two sockeye salmon populations, Tustumena Lake, Alaska. Transactions of the American Fisheries Society 129:1031–1043.

Wright, S. 1993. Fishery management of wild Pacific salmon to prevent extinctions. Fisheries 18(5):3–4.

Wunderlich, R. C., B. D. Winter, and J. H. Meyer. 1994. Restoration of the Elwha River ecosystem. Fisheries 19(8):11–19.

Sustaining North American Salmon: Perspectives Across Regions and Disciplines, 321–340

Chapter 15

Salmon, Society, and Politics: Moving Toward Ecosystem-Based Management for Pacific Salmon

Donald D. MacDonald, Megan Hanacek, and Laura Genn

Introduction

Salmon *Oncorhynchus* spp. and steelhead *O. mykiss* serve as powerful symbols of the quality of life that we enjoy in the Pacific Northwest and generate a wide range of economic, social, and cultural benefits in the region. Commercial fisheries contribute significantly to local, provincial/state, and national economies, both directly through the sale of fish and indirectly through the service and manufacturing industries. Likewise, sport fisheries provide diverse economic benefits through the tourist, manufacturing, and service sectors. First Nations (Canada) and tribal (U.S.) fisheries revitalize local economies, facilitate social stabilization and renewal, and help to maintain the rich cultural heritage that characterizes the Pacific Northwest. Because of the nature and extent of the benefits they bestow, ensuring the conservation and sustainability of Pacific salmonid resources should be both a national and an international priority.

Despite the vast resources that have been and continue to be spent on fisheries management each year, a widespread and marked decline in the size and number of salmonid populations in the Pacific Northwest has occurred during the past century. Numerous populations have been extirpated, and many of those remaining are at precariously low levels and continue to decline. Natural factors have undoubtedly played a role, but the primary causes of these declines are associated with human activities. The current status and continuing declines of many Pacific salmonid populations bear testimony to the ineffectiveness of past management practices and our inability to protect the fishery resource. The blame, as well as the responsibility for action, falls upon all of us—scientists, managers, resource users, and concerned citizens.

This book, *Sustaining North American Salmon: Perspectives Across Regions and Disciplines*, is intended to explore a variety of issues related to fisheries sustainability and to provide a series of policy recommendations for improving the prospects of North American salmon populations. This chapter is intended to support these objectives by discussing the social, political, and institutional aspects of Pacific salmon management and by offering a number of practical suggestions for restoring salmon and steelhead populations to sustainable levels. In developing these recommendations, we have relied heavily on the recommendations that were offered in the *Sustainable Fisheries Strategy* (SFF 1996) and the book *Sustainable Fisheries Management: Pacific Salmon* (Knudsen et al. 2000a). Both of these documents resulted from the Towards Sustainable Fisheries conference, which was convened by the Sustainable Fisheries Foundation and its partners in 1996 and attended by more than 500 delegates representing diverse salmon-related interest groups (e.g., First Nations, government agencies, stewardship groups, industry, fishers, landowners, and concerned citizens).

Overview of the Current Status of Pacific Salmon

Information from numerous sources indicates that Pacific salmon and steelhead populations have undergone widespread declines in recent years. In the landmark paper "Pacific Salmon at the Crossroads," Nehlsen et al. (1991) reviewed the status of salmon, steelhead, and sea-run cutthroat trout *O. clarki clarki* stocks in Washington, Idaho, Oregon, and California. The results of this review indicated that 106 stocks were extinct in these four states. A further 101 stocks were at high risk of extinction, while 58 stocks and 54 stocks were at moderate risk of extinction and of special concern, respectively. As the first broad-scale evaluation of the status of Pacific salmonid populations, this review confirmed what many biologists had feared for some time—that salmonid populations had undergone marked declines throughout the southern portions of their ranges.

"Pacific Salmon at the Crossroads" sparked a number of subsequent evaluations to further assess the status of Pacific salmon populations. Williams et al. (1992) reviewed the status of salmonid populations in the Columbia River Basin using the data presented by Nehlsen et al. (1991) and the Columbia Basin Fish and Wildlife Authority (CBFWA 1990). The results of that study revealed that, of the 192 anadromous salmonid stocks known to occur in the basin, 35% were extinct, 19% were at high risk of extinction, 7% were at moderate risk of extinction, and 13% were of special concern. Only 26% of the populations were considered to be secure. Overall, these results show that nearly three quarters of the stocks in the Columbia River Basin have been extirpated or are considered to be at some level of risk (Williams 2000). Similarly, Thurow et al. (2000) reported that chinook salmon *O. Tshawytscha* and steelhead stocks in the interior Columbia River Basin are extinct in more than 50% of their historic ranges.

Unfortunately, the situation does not appear to be much better in British Columbia. A review conducted by the American Fisheries Society (AFS) indicates that nearly 20% of the 5,487 stocks of salmon and steelhead for which data were available—including all large, commercially important stocks—were either at high risk of extinction (624), moderate risk of extinction (78), of special concern (230), or extirpated during the 20th century (142; Slaney et al. 1996). Insufficient data were available to evaluate the status of more than 40% of the stocks, which impaired AFS's ability to conduct an adequate review of the current diversity and overall status of salmonids in the province. Since these evaluations were completed, it has been determined that coho salmon *O. kisutch* populations in the Fraser and Skeena rivers and in the streams on the east coast of Vancouver Island are also severely depressed (Fisheries and Oceans Canada 1998a). This determination necessitated the elimination of all targeted fisheries on these stocks and the implementation of measures to minimize by-catch of this species in other fisheries (Fisheries and Oceans Canada 1998a).

Recent increases in the number of salmon populations afforded protection under the provisions of the U.S. Endangered Species Act (ESA) emphasize the fact that the situation has reached critical proportions. As of July 1999, 23 evolutionary significant units (ESUs) in California, Idaho, Oregon, and Washington had been listed under the ESA. Several other steelhead trout, chinook salmon, and coho salmon ESUs have been identified as candidates for future listing (Knudsen et al. 2000b). Following the passage of species-at-risk legislation in Canada (now pending), it is likely that a number of salmon and steelhead populations in British Columbia and/or the Yukon will also be listed for protection.

Overall, the most current information available strongly indicates that west coast salmonids are under a great deal of stress and that many stocks have already undergone serious declines as a result of this pressure. It is evident that the natural resiliency of salmonids is being overwhelmed by anthropogenic disturbances of a kind, magnitude, and frequency unlike any that have occurred during the species' evolutionary pasts. Without deliberate action to restore natural populations and their habitats to their former states, many salmon and steelhead populations will undoubtedly continue to advance toward extinction.

Factors Affecting the Sustainability of Pacific Salmon

A wide variety of factors have affected and continue to adversely affect the sustainability of Pacific salmon and steelhead populations. Habitat loss and degradation, harvest management, large-scale hatchery operations, and hydroelectric development are critically important factors that have had severe impacts on the health of Pacific salmon populations and their habitats. Several other factors must be considered, however, in the development of management strategies for protecting and restoring salmon and steelhead populations, including a variety of social, institutional, and political factors. Effectively addressing these factors will require a fundamental shift in how Pacific salmon and their habitats are managed in the Pacific Northwest.

Social Factors

The importance of Pacific salmon and steelhead to the people of the Northwest cannot be overestimated. For thousands of years, indigenous peoples have coexisted with Pacific salmonids and recognized the resource as intrinsic to the well-being of their communities. Traditionally, salmon provided an essential and primary source of food for native bands all along the western coastline of North America, from Monterey to the Aleutian Islands. Similarly, salmonids were critical to the survival of First Nations groups located in interior areas throughout the Northwest. These groups relied heavily upon smoked and dried salmon as an integral part of their food supply throughout the winter when other provisions were scarce. Because of this traditional reliance on salmonid resources, it is not surprising that salmon are revered by Canada's First Nations and Native American tribes. Even today, salmon continue to be a powerful cultural icon for these people and many others who reside in this region.

Pacific salmon were also of fundamental importance to North Americans of European descent during their colonization of the western regions of the continent over the past 200 years. According to Meggs (1991, p. 1–2):

> Europeans walked into British Columbia on the backs of salmon. Had the native people of the Interior not provided the fur trader and explorer, Simon Fraser, with fish, he likely would have died on his way to the Pacific. When the fur trade proved mediocre, the Hudson's Bay Company (HBC) turned to the salt salmon trade and reaped monopoly profits. When gold was discovered glittering in the spawning beds of the Fraser, the HBC purchased

> enormous supplies of salmon from native people to feed the miners whose rush to the gravel bars transformed the province. When those fortunes from the gold rush looked for another source of easy cash, the salmon canning trade offered itself as a ticket to wealth.

In many ways, the availability of the salmon resource made it possible for nonnative peoples to survive and prosper in the Northwest.

During the past 150 years, both native and nonnative peoples have enjoyed a tremendous array of economic, social, and cultural benefits due to the natural abundance of Pacific salmon in the region. While commercial fishing and canning interests have certainly reaped the greatest bounty, individuals and small corporations in coastal communities have also greatly benefited through their access to this resource. Unfortunately, the profligate harvest of salmonid resources in the latter decades of the 20th century, coupled with the degradation of freshwater and estuarine habitats, has generally left the resource in very poor shape. Not surprisingly, the people who have traditionally relied on this resource have been severely affected by reductions in the availability of Pacific salmon.

Today, numerous social factors influence the potential to effectively implement sustainable management strategies for Pacific Northwest salmonids. One of the most important of these factors is the value system upon which management decisions are based. The dominant value systems in place within any society are necessarily reflected in the cultural, social, economic, and political decisions that are made. Ultimately, our societies' value systems dictate how we choose to manage our natural resources and what we are trying to achieve through such management.

In the Pacific Northwest, the predominant societal values have largely been focused on promoting economic and population growth through the exploitation of natural resources. As we face the current situation in fisheries management, we must recognize that salmon recovery and long-term sustainability are unlikely to be achieved unless management strategies progress beyond a strictly utilitarian view of natural resources. This strategic shift will necessarily require enhanced public education and stakeholder involvement in resource management processes. The increased willingness of people to get involved in local management initiatives and stewardship activities strongly suggests that societal values are evolving in a way that is more likely to support fisheries sustainability. However, this transition will have the desired impact only if such values are ultimately reflected in the policies and programs that are developed and implemented by those that are entrusted with the management of our shared fisheries resources.

Political Factors

Political factors have considerable potential to influence fisheries sustainability, both positively and negatively. Anadromous fisheries resources, unlike forests or minerals, are migratory organisms that do not respect anthropogenically defined political boundaries. For this reason, the politics of fisheries management has become increasingly complicated as populations of salmon and other commercially important fish species are becoming depleted. The migratory behavior of Pacific salmonids represents one of the greatest political challenges affecting their management and, ultimately, their survival in the 21st century.

Both Canada and the United States have recognized the importance of developing cooperative agreements for managing transboundary fish stocks. Negotiations on a Pacific Salmon Treaty (PST) began in the 1960s, and an agreement was finally reached between Canada and the United States in 1985 (Huppert 1996). This agreement was based on two fundamental principles: conservation and equity. Although the agreement set limits on the number of fish caught, these limits were set too high to allow for rebuilding of those stocks that had already been depleted through overfishing. As a result, major declines in chinook and coho stocks occurred in Oregon, Washington, and southern British Columbia during the early 1990s (Trout Unlimited and Trout Unlimited Canada 1999). These declines were attributed to a variety of causes, including habitat destruction, overfishing, and changing ocean conditions.

The original PST expired in 1992. With the expiration of the agreement, both governments faced the prospect of managing transboundary fisheries resources in the absence of an effective binational framework. Inevitably, a dispute erupted between Canada and the United States in 1996 over fishing rights to the migratory salmon stocks of the Pacific Northwest (Strangway and Ruckelshaus 1998). Canadian fishers found that they were catching significantly fewer fish of American origin than they had when the original PST was signed. In addition, U.S. fishers were catching relatively more Canadian fish, particularly in Alaska. In an attempt to pressure the United States to make concessions on harvests of Canadian fish, Canadian fishers continued to overharvest the declining coho and chinook stocks of U.S. origin. Unfortunately, this politically motivated overharvesting became a significant factor contributing to the eventual collapse of certain stocks (Trout Unlimited and Trout Unlimited Canada 1999).

After many failed attempts to resolve the on-going dispute between Canada and the United States, a new agreement was reached in 1999. This comprehensive agreement includes four important elements (Fisheries and Oceans Canada 1999):

- The establishment of long-term fishing arrangements governing transboundary fisheries and waterways that are founded on an abundance-based management framework that increases sensitivity to conservation requirements.
- The establishment of two Pacific Salmon Treaty endowment funds (one for the north and one for the south) to be administered jointly by the two countries for investment in stock enhancement, habitat, science, and salmon initiatives.
- Strengthened institutional arrangements for cooperation among U.S. and Canadian fisheries managers and scientists.
- A formal commitment by both countries to protect and restore salmon habitat.

The new agreement has been seen as a cooperative, conservation-based approach to the management of Pacific salmon fisheries and one that will provide a more equitable sharing of the resource between Canada and the United States (Fisheries and Oceans Canada 1999). Attaining such an agreement definitely signals a much-needed political shift toward conservation-based management and perhaps reflects a shift in social values that is starting to infiltrate our political institutions. Unfortunately, this shift took place under the threat of decimation of many Pacific salmon stocks and not as a precautionary measure to assure fisheries sustainability.

Trade agreements and policies, such as those implemented under the World Trade Organization (WTO) and the Canada-U.S. Free Trade Agreement, represent another important example of the negative impacts that political institutions and their decision makers can have on the management of natural resources. By design, trade agreements serve to define the fundamental rights of their constituents. These agreements are most often negotiated behind closed doors by government and business interests (West Coast Environmental Law 1999). Such negotiations often result in agreements that satisfy the economic interests of corporations but provide few provisions for addressing other high priority societal needs, such as pollution abatement, maintenance of global biodiversity, and ensuring the sustainability of natural resources.

Because 135 countries currently belong to the WTO, this organization represents a major step toward the establishment of a global economic order (West Coast Environmental Law 1999). The WTO operates under general definitions of trade, encompassing not only traditional goods but also services, property rights, and domestic regulations. The WTO has strong regulatory powers with which all participating nations must comply or face retaliatory trade sanctions. For example, one of the first trade complaints to be resolved under the WTO ended with the conclusion that U.S. Clean Air Act regulations violated WTO rules, and the United States was left to choose between removing the offending provisions of the regulations or face sanctions on the order of US$150 million a year (West Coast Environmental Law 1999). When faced with such choices, it will be difficult for most countries to maintain their commitment to natural resource sustainability.

Future decisions by the WTO have the potential to adversely affect both fishing communities and fish stocks (Menotti and Grader 1999). For example, the WTO is currently considering a proposal to eliminate all tariffs on fish products by 2005 (Menotti and Grader 1999). Adoption of this proposal could seriously affect fishing communities by letting cheap imported seafood products flood domestic markets. This, in turn, may create pressure to lower existing conservation standards to compete in the marketplace. Additionally, lower market prices for certain fish and fish products could have the effect of stimulating consumption of already declining stocks and populations.

Elimination of unfair subsidies is another issue that is likely to be addressed by the WTO and warrants the attention of fishing communities. Elimination of certain subsidies, such as those that are currently being offered to fishers under the Pacific Fisheries Adjustment and Restructuring Program (i.e., the Community Economic Development and Adjustment Initiative: Vessel Tie-Up), could lead to a reduction in fishing pressure and, hence, a temporary respite for threatened fish stocks. However, government-funded restoration and enhancement programs are also commonly viewed as subsidies and, as such, could be eliminated to comply with WTO decisions (Menotti and Grader 1999). WTO decisions to address such subsidies compromise the long-term sustainability of Pacific salmon. Unfortunately, these issues receive little media coverage and, as a result, those individuals and communities that could be affected by pending resolutions have limited knowledge of the potential impacts of such decisions (Menotti and Grader 1999).

Another example of how political factors can influence the management of natural resources is the salvage logging rider that was signed into law by President Clinton in 1995 as an adjunct to an emergency appropriations bill that provided funds for the victims of the Oklahoma City bombing (Gorte 1996a). The rider was ostensibly written to permit salvage logging of diseased timber in the interest of maintaining forest health. In reality, the legislation is intended to facilitate increased logging, including clearcutting, in old-growth forests that had been set aside to protect threatened and endangered species such as the spotted owl. Many people have called for the repeal of this rider because they believe that it overrides environmental laws, destroys wildlife habitat, increases erosion, severely affects rivers and streams, and accelerates the logging of healthy, ancient forests (Gorte 1996a). The incremental impacts of salvage logging are likely to have serious implications relative to the recovery of salmon and other endangered species (Gorte 1996b).

Political forces can seriously affect our ability to effectively implement conservation-based strategies, but they can also serve to support the implementation of such strategies when policies are developed that reflect ecosystem-based goals and objectives. Within the past five years, overwhelming evidence has come to light that coho salmon stocks in British Columbia are declining in abundance at alarming rates and that several stocks are at extreme risk (Fisheries and Oceans Canada 1998b). In response to this management challenge, David Anderson, then minister of fisheries and oceans, demonstrated something that is uncommonly exhibited by politicians—leadership in the environmental management arena. In 1998, Minister Anderson announced a comprehensive plan to rebuild the fishery resource in British Columbia (Fisheries and Oceans Canada 1998b). This management plan, which is supported by Can$400 million in federal funding, was intended to introduce a fundamentally new direction for the management of the Pacific salmon fishery, including the introduction of selective, conservation-based fishing techniques to conserve coho and other stocks at risk. A significant portion of the funding (25%) was to go toward measures for protecting and rebuilding habitat (Fisheries and Oceans Canada 1998b). At the same time and in the absence of a PST, Mr. Anderson negotiated an agreement-in-principle between Canada and Washington State to conserve threatened salmon stocks in both regions, with Washington agreeing to implement measures that would achieve a 22% reduction in its 1998 catch of coho salmon bound for the Upper Thompson River in British Columbia (Fisheries and Oceans Canada 1998c).

Another positive example of political intervention occurred on the U.S. side of the border in 1999. In this case, the U.S. Forest Service implemented a temporary moratorium on roadbuilding in roadless areas (i.e., federal lands with no roads in or across them; Gorte 1999). Debates on this issue have largely focused on national forests, which have been traditionally managed for multiple uses. Although the legislation was estimated to affect only 2% to 6% of the annual Forest Service timber program, timber interests vigorously opposed it. Opponents cited the inevitability of roadbuilding in these areas and the near-term loss of timber revenues as the primary reasons for their antithetical view of this legislation. Nevertheless, the Clinton administration pressed forward with this initiative, which has the potential to provide important protection for salmonid habitats, particularly in Alaska.

These examples demonstrate that political decisions can have dramatic effects on the management of Pacific salmon and the habitats upon which they depend. Although political solutions could, potentially, solve the challenges that we are currently facing, trust in politicians and governments has eroded to such a degree that even good policies are unlikely to receive the public support that is needed to move toward implementation. When trust in government institutions erodes to this extent, two solutions are possible: one is dictatorship; the other is more democracy. We are suggesting the second alternative, more democracy. This solution requires that society and the government agencies charged with the management of natural resources work together in shared decision-making processes (i.e., developing a shared vision for the future, establishing common policies, and building collaborative management systems). These types of processes are best implemented at the watershed level by moving toward ecosystem-based management. Again, education and involvement of communities in natural resource management processes will greatly increase the potential to realize the long-term conservation and sustainable management of Pacific salmonids.

Institutional Factors

A wide variety of institutions and organizations are involved in fisheries and environmental management processes. In fact, the existing management framework is so complicated that even those involved in the process often find it difficult to fully understand. This complexity in the management of Pacific salmonids and their associated habitats has arisen for several reasons. Our existing management structures have evolved during a period of more than a hundred years, over which time our understanding of the resource and its interactions with people and the environment has increased dramatically. However, rather than implementing a holistic, ecosystem-based approach to management, we have tended to compartmentalize the environment and establish agencies that assume the primary responsibility for managing one or more compartments (i.e., fish, water, forests, urban developments, etc.).

Although most of the existing institutions have adequately fulfilled their stated mandates, the current status of salmon and steelhead populations and their associated habitats suggests that the existing top-down management structure does not adequately respond to the challenges that we are currently facing. One need look only at the precipitous and widespread declines of salmon populations in the Northwest to conclude that, together, those institutions responsible for their management have failed to protect them. Although many factors have contributed to the decline (e.g., overfishing, habitat degradation, water quality degradation, etc.), the overriding cause has been our failure to effectively manage human activities and their impacts on the natural environment.

Resource agencies have made valiant attempts to rectify the shortcomings of the existing management structure, but they are often constrained by the lack of a clear mandate, insufficient funding, and chronic understaffing.

When changes in policy and management have occurred, they have often been incremental, rarely deviating sufficiently from the status quo to correct the problem at hand. There has been a lack of consistency, over time and between agencies, in setting and implementing management priorities. Furthermore, most agencies are funded separately and operate with separate agendas. The result is uncoordinated and piecemeal management. Lacking clear policy direction and incentives for meaningful change, agencies have often found it difficult to establish and enforce regulations.

The lack of consistency and effectiveness among the institutions governing natural resource management in the Pacific Northwest stems in part from the large number of jurisdictions involved. Two federal governments, each represented by a host of agencies; at least six state and provincial governments; a plethora of local jurisdictions; and numerous Indian tribes and First Nations all have some responsibility for the management of fisheries resources, and many have unique approaches. Given the number and diversity of jurisdictions and institutions involved, there is little wonder that efforts to protect salmon have been thwarted. Despite efforts to cooperate and achieve consensus among themselves, our institutions have not developed the collective ability to manage these resources effectively.

Technological and financial fixes have been applied successfully to many other societal problems, but this approach has failed to solve difficult environmental management problems. Already, substantial effort and untold billions of dollars have been devoted to salmon recovery. For example, it has been estimated that during the period 1996–2001, the United States will have spent US$535 million to $700 million annually on salmon recovery projects in the Columbia River basin alone (Marvin Shaffer and Associates 1998). A lot of these resources have been applied to short-term fixes without treating the root causes of the problem. Not surprisingly, this investment of time and money has paid poor dividends—the stocks continue to decline. Therefore, institutional and regulatory structures must be reformed if we are to reverse the current trends and restore populations of Pacific salmonids to historic levels.

In the past, there have been numerous attempts to redefine the mandates and activities of our public institutions. These attempts have been largely ineffective because they failed to create effective resource management structures. In recognition that the watershed is the basic biological production unit for Pacific salmon, future institutional structures must be established to facilitate the management of human activities at this scale. For this reason, sustainable fisheries and watershed management will necessitate the development of institutional structures that make room for government agencies, First Nations, nongovernmental organizations, and other stakeholders in decision-making processes.

Moving Toward Ecosystem-Based Management

Humans have maintained a bond with Pacific salmon for thousands of years. Since the earliest residents of the Northwest crossed the Siberian land bridge, salmon have been an integral element of the economy, society, and culture that developed on the west coast of North America. Significant social and economic changes have occurred since Europeans arrived some 200 years ago, but the importance of salmon to the region remains essentially undiminished. In many ways, salmon continue to define the Pacific Northwest and the people who inhabit it, both in culture and societal values. One has difficulty considering the plight of the salmon without experiencing a personal sense of loss; our fate seems inextricably linked to theirs. The extinction of even one stock of salmon or steelhead diminishes our stature as humans and stewards of the resource.

People throughout the Northwest share a common desire to see salmon and steelhead populations return to health. This overwhelming concern over the fate of the salmon has given rise to a plethora of salmon protection, enhancement, and recovery plans and initiatives. In spite of our best efforts, we have not succeeded and are unlikely to succeed until these disparate efforts are effectively integrated into an ecosystem-based framework that encompasses the entire west coast. Implementation of such an approach requires the development of an integrated set of policies and practices that place human activities in an ecosystem context, instead of the external resources or environments with which they interact (Christie et al. 1986; Vallentyne and Beeton 1988). That is, the preferred approach is one that views salmon and people as inextricably linked and recognizes that decisions affecting the well-being of the salmon will also influence our economy, culture, and communities, both now and in the future.

The ecosystem approach is not a new concept and does not hinge on any one program, definition, or action. The Canadian Council of Ministers of the Environment (CCME) has defined the ecosystem approach as a geographically comprehensive approach to environmental planning and management that recognizes the interrelated nature of environmental media and humans as key components of ecological systems (CCME 1996). It places equal emphasis on concerns related to the environment, the economy, and the community. This approach recognizes that it is human activities rather than the ecosystem itself that must be managed (Environment Canada 1996; Thomas et al. 1988). Adopting the ecosystem approach means viewing the basic components of the ecosystem (i.e., air, water, land, and biota) and its functions in a broad context—one that effectively integrates environmental, social, and economic interests into a decision-making framework that embraces the concept of sustainability (Figure 1). In turn, this expanded view of the ecosystem helps shape the planning, research, and

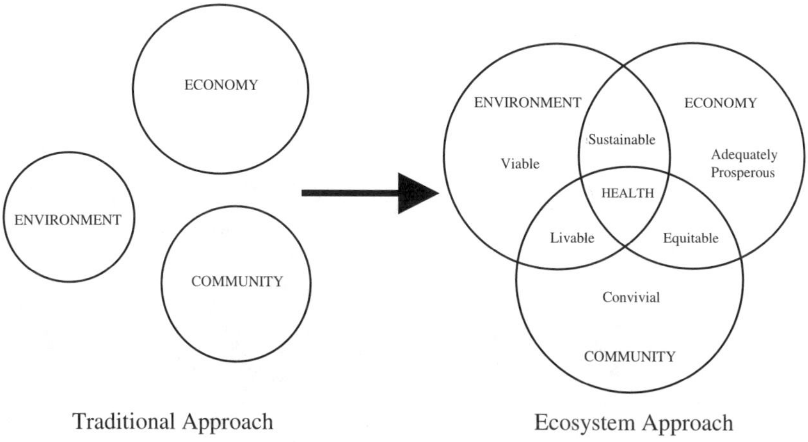

Figure 1. The shift from traditional to ecosystem-based decision making (from CCME 1996). Relationships within ecosystems can best be visualized as three interlocking circles: environment, economy, and community. Traditionally, most decision making separates these three components, with little understanding (or even heed), for example, of the effects of economic decisions on community needs or the environment. The challenge now is twofold: to understand the links between these components and to redress the balance among them. The ecosystem approach requires an equal and integrated consideration of these elements.

management decisions that are made collectively by government agencies, First Nations, industry, and other stakeholders.

The CCME has developed a framework for ecosystem-based management based on the lessons that have been learned under a number of ecosystem initiatives (CCME 1996). This framework describes a four-step process to support informed decision making (Figure 2; Environment Canada 1996). The first step in the process involves the identification and assessment of the key issues and collation of the existing ecosystem knowledge base, with a focus on ecosystem science, socioeconomic information, historical resource use patterns, and traditional knowledge. In the second step of the process, stakeholders utilize the common knowledge base to support the formulation of a common vision for the future of the ecosystem. In turn, this long-term vision supports the development of ecosystem goals and more specific ecosystem objectives that clearly articulate societal values relative to the management of natural resources (e.g., through consultation with competing users of ecosystem resources). The third step is to select a suite of ecosystem health indicators that provide information on the status of the ecosystem as a whole and can be used to track progress toward the

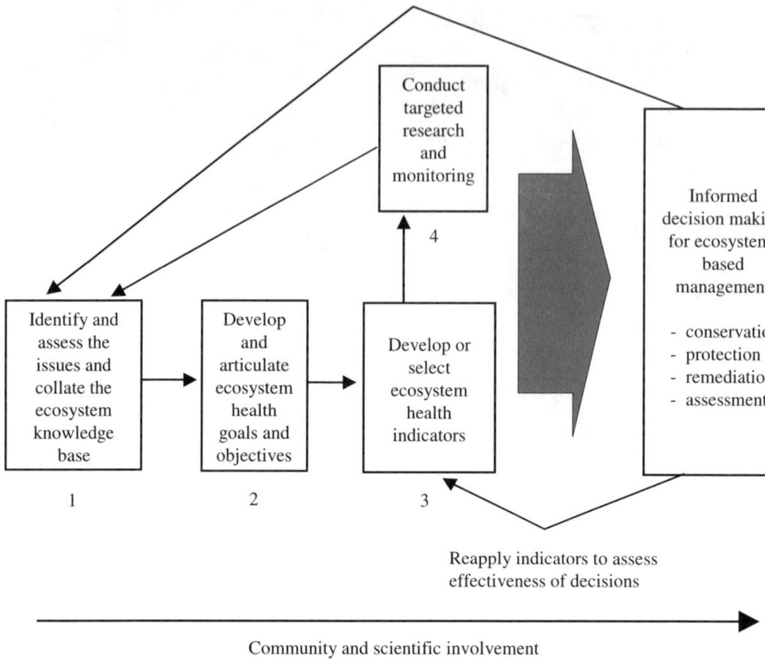

Figure 2. A framework for ecosystem-based management (from Environment Canada 1996).

ecosystem goals and objectives. Because salmon are keystone species in many ecosystems throughout their range, the key indicators of ecosystem health are likely to provide information that is directly relevant for assessing the status of salmon populations. Such indicators are supported by metrics, which identify the variables that need to be measured to assess the status of each indicator, and targets, which define the acceptable ranges for each metric. In the final step of the process, the data gaps and information requirements that are identified during steps 1–3 are addressed through the design and implementation of directed research and monitoring programs. Implementation of these four basic steps provides the information and guidance needed to make informed decisions regarding the management of natural resources.

This framework is intended to support sound management decisions that help to protect, maintain, restore, and enhance ecosystem health. To be effective, however, decision-making processes must be accessible to all stakeholders, including communities of place (i.e., encompassing the residents within a watershed, including individual First Nations, community stewardship groups, and the public) and communities of interest (i.e., encompassing those groups with a legitimate interest in the watershed, including government agencies, nongovernment organizations, and others). Indeed, full implementation of ecosystem-based management necessitates the development of cooperative decision-making processes in which all interested stakeholders participate as partners. The formation and empowerment of watershed and basin councils can facilitate the transition to ecosystem-based management by creating structures within which stakeholders can identify and pursue common interests. The development of watershed management plans and salmon recovery plans represent two important vehicles for sharing authority, responsibility, and accountability for decision making among the partners. Importantly, these processes provide a basis for clearly describing the actions that should be taken to achieve the ecosystem goals and objectives and identifying the participants who are responsible for implementing each initiative.

The framework for ecosystem-based management should be viewed as a dynamic process in which new information from research and monitoring programs continually feeds into an expanding knowledge base. Ecosystem goals, objectives, and indicators are refined as societal values change or new information becomes available. Importantly, the status of the indicators needs to be reevaluated on a regular basis following the implementation of management actions to assess their efficacy in advancing the prospects for fisheries sustainability (i.e., achieving the ecosystem goals and objectives).

The ecosystem approach provides an effective basis for managing human activities in watersheds that support Pacific salmon. However, it should not be viewed as the ultimate solution to the challenges that we are currently facing in the Pacific Northwest. The life history patterns of anadromous salmon and steelhead and the existence of ocean fisheries necessitate coordination of management efforts at the basin- and coast-wide levels. The formation and empowerment of major basin councils and full implementation of the Pacific Salmon Treaty will help support the transition to ecosystem-based management of Pacific salmon. It is likely that the strategies and principles that underlie the ecosystem approach can also be applied to these broader processes to enhance their prospects for success. (For more information on the implementation of ecosystem-based management at the watershed and basin scales, the reader is directed to Harris et al. [1987], RCFTW [1991, 1992], MacDonald [1995, 1997, 1999], SFF [1996], Crane et al. [2000] and Knudsen et al. [2000a].)

Conclusions and Recommendations

The participants of the Towards Sustainable Fisheries conference developed a long-term vision for the future that can be shared by residents throughout the Northwest. This vision includes healthy, diverse, and productive ecosystems; viable aboriginal, sport, and commercial fisheries; and vital and stable communities throughout the historical range of Pacific salmonids. Realization of this vision will necessarily require a transition to a sustainable fisheries management regime in the Pacific Northwest—one that effectively integrates social, cultural, economic, and ecological values within a decision-making framework that is accessible to all stakeholders.

As new approaches to managing fisheries resources are being explored to address the issues of declining populations and deteriorating habitats, it is essential that the roles of communities of place and communities of interest are redefined so that all stakeholders are afforded meaningful opportunities to participate in the Pacific salmon management process. Ecosystem-based management provides a framework for developing and implementing cooperative management strategies by fostering the development of a shared vision for the future, the establishment of long-term ecosystem goals and objectives to guide management decisions, and the selection of ecosystem health indicators that can be used to track progress toward fisheries sustainability. Because these tools are developed collaboratively by all interested stakeholders and are integrally linked to ecosystem science, they provide both the moral authority and the scientific credibility to effectively guide resource management initiatives. Furthermore, incorporation of these tools into watershed management plans and salmon recovery programs provides relevant mechanisms for translating societal values into effective management actions.

Clearly, we will face a number of challenges in our journey toward fisheries sustainability. There are innumerable social, cultural, economic, and political factors that will conspire to thwart such efforts. But, in the words of William D. Ruckelshaus, former USEPA administrator, "While our work is cut out for us, we cannot give up. After all, we human beings are supposed to be the most intelligent species on the planet. Collectively, we have put in jeopardy other species that are completely dependent on us for survival. Shame on us. Maybe we are not as intelligent as we thought. Now is the time for us to get on with proving that we are." In our opinion, ecosystem-based management is likely to provide our best chance at success.

References

CBFWA (Columbia Basin Fish and Wildlife Authority). 1990. Integrated system plan for salmon and steelhead production in the Columbia River Basin. Columbia Basin System Planning Report 91–16, Portland, Oregon.

CCME (Canadian Council of Ministers of the Environment). 1996. A framework for developing ecosystem health goals, objectives, and indicators: tools for ecosystem-based management. Produced by the Water Quality Guidelines Task Force (WQGTG) of the CCME, Winnipeg, Manitoba.

Christie, W. J., M. Becker, J. W. Cowden, and J. R. Vallentyne. 1986. Managing the Great Lakes Basin as a home. Journal of Great Lakes Research 12:2–17.

Crane, J. L., D. D. MacDonald, C. G. Ingersoll, D. E. Smorong, R. A. Lindskoog, C. G. Severn, T. A. Berger, and L. J. Field. 2000. Development of a framework for evaluating numerical sediment quality targets and sediment contamination in the St. Louis River area of concern. Final report. EPA 905-R-00-008. United States Environmental Protection Agency, Chicago, Illinois.

Environment Canada. 1996. The ecosystem approach: getting beyond the rhetoric. ISBN: 0-662-62473-4. Prepared for the Environment Canada Long Term Strategic Plan for Ecosystem Initiatives, Ottawa, Ontario.

Fisheries and Oceans Canada. 1998a. Fisheries minister urges strong coho conservation measures. Fisheries and Oceans Canada news release, May 11, 1998. NR-HQ-98-22E. Ottawa, Ontario.

Fisheries and Oceans Canada. 1998b. Ministers announce Canada's coho recovery plan and $400 million for Pacific salmon fishery. Fisheries and Oceans Canada news release, June 19, 1998. Vancouver, British Columbia.

Fisheries and Oceans Canada. 1998c. Agreement reached on southern coho and chinook. Fisheries and Oceans Canada news release, June 26, 1998. NR-HQ-98-36E. Vancouver, British Columbia.

Fisheries and Oceans Canada. 1999. Canada and the U.S. reach a comprehensive agreement under the Pacific Salmon Treaty. Fisheries and Oceans Canada news release, June 3, 1999. NR-HQ099-29. Ottawa, Ontario.

Gorte, R. W. 1996a. The salvage timber sale rider: overview and policy issues. Congressional Research Service Report for Congress, No. 95-569 ENR. The Committee for the National Institute for the Environment, Washington D.C.

Gorte, R. W. 1996b. Salvage timber sales and forest health. Congressional Research Services Report for Congress, No. 95-364 ENR. The Committee for the National Institute for the Environment, Washington, D.C.

Gorte, R. W. 1999. Roadless Areas: the administration's moratorium. CRS Issue Brief for Congress. Available through the Committee for the National Institute for the Environment, 1725 K Street, NW, Suite 212, Washington, D.C. 20006.

Harris, H. J., P. E. Sager, S. Richman, V. A. Harris, and C. J. Yarborough. 1987. Coupling ecosystem science with management: a Great Lakes perspective from Green Bay, Lake Michigan, USA. Environmental Management 11:619–625.

Huppert, D. 1996. U.S./Canada salmon wars: why the Pacific Salmon Treaty has not brought peace. New directions in marine affairs. Report No. 1. January 1996. School of Maine Affairs, University of Washington, Seattle.

Knudsen, E. E., C. R. Steward, D. D. MacDonald, D. W. Reiser, and J. E. Williams, editors. 2000a. Sustainable fisheries management: Pacific salmon. Lewis Publishers, Boca Raton, Florida.

Knudsen, E. E., D. D. MacDonald, and C. R. Steward. 2000b. Setting the stage for a sustainable Pacific salmon fisheries strategy. Pages 3–13 *in* E. E. Knudsen, C. R. Steward, D. D. MacDonald, D. W. Reiser, and J. E. Williams, editors. Sustainable fisheries management: Pacific salmon. Lewis Publishers, Boca Raton, Florida.

MacDonald, D. D. 1995. Science advisory group on sediment assessment in Tampa Bay. Summary report. Technical Publication #06-95. Tampa Bay National Estuary Program, St. Petersburg, Florida.

MacDonald, D. D. 1997. Tampa Bay sediment quality workshop: setting targets and defining management strategies. Final summary report. Tampa Bay National Estuary Program, St. Petersburg, Florida.

MacDonald, D. D. 1999. Tampa Bay sediment quality workshop: establishing impact levels and setting sediment quality targets. Workshop summary report. Tampa Bay Estuary Program, St. Petersburg, Florida.

Marvin Shaffer and Associates Ltd. 1998. Pacific Northwest Salmon Recovery Efforts and the Pacific Salmon Treaty. Prepared for the Departments of Fisheries and Oceans and Foreign Affairs and International Trade of the Government of Canada. White Rock, British Columbia.

Meggs, G. 1991. Salmon: the decline of the British Columbia fishery. Douglas & McIntyre, Vancouver, British Columbia.

Menotti, V., and Z. Grader. 1999. The World Trade Organization (WTO): flying under fishermen's radar. The fishing industry confronts world trade issues. The Pacific Coast Federation of Fishermen's Associations. Fishermen's News. November 1999.

Nehlsen, W., J. E. Williams, and J. A. Lichatowich. 1991. Pacific salmon at the crossroads: stocks at risk from California, Oregon, Idaho, and Washington. Fisheries 16(2):4–21.

RCFTW (Royal Commission on the Future of the Toronto Waterfront). 1991. East Bay front and Port Industrial Area: pathways toward an ecosystem approach. Royal Commission on the Future of Toronto Waterfront, Toronto, Ontario.

RCFTW (Royal Commission on the Future of the Toronto Waterfront). 1992. Regeneration: Toronto's waterfront and the sustainable city. Final report. Royal Commission on the Future of Toronto Waterfront, Toronto, Ontario.

SFF (Sustainable Fisheries Foundation). 1996. Towards sustainable fisheries: building a cooperative strategy for balancing the conservation and use of west coast salmon and steelhead populations. Sustainable Fisheries Foundation, Nanaimo, British Columbia.

Slaney, T. L., K. D. Hyatt, T. G. Northcote, and R. J. Fielden. 1996. Status of anadromous salmon and trout in British Columbia and Yukon. Fisheries: 21(10):20–35.

Strangway, D. W., and W. D. Ruckelshaus. 1998. Pacific salmon: report to the President of the United States and the Prime Minister of Canada. January 12, 1998. Department of Fisheries and Oceans, Ottawa, Ontario.

Thomas, R. L., J. R. Vallentyne, K. Olgilvie, and J. D. Kingman. 1988. The ecosystems approach: a strategy for the management of renewable resources in the Great Lakes Basin. Pages 31–59 *in* L. Caldwell, editor. Perspectives on ecosystem management for the Great Lakes. State University of New York Press, Albany, New York.

Thurow, R. F., D. C. Lee, and B. E. Rieman. 2000. Status and distribution of chinook salmon and steelhead in the interior Columbia River Basin, and portions of the Klamath River Basin. Pages 133–160 *in* E. E. Knudsen, C. R. Steward, D. D. MacDonald, D. W. Reiser, and J. E. Williams, editors. Sustainable fisheries management: Pacific salmon. Lewis Publishers, Boca Raton, Florida.

Trout Unlimited and Trout Unlimited Canada. 1999. Resolving the Pacific Salmon Treaty stalemate. Trout Unlimited USA, Seattle, Washington.

Vallentyne, J. R., and A. M. Beeton. 1988. The "ecosystem" approach to managing human uses and abuses of natural resources in the Great Lakes Basin. Environmental Conservation 15:58–62.

West Coast Environmental Law. 1999. The World Trade Organization: a guide for environmentalists. Common Front on the World Trade Organization, Vancouver, British Columbia.

Williams, J. E., J. A. Lichatowich, and W. Nehlsen. 1992. Declining salmon and steelhead populations: new endangered species concerns for the West. Endangered Species Update 9(4):1–8.

Williams, J. E. 2000. The status of anadromous salmonids: lessons in our search for sustainability. Pages 95–102 *in* E. E. Knudsen, C. R. Steward, D. D. MacDonald, D. W. Reiser, and J. E. Williams, editors. Sustainable fisheries management: Pacific salmon. Lewis Publishers, Boca Raton, Florida.

Sustaining North American Salmon: Perspectives Across Regions and Disciplines, 341–368

Chapter 16

The Pacific Salmon Dispute: Rationalizing a Dysfunctional Joint Venture

Robert McKelvey and Kathleen A. Miller

Introduction

On 30 June 1999, the Canadian and U.S. governments signed a new Pacific salmon treaty that they hope will end a period of crisis and disarray in the efforts of the two nations to coordinate management of their North Pacific salmon fisheries. The recent turmoil was part of the long and rocky history of alternation between cooperation and discord in this arena. These fisheries are inherently binational because anadromous Pacific salmon migrate across state and international boundaries during their ocean phase. The need for cooperative management arises because harvesters in one jurisdiction can intercept salmon heading to spawn in the rivers of another jurisdiction. As is so often the case with such fisheries, cooperation has proved to be elusive and difficult to maintain. In this paper, we explore the sources of this instability, using game theoretic concepts to describe the negotiation process, and discuss the new agreement in light of this analysis.

The difficulty of working out a mutually satisfactory division of the harvest has been compounded, in this case, by major shifts in the spatial pattern of salmon abundance. In particular, during the past two decades, salmon abundance has increased dramatically in Alaskan waters while more southerly salmon stocks have fared poorly. As a result, Alaskan interceptions of intermingled salmon stocks originating in British Columbia have increased, while B.C. interceptions of U.S.-origin salmon have declined. The extent to which these changes in abundance are natural or human-caused is subject to debate, though climate-related changes in the marine environment are widely identified as playing a role (Downton and Miller 1998; Gargett 1997; Mantua et al. 1997; Francis and Mantua 1996; Hare 1996; Francis and Hare 1994; Pearcy 1992; Nickelson 1986). Our analysis suggests that the shift in abundance destabilized cooperation by interfering with efforts to implement a newly articulated principle of equitable allocation as stipulated in the Pacific salmon treaty.

The transboundary Pacific salmon dispute illustrates a common pattern in the history of cooperative arrangements among jurisdictional entities. Agreements worked out painfully over extended periods often have been disrupted and finally destroyed by their inflexibility in the face of changed environmental and sociopolitical conditions. Such inflexibility might arise from a rigid allocation rule embodied in the agreement or merely from overly narrow interpretations of rules that *could* be interpreted more broadly. The analysis presented here is part of an ongoing research effort focused on understanding the extent to which more flexible institutions, combined with more accurate and timely environmental information, might overcome these recurring impediments to stable, sustainable joint fisheries management.

Game Theoretic Concepts

Game theoretic representations of international fishery negotiations can help to identify the root causes of disputes and suggest principles that, if properly applied, might help to stabilize cooperation. The regime-setting process under the Pacific salmon treaty could be modeled as a dynamic cooperative game in which any agreement on harvest allocations results from hard bargaining among the players rather than from an a priori definition of entitlements.

Game theory is designed to analyze strategic interactions among independent, self-interested players. Each player is modeled as rationally choosing a strategy on the basis of expected payoffs, given the likely actions of the other players over the entire sequence of play. A game is said to have a "solution" if the players' interactions result in a stable outcome—that is, if each player's strategy is the best possible response to the strategies of the others (Owen 1982; Osborne and Rubinstein 1994).

In a cooperative game, the players attempt to coordinate their strategies to mutual advantage. If cooperation is achieved, it is not motivated by altruism but by the possibility that all parties can gain by avoiding the destructive consequences of noncooperation.

Such a game may or may not result in a cooperative agreement regarding division of the payoff. And the agreement arrived at may or may not prove to be stable. It is the role of game theory to examine the strategic aspects of the game and the existence and stability of possible cooperative solutions.

To be stable and efficient, a cooperative solution must satisfy the following conditions: the solution must be "Pareto optimal," which means that it must not be possible to make one player better off without harming the other(s); and it must be individually rational, which means that it must not be possible for any player to do better by refusing to cooperate (Munro et al. 1998). These conditions cannot always be met under particular rules of the game but perhaps may be achievable by modifying the game structure.

The Bargaining Set—Individual Rationality

The central puzzle is why cooperation has been so difficult to achieve and to maintain in the case of the Pacific salmon treaty. Cooperation can pay in an international fishery because it can dramatically expand the overall size and sustainability of the harvest. This year's harvesting activities affect both net returns this year and the size of potential harvests in future years. If cooperation leads to better conservation and/or lower harvesting costs, all parties can benefit. Certainly both countries' management agencies understand that uncoordinated, purely competitive harvesting is destructive to the long-term productivity of the stocks, so that both fisheries ultimately will suffer. They are also well aware that cooperative management can build and maintain healthy, productive stocks capable of sustaining high annual harvests. Efforts to achieve such cooperation have repeatedly foundered, however, primarily over the issue of allocation of the total harvest between the fleets. Typically, both countries regard this allocation decision to be a zero-sum game—one party's gain in harvest share, it is thought, is inevitably the other party's loss.

We believe this view to be incomplete and also misleading in its inadequacy for portraying the full complexity—and potential trade-off benefits—that cooperative exploitation can offer. We shall return to this theme later in our discussion. First, however, we need to look more closely at the strategic situation that bargaining entails.

Figure 1 illustrates the usual fisheries perspective in a format commonly employed in *decision theory.* In the figure the stippled area represents the set of feasible management and allocation arrangements, with the coordinates of a particular feasible point representing the size of respective harvests, H_1 and H_2 by the two fleets, labeled, respectively, 1 and 2. Along any 45-degree line, the sum H_1+H_2 is constant. Therefore all of the points along such a line represent the same total sustained harvest, with each point on the line corresponding to a particular harvest allocation between the fleets.

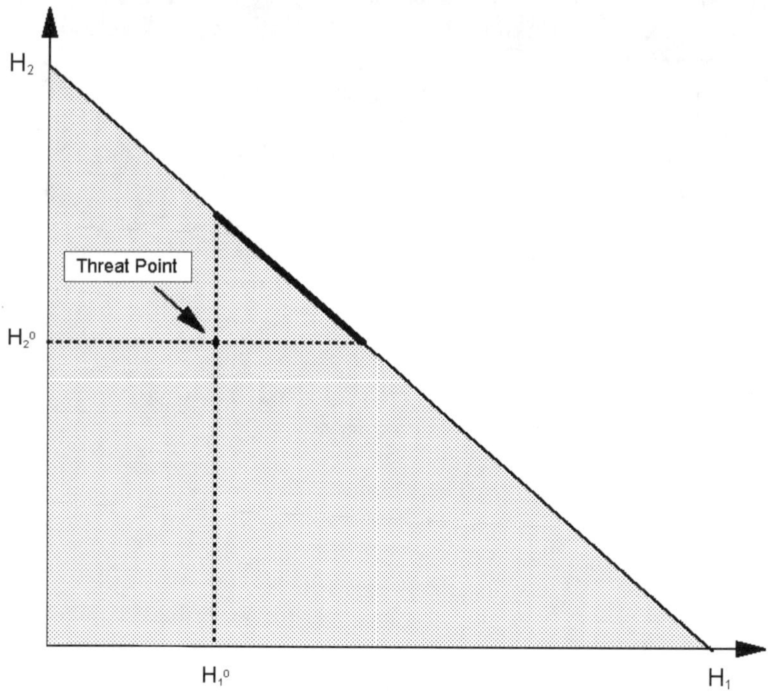

Figure 1. Cooperative management: a naive view. H_1; H_2 = harvests of parties 1 and 2, respectively. H°_1; H°_2 = "threat-point" harvests attainable without cooperation.

In this conception, the value of cooperation arises solely from its potential for achieving an increase in sustained total harvest. In Figure 1, the gain from cooperation would be depicted as a shift of the 45-degree upper boundary of the feasibility set (the so-called "Pareto boundary") upward and toward the right. But then negotiations must still be undertaken to determine the allocation point on the boundary line of this enhanced harvest—the result of a bargaining process between the fleets. The figure portrays the trade-offs in this bargaining as being zero-sum.

This decision diagram also includes a "threat-point" [H°_1, H°_2] within the set of feasible outcomes. This point corresponds to the harvest size and distribution to be expected in the *absence* of cooperation. It represents the outcome when the fleets fall back to a narrowly self-interested (and mutually destructive) competition. Because neither fleet will agree to accept less from a cooperative arrangement than it could achieve unilaterally (the "principle of *individual rationality*"), the "bargaining set" is effectively confined to that (darkened) segment of the Pareto boundary which lies between the horizontal and vertical lines that pass through the threat point.

The Fraser River Treaty: A Period of Stability

Obvious though the individual rationality principle may seem, it has nevertheless been ignored repeatedly in the history of Pacific salmon negotiations (see Munro et al. 1998). Typically, one or another of the parties has been motivated—through miscalculation or simple disregard of relative bargaining strengths—to insist upon a harvest allocation with which some other party will feel it cannot live. Inevitably, the result has been a breakdown in cooperation and a return to a period of unbridled "fish wars."

There was a period of time, however, extending over at least a quarter century, in which a mutually satisfactory cooperative regime did prevail—in large part because all parties' individual rationality was then respected.

This was during the period from 1937 until the early 1960s, during which the original Fraser River salmon convention was in place. This treaty was restricted to the management of Fraser River sockeye salmon *Oncorhynchus nerka* and, eventually, pink salmon *Oncorhynchus gorbuscha* stocks. The Fraser, said to be North America's most important salmon river, lies entirely in Canada, draining a vast area of interior British Columbia. Its mouth is close to the U.S. border, however, and a large proportion of Fraser River salmon returning from the north to spawn travel around the outside of Vancouver Island and through the Strait of Juan de Fuca, where they are accessible to harvest by U.S. fishermen (see the map in Figure 2). Indeed, prior to implementation of the treaty, a very large portion of the harvest had historically been taken, entirely legally, by U.S.-based vessels.

When no agreement was in place to constrain interceptions by the U.S. fleet, Canada had little incentive to invest in enhancing the Fraser River runs. Rockslides in 1913–1914 decimated the Fraser's salmon stocks by blocking access to a major part of their spawning habitat. The Canadians, however, were unwilling to undertake the enormous costs of clearing the blockages and restoring the runs on their own because they realized that the U.S. fleet was in a position to capture a large share of the benefits of the restoration effort. The possibility of mutual gain arising from this situation created an opening for negotiations, while the competitive power of the two nations at the time certainly affected the outcome. After many years of bargaining, the United States and Canada developed the Convention for the Protection, Preservation and Extension of the Sockeye Salmon Fisheries in the Fraser River System. The convention established a binational body, the International Pacific Salmon Fisheries Commission (IPSFC) to manage harvests within the convention area.

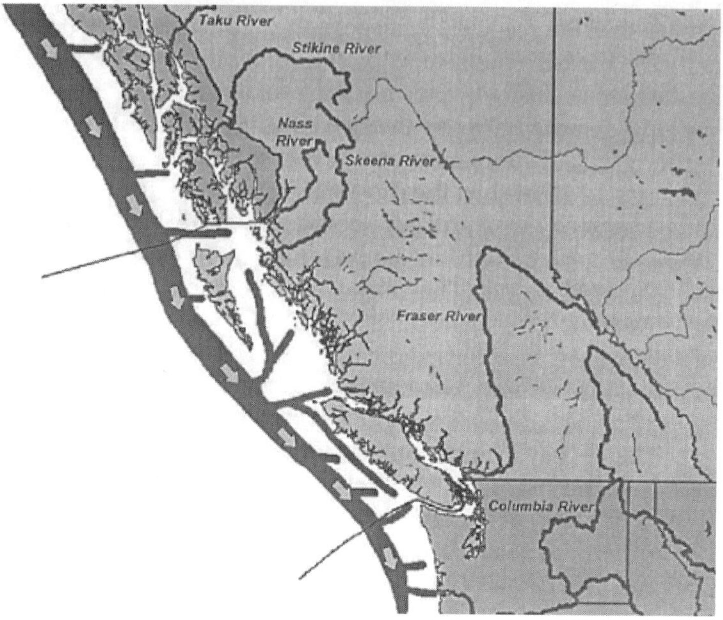

Figure 2. Generalized return migration routes. Source: Department of Fisheries and Oceans, Canada (1997). Pacific Salmon Treaty: Moving Towards Equity and Conservation. Paper prepared by Bud Graham, director of Fisheries Management, Department of Fisheries and Oceans, Pacific Region.

The agreement called for restriction of the overall harvest and, in recognition of the strong competitive position of the United States, a 50–50 split between the two countries' fleets of the total harvest within this terminal area. This arrangement was supported by a 50–50 sharing of the direct costs of management and of on-river habitat enhancement.

The Fraser River salmon harvest agreement began to destabilize in the 1960s when the Canadians came to believe that the balance originally agreed to was no longer equitable. In contrast to what had happened on the Columbia River in the U.S. Pacific Northwest, Canada had built no dams on the Fraser and thus had foregone potential benefits from hydropower development there. Also, Canada was unilaterally bearing the costs of off-river pollution control measures that served to enhance the productivity of the jointly harvested fishery.

Furthermore, there had been recent advances in the International Law of the Sea, incorporating a concept supported by both Canada and the United States that asserted a (vaguely defined) management "preeminence" for the country-of-origin of an anadromous stock. Thereby a new issue of fairness or "equity" was injected into the Fraser River allocation question, a perspective seemingly at odds with the previously agreed acceptance of a neutral 50–50 split of the harvest. In particular, Canada was now seeking a new, more favorable share of harvest rights to those stocks that spawn in the Fraser River. In an effort to press this claim, British Columbia intentionally redirected its fisheries to harvest a larger share of the U.S. chinook *Oncorhynchus tshawytscha* and coho *Oncorhynchus kisutch* migrating southward along the west coast of Vancouver Island to spawn in the Columbia and other U.S. rivers. B.C.'s harvest of those fish increased by 150% between the early 1960s and the early 1970s (Munro et al. 1998).

At the same time, Alaskan interceptions of chinook spawned in the rivers of Washington and Oregon were creating tensions among the states, but increasing Canadian troll harvests of those stocks precluded an effective internal U.S. solution. In addition, mutual interceptions of salmon of Canadian and Alaskan origin were seen as a barrier to effective management in the northern area (Yanagida 1987). Negotiations for a new agreement began in 1971. The talks dragged on for 14 years. During this time, British Columbia maintained intense pressure on the southward-bound U.S. stocks. As Washington and Oregon's coho and chinook harvests declined, particularly in the wake of the 1982–1983 El Niño, Canada's "fish war" strategy succeeded in convincing the southern U.S. parties to support the proposed treaty (Munro and Stokes 1989; Munro et al. 1998).

Alaska, however, showed little interest in reaching an agreement until the Pacific Northwest Treaty Tribes filed suit to extend the 50–50 sharing rule established by the landmark Boldt decision (United States v. Washington [W.D. Wash. 1974]) to restrict Alaskan harvests of chinook originating in the rivers of the Pacific Northwest (Yanagida 1987; Munro et al. 1998). Language in the U.S. Senate's implementation act (1985) and a side agreement between Alaska and the tribes in Confederated Tribes and Bands v. Baldridge (W.D. Wash. 1985) broke the impasse. Under the settlement, Alaska, the northwestern states, and the tribes obtained representation and a veto within the U.S. delegation to the joint commission, and the tribes agreed to give up their right to litigate north/south chinook allocations (Stevens 1986; Yanagida 1987).

The 1985 Treaty

A new comprehensive Pacific salmon treaty signed in 1985 intended to govern the overall harvest and allocation of landings of Pacific transboundary salmon stocks along the entire northern Pacific coast that the two countries share. The treaty calls for joint management through the Pacific salmon commission, with the primary goals:

a) to prevent overfishing and to provide for optimal production, and

b) provide for each party to receive benefits equivalent to the production of salmon originating in its waters.

The second clause formulates an equity principle that established a formula for sharing the overall benefits of the fishery. In practice, the parties have disagreed regarding the steps needed to implement the principle, but many observers have construed it to mean that *the net landings values of fisheries interceptions should balance.* This concept appears to go well beyond the country-of-origin priorities incorporated into the Law of the Sea (LOS). Its categorical language is somewhat mitigated by a subsequent passage in the treaty, which directs that *the parties shall take into account:*

a) the desirability in most cases of reducing interceptions;
b) the desirability in most cases of avoiding undue disruptions of existing fisheries; and
c) annual variations in abundance of the stocks.

Though the treaty was initially hailed by many in both countries as having brought about the end of the salmon wars, close observers of the negotiations (e.g., Yanagida 1987) immediately pointed out the ultimately irreconcilable conflicts among its provisions—conflicts that, soon enough, became evident to all.

The interim arrangements decided upon for managing Fraser River stocks demonstrate how the apparently conflicting goals of the treaty were expected to be reconciled in practice: it was anticipated that Canada would undertake major river habitat enhancement on the Fraser. This was expected to substantially increase stock productivity, and it seemed fair to both parties that Canada should reap the benefits from its investment.

With this in mind, it was agreed to cap the U.S. harvest quota, holding it at a level approximately unchanged from that realized over the previous decade. All additional harvest would be allocated to Canada. This arrangement did indeed "avoid disruption" of the existing U.S. Fraser harvest while protecting the benefits to Canada from its investments in enhancement.

In conformity with the equity clause, however, the continuation of U.S. harvesting of Fraser stocks was now contingent upon a compensating harvest of Pacific Northwest stocks by Canadian vessels operating off the west coast of Vancouver Island. This arrangement acknowledged the fact that B.C.'s troll fishery had developed into a serious contender for a share of those stocks.

Initially this arrangement worked well, but within a very few years, as Northwest stocks declined, it was no longer possible for the United States to pay for its allocation of the Fraser harvest in this way. Canada then demanded that the U.S. allocation be cut, but the Americans refused.

Furthermore, there was no possible way to expand interceptions in British Columbia of U.S.-origin stocks to balance the harvest in Alaska of B.C.-

originated stocks, which Canadian officials argued were large and increasing (Shepard and Argue 1998). Any balancing of harvest interceptions negotiated within the binational salmon commission would have required a very substantial reduction in Alaskan catches of B.C.-originated stocks. It is thus evident that no cooperative agreement relying on a balancing of interception landings values could possibly have satisfied Alaska's individual rationality—Alaska could only lose from such an agreement.

The upshot was that Alaska, in 1993 and subsequently, used its effective veto power over the U.S. position at the binational commission to block any agreement that would substantially interfere with Alaskan harvests. The bargaining framework implemented in 1985 called for frequent renegotiation of the fishing regimes that guided harvest management for the stocks covered by the treaty. Negotiations for new regimes were to follow a consensus rule. On the U.S. side, this rule gave the three voting U.S. commissioners—representing Alaska, Washington/Oregon, and the Treaty Indian Nations—an effective veto over proposed fishing regimes.

When the parties failed to agree on fishing regimes, management reverted to the individual national authorities, and in particular the U.S. harvest quota decisions fell to the regional management commissions and the individual states. Thus the treaty's equity clause was effectively blocked, and management retreated from the Pareto-optimal bargaining set back to the noncooperative threat point.

June 1999 Agreement

On June 30, 1999, the United States and Canada formally signed a new agreement amending the Pacific salmon treaty. The agreement represents a departure from the previous approach in that it replaces short-lived ceiling-based regimes with a longer-term allocation framework in which harvest shares are to be defined as a function of the abundance of each salmon species in the areas covered by the treaty. For example, until 2011, the U.S. share of Fraser River sockeye will be fixed at 16.5% of the annual harvest. This represents a decrease from the post-1985 average U.S. share of 20.5% but an increase relative to the share actually attained by the U.S. fleet during the 1992–1997 salmon war period (DFO 1999; O'Neil 1999a). This percentage approach allows the number of Fraser River sockeye harvested by the U.S. fleet to increase in years of high sockeye abundance while requiring reduced harvests when abundance is depressed. In contrast, in the 1985 treaty, U.S. harvests of Fraser sockeye were to be held to a cap of 7 million fish over each of two successive four-year periods (Pacific Salmon Treaty, Annex 4).

The new arrangements for chinook, which will be in effect until 2009, take account of the fact that the various fisheries along the coast differ considerably in the extent to which they rely on healthy or depressed chinook stocks (U.S. Department of State 1999). Accordingly, the agreement designates two types of fisheries: abundance-based management (AABM) fisheries will be managed based on indices of the aggregate abundance of chinook present in the fishery; and individual stock-based management (ISBM) fisheries, which are primarily located in inside fishing areas, will be managed based on the status of individual stocks or groups of stocks (i.e., on the basis of the evolving status of currently endangered or threatened stocks). Abundance-based allocation rules for coho have not yet been developed, but the agreement instructs the parties to jointly develop such a management approach and specifies various deadlines for the accomplishment of particular tasks.

The agreement also provides for two endowment funds, the proceeds of which are to be used to support scientific research, habitat restoration and enhancement of wild stock production in their respective (northern and southern) areas. Initially, the funding is to be provided entirely by the United States, and the entire agreement is contingent on U.S. Congressional approval of US$75 million for the northern fund and US$65 million for the southern fund. In the future, either party may make additional contributions, and even third parties may contribute, with the consent of the parties.

It is not yet clear if the new approach will provide a path to sustainable cooperation. The focus on conservation will tend to protect some of the weak stocks that were jeopardized by the recent turmoil, but the new agreement does little to resolve long-standing differences over the division of benefits. In particular, many Canadians remain convinced that Canada will come out short under this agreement, and they have labeled it "profoundly disappointing" (Culbert and Beatty 1999). They are also unhappy about the long-term implications of the agreement because they feel that it risks committing them to an unsatisfactory arrangement for many years. In fact, the Canadian delegation had unsuccessfully argued for a shorter term. Although the parties have formally stipulated that compliance with the terms of the new agreement shall be deemed to fulfill the requirements of Article III of the treaty, the stipulation applies only for the duration of the current agreement. If Canadians continue to feel that their interests have been compromised, there may be renewed turmoil when this agreement expires.

Fairness or Competitive Power: Conflicting Principles for Utility Allocation

The salmon dispute may be characterized as a quarrel over both control and allocation: over the right to manage and the right to exploit the fish stocks.

Throughout the modern history of marine fisheries, there has been a slow evolution in shared perceptions of who owns the fish. Moving away from the traditional view of the "rule of capture" (which still prevails on the high seas), law and custom came, in the mid-20th century, to embrace the notion that each coastal nation should enjoy exclusive fisheries harvest rights in the seas near to its shores. Initially asserted in a power play by a certain few coastal nations, this concept came to be widely accepted and was then formalized in international law as the 200-mi Exclusive Economic Zone (EEZ).

As noted above, in the 1937 Fraser River treaty, Canada and the United States agreed to a 50–50 split of the terminal harvest of Fraser sockeye stocks and matched contributions to management and enhancement. This allocation in the treaty can be interpreted as a tacit recognition and pragmatic acceptance by both parties of their effectively equivalent competitive power in any contest over the harvest of these stocks.

By contrast, the 1985 Pacific salmon treaty, through its equity clause, articulates a sharply different principle. Taken literally, the equity clause can be interpreted as vesting fishery *benefit* rights primarily in the country-of-spawn. This concept may be thought of as a *principle of fairness*—a recognition of the paramount benefit to the joint fishery of proper stewardship on the river. The primacy of country-of-origin is modified in the treaty—again, presumably as a matter of fairness—by the admonition to "take into account the desirability...of minimizing the disruption of existing fisheries."

The country-of-origin priority is reiterated repeatedly in various memoranda of understanding between the parties subsequent to the enactment of the treaty. However, the two countries do not agree on precisely how far this country-of-origin priority should extend. Canada argues that location of spawning is the primary consideration in accounting for the entitlements of each party; the United States remains unwilling to embrace this position. Envoys Strangway and Ruckelshaus (1998, p. 4) describe the U.S. position as follows: "The U.S. contends that all the principles in Article III must be read together, and that whether an equity imbalance exists or interceptions need to be reduced in any particular situation, require consideration of a number of factors, including annual variations in the abundance of stocks, conservation and avoiding undue disruption of existing fisheries."

The parties to the treaty may not agree on precisely what *constitutes* an equitable allocation of benefits, but they both seem to have accepted the doctrine that *fairness* should govern this allocation. Furthermore, all recognize that, in view of the dominant pattern of stock migration and intermixing across jurisdictional boundaries and the desire to avoid undue disruption of existing fisheries, some stock interceptions are inevitable. Hence achieving equity will require a trade-off of benefits, including those resulting from stock interceptions. The equity principle includes the notion that an overall balancing of benefits must be achieved, a balancing consistent with the goal of fairness. In the literal language of the treaty (and many subsequent memoranda), this balance is to "provide for each party to receive benefits equivalent to the production of salmon originating in its waters."

However, the requirement of achieving equity clashes directly with the interests of several regional and political entities in the United States—specifically, the states of Alaska, Washington and Oregon, and the Treaty Indian Nations. Until the June 1999 amendments to the treaty, these entities had frequent opportunities to exercise their veto power over the U.S. vote within the Pacific salmon commission when fishing regimes came up for renegotiation. It follows that the treaty's equity clause could be—and routinely was—suspended by unilateral objection by one or another of these regional entities. When this occurred, management and establishment of harvest quotas reverted to national control and, in the United States, to control by the regional authorities. Their decisions were made on the basis of self-interest, and thus the harvest allocations once again tended to reflect relative competitive power rather than being consistent with the treaty's characterization of fairness.

Though it is recognized that parties operate on the basis of individual rationality, their relative competitive strengths are determined by the institutional structures within which they operate. Thus in particular, though accepted international law gives full authority to each coastal nation over fisheries within its own EEZ, this authority can nevertheless be contracted away, partially or fully, through treaty agreement. A fundamental flaw in the 1985 treaty, as well as in the new agreement, is that harvest authority has *not* been ceded to an independent binational salmon commission but must be renegotiated periodically. The new agreement calls for less frequent renegotiations, in that the newly articulated abundance-based regimes are to be in place for 10 to 12 years. Negotiations will again be required when they expire.

Thus, the opposing principles of competitive power and fairness remain unresolved—and mutually inconsistent—in the current Pacific salmon treaty. Reconciliation of their conflicting requirements will require that each must give some ground.

The Transboundary Fishery as a Cooperative Joint Venture

One interpretation of the equity principle to which many Canadians adhere is that it requires that interceptions be minimized and their landings values balanced. Authority to recommend management regimes to achieve the equitable balance is delegated to a binational commission over whose deliberations both countries retain veto power. Though the new agreement provides less frequent opportunities to exercise this veto power, the fundamental institutional structure of the commission remains unchanged (Waldeck and Buck 1999). Implementation of the agreed-upon regimes is the responsibility of each nation.

It can be argued that a narrow interpretation of the equity principle—as requiring interceptions balancing—is a recipe for failure. Restricting in this way the options open for distributing benefits diminishes the likelihood of finding a solution that will satisfy individual rationality for all of the players. Ignoring individual rationality accentuates the potential for instability in the functioning of the treaty. It is unrealistic to expect that a balance can always be struck merely by a transfer in fish landings. Rather, success in negotiations may very well depend on enlarging the mix of benefits to be traded off and balancing them by incorporating components that fall outside of the fishery harvesting sector.

In searching for a way out of the impasse, Strangway and Ruckelshaus (1998) urged that repairing the "dysfunctional" salmon treaty would require that the parties to the treaty must first achieve a consensus on a precise operational meaning of the equity clause. At first glance, given Alaska's dominant strategic position, consensus would seem to require that Canada must settle for very much less than "benefits equivalent to the production of salmon originating in its waters."

But this view may be unduly pessimistic. Indeed, a simple game theoretic analysis raises the possibility that expanding the framework for negotiation might create new advantageous cooperative arrangements permitting the reasonable goals of all parties to be satisfied. The analysis also suggests a conceptual approach for mediating the conflict between fairness and competitive power.

The wider framework for negotiation would be achieved, as several observers have urged, by explicitly rejecting the narrow interpretation that equity must be achieved exclusively through a balancing of interceptions. Instead, imbalances in interceptions could be compensated for by transfer payments ("side payments") *in currency other than fish* (Munro et al. 1998). Such an expansion of options, it has been suggested, could make possible an accommodation that would respect the vital interests of all.

Enlarging on this point of view, we would argue that significant collateral benefits, other than direct landings values, accrue to each of the parties from participating in cooperative management of the transboundary salmon fishery. These values are substantial and are not zero-sum. By explicitly recognizing these broader benefits of cooperation and balancing them through appropriate side payments, it might well be possible to fashion feasible cooperative arrangements that are consistent with mutually acceptable principles of fairness and still satisfy all parties' individual rationality.

Such collateral benefits fall into several broad categories. First we note that there are additional *direct* market values that may be realized by relaxing cross-jurisdictional constraints to permit greater flexibility in the choice of harvest sites (e.g., to optimize catchability or market price) or of timing (e.g., to take advantage of in-season information about run size and temporal distribution). These potential benefits arise most often when stocks are intermingled. An oft cited example is the potential benefit from relaxing current restrictions on harvest of Alaskan pinks in Alaskan waters, restrictions due to intermixing of these stocks with migratory sockeye from northern B.C. rivers.

Less easily measurable but nonetheless highly valuable are nonmarket socioeconomic and political benefits, such as enhanced employment in currently depressed areas. These benefits are especially important in outport areas of B.C. and Alaska, where the fishery has been a major—perhaps the only—seasonal source of cash employment. Another example concerns the cultural values attached to traditional harvest by many indigenous peoples.

Finally, we emphasize the potentially enormous cost reductions in current on-river mitigation measures that might be realized by limiting the ocean harvest of threatened or endangered Northwest salmon stocks. The prime application is to the Columbia River, where hugely costly—and eventually politically unsustainable—measures are now being taken to maintain residual populations of endangered stocks.

In addition to direct costs, such as the cost of barging smolts through slackwater pools, salmon conservation measures on the Columbia incur major opportunity costs in foregone hydroelectric production, in diminished irrigation and navigation capacity, and in upstream ecosystem degradation resulting from flow surges and reservoir drawdowns. Clearly fish from highly endangered salmon stocks have their highest value when swimming in the sea or the river, not in the hold of anyone's fishing vessel.

There may well be no single common currency with which to pay off a perceived imbalance in these multiple interests. It is the art of the mediator, and of the diplomat, to invent mixtures of concessions and incorporate forms of benefit that the parties may agree upon as achieving a balance. What this requires depends on the subjective values that the various stakeholders to the fishery may place on these benefits, as well as the strength of their bargaining powers in the negotiations.

Recognizing these pervasive internal linkages and the currently unrealized potential for exploiting them, one is led to regard the North American Pacific salmon fishery as a natural *multiregional joint venture*—one that is currently hobbled by divided management authority and jurisdictional constraints that limit its ability to exploit the full range of its options.

The Bargaining Game: Capturing the Competitive Surplus

Recognizing the full extent of potential gains from cooperation leads us to significant elaboration of our conceptual model of the bargaining framework. Obviously, the actual bargaining problem under the Pacific salmon treaty is highly complex, involving a bewildering array of variously intermixed transboundary stocks. There are also multiple interests to be satisfied within each nation, including regional and indigenous groups, some of which are directly represented in the binational negotiations.

Nevertheless, and for orientation, we shall initially maintain the perspective of bargaining over a single jointly harvested stock and then will examine the possible trade-off of benefits between two separately targeted but jointly harvested stocks. Later we shall return to a discussion of the potential problems and benefits of binational coordinated management of still larger aggregates of transboundary stocks.

The special "atomic" cases will be illustrated with the aid of Figures 3 and 4, respectively. These, as before, are schematic representations of the decision framework for joint management of an isolated stock. For representational simplicity, the decision context is again portrayed as two-dimensional—involving only two separate jurisdictional entities. (Actually, as we know, there are several independent regional entities, each competing to maximize its share of the gains from the common enterprise.)

As in Figure 1, alternative feasible management arrangements are shown as stippled areas. The axes in Figures 3 and 4 have been relabeled because the coordinates $[U_1, U_2]$ of a representative point no longer represent only landings values. Rather, they incorporate both market and nonmarket values and thus the total utility accruing to each jurisdictional entity as a result of its participation in the single-stock fishery.

We hypothesize a feasibility set that is convex—bulging upward at its center. It is reasonable to suppose that the Pareto frontier would have such a shape because the marginal utility to either party of a change in its share of the total harvest is not likely to be constant. For example, near the upper left-hand corner of the frontier, the value to party 1 of an increment in its share of the run may be very high because the first increments are likely to be used to serve important conservation, cultural, and recreational uses.

The principle of diminishing marginal utility implies that party 1's valuation of further increments would decline as one moves downward and to the right along the frontier, but simultaneously the marginal utility to party 2 grows, more than offsetting the former's decline. The convexity feature affirms that there is a substantial added utility to be achieved through cooperation—a bonus beyond the purely biological consequence of enhanced

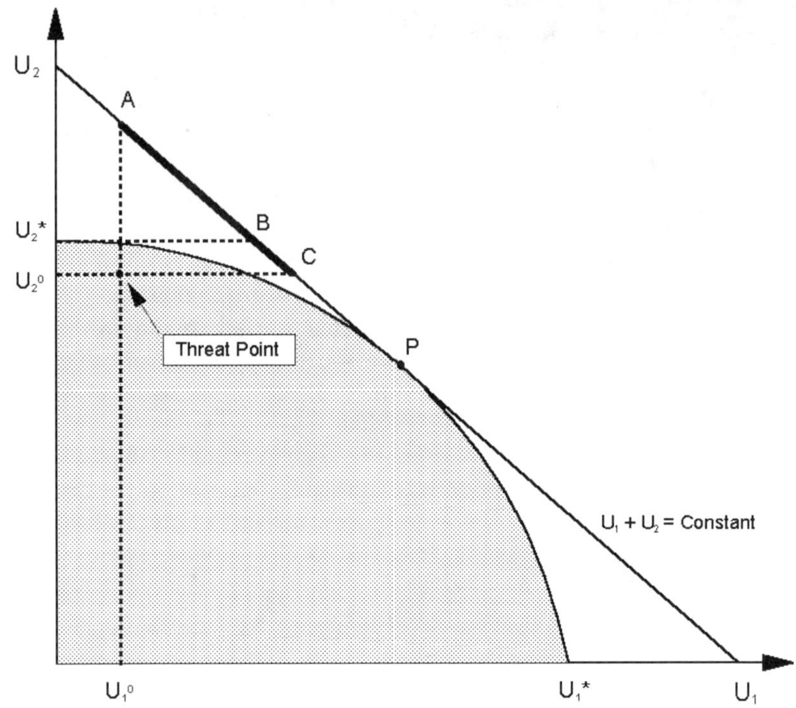

Figure 3. Bargaining framework with convex pareto frontier. U_1; U_2 = utility levels of parties 1 and 2, respectively. U^o_1; U^o_2 = "threat-point" utility levels without cooperation. U^*_2 = utility attainable by party 2 as sole owner of stock. P = coordinated arrangement maximizing joint utility.

stock productivity. It affirms that the total payoff, U_1+U_2, to the binational community will be maximized at some cooperative mix of shared landings.

If the parties' utilities are given equal weight, then the coordinated management arrangement that maximizes total community utility corresponds to the point *P*, shown in the figures. *P* is the point of tangency of the Pareto boundary of the feasibility set with a 45-degree tangent line.

An immediate conflict, clearly evident in these diagrams, is that the *distribution* of utility at point *P* is not likely to be acceptable to both parties. For example, it may, as illustrated, violate the individual rationality of one of the parties (here, party 2). The most stringent interpretation of equity would require, if country 2 contains the river-of-spawn, that the utility level for fleet 2 be even higher, matching that at the point where the upper boundary curve intersects the vertical axis.

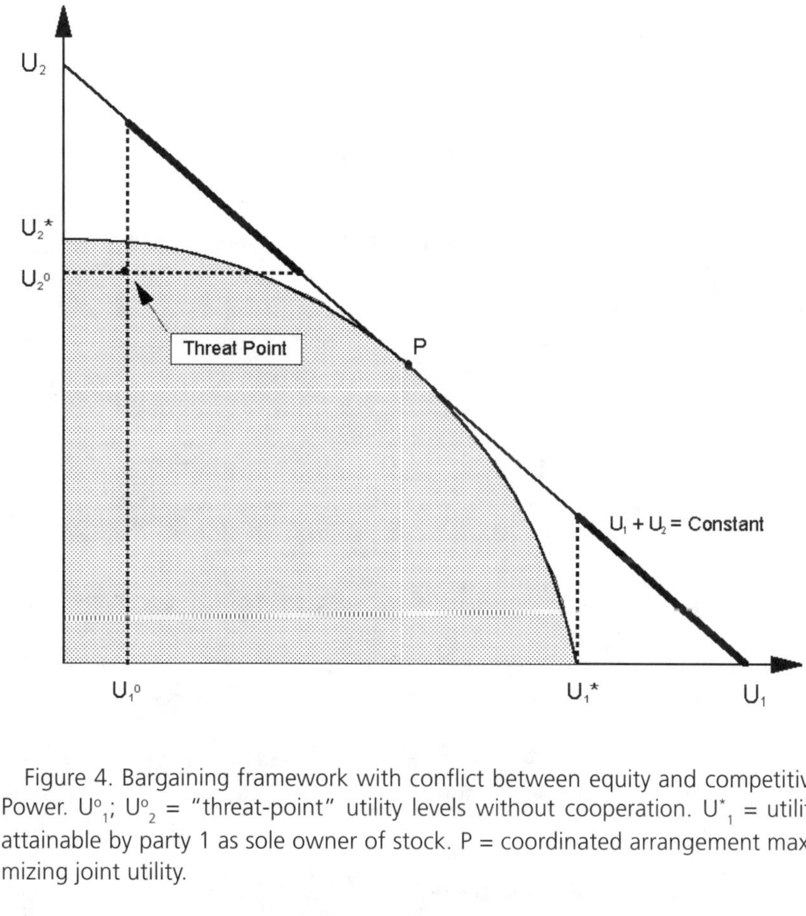

Figure 4. Bargaining framework with conflict between equity and competitive Power. U^o_1; U^o_2 = "threat-point" utility levels without cooperation. U^*_1 = utility attainable by party 1 as sole owner of stock. P = coordinated arrangement maximizing joint utility.

The notion of a side payment, not necessarily in fish, is captured in our decision-theoretic diagram as an enlargement of the set of feasible outcomes. Harvesting at *P* and then transferring utility by a side payment from fleet 1 to fleet 2 will not change the harvest distribution but will move the *utility* distribution upward and to the left along the tangent line. In the example case shown in the figures, the new bargaining set—line segment A-C (still limited by individual rationality)— is nonempty.

Now, in Figure 3, let us assume that country 2 contains the river-of-spawn and that, on this basis, claims full equity rights in the stock. The schematic demonstrates that, with a side payment, it may be possible *in principle* simultaneously to satisfy both individual rationality and equity. In this example, it holds throughout a constrained bargaining set the line segment A-B. Notice that this is possible *only because of the convexity of the original feasibility set*—i.e., the presence of a cooperative surplus, due mainly to the indirect and nonmarket values in the fishery.

Note that in this particular example, recognition of party 2's sole equity right to the fish is not accompanied by an increase in that party's harvest. Rather, cooperation in this case results in increased harvests by party 1 (relative to the threat point), followed by a side payment to party 2. Such a result is possible only if there is a distinct utility advantage to harvesting the fish in the waters of jurisdiction 1.

For example, the figure might be applied to the previously mentioned example of harvest, by Alaska (as player 1), of B.C. sockeye intermixed with Alaskan pinks. The diagram shows explicitly only the distribution of harvest of the *single transboundary* stock—the sockeye. The presence of intermixed domestic Alaskan pinks in Alaskan waters is reflected only by the increased utility attributable to the Alaskan portion of sockeye harvest. This, however, demonstrates how an Alaskan harvest of the sockeye might conceivably (with suitable side payment) be consistent with B.C.'s equity condition while simultaneously satisfying Alaska's individual rationality. (This might also be achieved by giving B.C. fishermen a quota of the sockeye taken in Alaskan waters.)

Now, of course, it is not usually the case that there is enough added utility from cooperative joint harvest to satisfy transboundary equity claims on a stock-by-stock basis. For example, this could not be achieved entirely within the Fraser River basin, where equity clashes strongly with individual rationality.

This is illustrated by Figure 4, with the same feasibility set as in Figure 3 but with player 1 being the country-of-spawn and player 2 having the intercepting fleet. As shown in Figure 4, the added value in the fishery from operating at P might be redistributed to achieve *either* fleet 2's individual rationality *or* fleet 1's full equity rights—*but not both simultaneously*. The only way to accomplish this is to inject value from outside of this single-stock fishery. But both countries seem to have recognized this problem and agree that equity claims need to be balanced only in the aggregate. To this we add that individual rationality claims need also to be adjusted but that this, too, needs to be achieved only in the aggregate.

As an example, since signing the current treaty, the two countries have attempted to balance Canadian harvest of Pacific Northwest chinook and coho against the U.S. harvest of Fraser sockeye. That is, they have attempted to meet equity and individual rationality requirements simultaneously by pairing two groups of transboundary stocks that are separately targeted and have different countries of origin.

If those two groups of stocks could, in fact, be considered in isolation from the issues of Alaskan interceptions and perceived imbalances elsewhere, then it might be possible, in principle, to achieve such a balancing by basing negotiations on a comparison of the benefits attainable under three distinct harvesting regimes. The first, defining each country's individual equity position, would restrict harvest of each stock to country-of-origin fleets. The second, a competitive harvest of all stocks, would determine each country's individual rationality constraint. The third regime, a cooperative joint harvest, would determine that economically efficient regime which would generate maximal societal utility. Note that, so long as the stocks from different countries of origin are not intermixed, all three of these harvesting regimes are obtained by simple (additive) aggregation over the single-stock component fisheries.

The actual negotiations have been complicated by the fact that Alaskan harvesters intercept both chinook from the Pacific Northwest and sockeye from the Skeena, Nass, and other B.C. rivers. Figure 4 might serve as a useful representation of the current north/south chinook allocation problem. The precipitous decline of southern coho and chinook stocks has made their harvest in B.C. an increasingly inadequate trade-off for the U.S. take of Fraser stocks. Furthermore, there is an enormous disparity between the value these harvested chinook stocks have to Canada and Alaska (jointly, the intercepting player 2) compared with the potential value to the Northwest (the region of origin, player 1) of lowering the huge current costs of maintaining endangered Northwest chinook stocks by riverine enhancement measures. In this case, the hope for equity balance between the two interception fisheries surely rests on the sizeable utility asymmetry.

In the above example cases, transboundary stocks originating in different nations or regions are assumed to be targeted separately. Consequently, definition of equity rights to these stocks is straightforward, and overall national or regional equity can be arrived at by simple aggregation over individual stocks.

The reality, of course, is that transnational Pacific salmon will likely continue to be harvested in areas where stocks are intermixed. In principle, such interceptions might be avoided by confining all harvests to national waters close to the rivers-of-spawn. Though many observers have extolled the merits of such an approach—focusing attention, for example, on the advantages of fish wheels and other fixed gear (Crutchfield and Pontecorvo 1969; Higgs 1982; Jaeger 1997)—others have pressed both economic and management efficiency arguments in favor of the current mixed-stock approach (e.g., ADFG 1994). Sands and Hartman (2000) estimate, for example, that eliminating the Alaskan–B.C. interceptions imbalance by harvesting only in nonintercepting fisheries would diminish the net value of the binational harvest by approximately one-third. One of the factors driving this result is their estimate that the market price of terminally caught pink and sockeye salmon in northern B.C. and southeastern Alaska is only half that of ocean-caught fish. Therefore, attempts to implement significant constraints on mixed-stock fisheries would certainly trigger defensive responses based on regional self-interest, such as we have seen throughout the history of the Pacific salmon fishery.

If the estimates of Sands and Hartman are even approximately correct, such constraints would impose major economic inefficiencies in the overall binational fishery. If the parties attempted to calculate fair-share national and regional equity valuations on the basis of such a hypothetical fishery, the result would understate the values that could be achieved in a rationally managed mixed-stock fishery. Stock-by-stock valuation would necessarily be based on a conjectural harvesting pattern, one that would be far from actual practice and far from any socially optimal management regime. Such an approach to valuation would in fact misrepresent the actual contribution of particular stocks to the overall benefits of the fishery and thus be subject to challenge on the basis of individual rationality. In summary, a stock-by-stock approach to defining equity is neither desirable nor even feasible.

Rather, the proper fisheries unit for management and equity valuation is not an individual stock but is rather the total coast-wide multistock fishery. That is, *the entire binational salmon fishery ought to be viewed as a single joint venture*, a venture in which all of the partners have agreed equity shares of overall benefits. Management would then start with identification of an overall optimal pattern of harvesting, followed by a sharing of the gains from optimal management.

Regarding the entire Pacific salmon fishery as a common venture would lead to consideration of the indirect and nonmarket values to be gained by cooperative management of the industry and elimination of some of the jurisdictional constraints that now force second-best harvesting options. It would also require prior negotiation and a mechanism for periodic renegotiation to determine stable equity shares to be allocated among the parties to the treaty.

Once again, one might represent this situation as a bargaining game among the several players. A two-dimensional schematic representation would look much the same as our schematic Figures 3 and 4, except that equity rights to the entire fishery would now be shared between the parties as the result of prenegotiation and might or might not be based on rivers-of-origin alone.

Meeting individual rationalities would still be necessary and would impose constraints on the geographical trade-offs in harvests among various groups of fishermen. It might also still be necessary to inject value from outside of the fisheries sector. In the Columbia River case, for example, this might involve monetary payments by the Bonneville Power Administration and Northwest electric power consumers to Alaska and Canada to secure reduced harvesting pressure on Columbia River salmon with less disruption to the regional power supply industry.

Adopting such an integrated decision process will not automatically eliminate the possibility of deadlock within the joint management commission, which might trigger a reversion to noncooperative subregional management. It does, however, provide a much broader range of options to work with and the promise of enhanced returns for all.

Cooperative Management in an Unstable World: Adapting to Change

The models we have shown describe a deterministic world. Yet we know that the Pacific salmon fishery has, throughout its history, been buffeted by change, and that it is precisely the onset of change, especially unexpected change, that has repeatedly destabilized efforts at binational cooperation in salmon management.

Historically, change and uncertainty—short-term stochastic fluctuations, longer-term temporal trends, unanticipated surprises, and environmental regime shifts—have been ubiquitous features of marine fisheries worldwide. In the transnational context, change—especially unanticipated change—can be expected to have destabilizing effects on any rigid arrangement for coordinating stock management across national jurisdictions.

In responding to such changes, two simple principles are relevant. First, responses to change should be keyed to the time scale on which the change occurs. Second, the chosen response mechanism needs to be simple, transparent, and cost effective, given the technology currently available for its implementation.

Plainly the 1985 Pacific salmon treaty behaved poorly on both counts. We examine first the problem of time scales, then, in the next section, the problem of mechanism design.

As the treaty operated, a disagreement over a single season's allocation of harvest could lead to immediate breakdown of cooperative management, and thus to an abrupt shift in patterns of stock exploitation. In this way, the allocation of fishery benefits specified in the treaty's equity clause could be immediately nullified.

It is important to break the link between the allocation of current harvest landings and the determination of benefit shares in the joint venture. Both will reasonably be expected to undergo changes over time, but normally on quite different time scales.

In principle, benefit shares ought to evolve slowly in response to evolving perceptions of fairness and of shifts in relative competitive power. By contrast, harvest levels and allocations need to be adjusted more often: seasonally, in response to short-term fluctuations in stock demography or in market conditions, and cyclically, in response to successive El Niño events. This is necessary to optimize overall societal benefits from the fishery.

It is instructive to compare once again the 1985 salmon treaty with its predecessor, which dealt only with stocks on the Fraser. Both agreements eventually came apart because of their inability to respond flexibly and creatively to change. The Fraser treaty unraveled over a period of decades; its successor's unraveling took only a brief few seasons.

Under terms of the Fraser treaty, short-term harvest allocation was directly determined through the institution of a 50–50 joint fishery. The treaty remained viable so long as both parties were satisfied with this balanced allocation of both management costs and benefits. A renegotiation of the treaty became necessary when a new concept of benefit distribution, focusing on nation-of-spawn, had gained ascendancy.

Had only Fraser and U.S. Northwest stocks been at issue, there is little doubt that the 1937 treaty could have been renegotiated successfully along the lines actually adopted on an interim basis. In fact, even during the recent years of turmoil, partial interim agreements involving only those stocks were successfully negotiated in 1995, 1996, and 1998. Those partial agreements, however, were fragile given the on-going disagreement over coast-wide balancing. What could not be reconciled under the 1985 treaty was the imbalance between harvest returns and perceived benefit rights due to subsequent decline of Northwest stocks and expansion of Alaskan interceptions.

The 1999 agreement represents a step in the right direction, but it does not actually fulfill the recommendation of envoys Strangway and Ruckelshaus that "governments from both sides should develop a practical framework for implementing Article III..." (Strangway and Ruckelshaus, p. 8). The division of benefits is still largely tied to the division of the catch. Depending on the vagaries of nature, the parties may or may not find this division to be satisfactory. If they do not, and if they fail to develop other methods to ensure a fair division of benefits, then another breakdown in cooperation will likely occur when this agreement expires. However, the arrangements should help to protect stocks that had been imperiled by recent fish war tactics and may thus buy time to work out a longer-term agreement on fair division of the benefits of the fishery.

We have seen that the introduction of side payments into transboundary fishery management agreements will expand the bargaining set, creating new options for negotiation. But it does something else, too. The mechanism of utility transfer through side payments effectively decouples the allocation of current harvest from the allocation of the benefits from the fishery. Thus benefits (which will fluctuate annually) can be allocated on the basis of a stable, agreed distribution of equity rights and realized through a combination of direct harvest returns and side payments.

We recognize that the effectiveness of the transfer payment concept depends on the willingness of the parties to trade off fishing rights for other valuable assets or considerations. In particular, fishermen generally prefer catching fish to being paid off through a monthly welfare check. It will require ingenuity and imagination on the part of negotiators and mediators to arrive at mutually acceptable arrangements. In general, all that can be claimed is that introducing the concept of side payments can greatly expand the range of available options.

The endowment funds that are to be established under the new agreement might provide a vehicle for providing such side payments. However, their yield will be far smaller than the debt claimed by Canada from the United States for the accumulated harvest imbalance. A portion of the money available from the northern fund will support Alaskan research and enhancement (O'Neil 1999b), which, together with other payments, might elicit greater cooperation from Alaska. The funds are to be managed by committees composed of representatives appointed by the federal governments of Canada and the United States. If these committees are able to use the funds creatively to offset perceived harvest imbalances and maintain goodwill, this approach might constitute a step towards the international joint venture that we advocate. It remains to be seen, however, if this tool will be put to good use or if its promise will be squandered in quarreling over distribution of the proceeds.

Measuring Benefits

Establishing an equitable return to productive use of an asset requires a determination of the value to be imputed to this asset. In the case of fish stock assets in the binational Pacific salmon fishery, efforts to carry out such an evaluation have been bedeviled by conceptual and practical difficulties.

Mainly at the insistence of Canada, elaborate accounting schemes have been developed to measure, in particular, the value of interceptions. The methods for carrying this out are complex and controversial, and the two nations' separate accounting processes typically have yielded radically divergent results, particularly when they have applied different standards for valuation. Even the biological estimates of numbers of fish intercepted in the various fisheries have been subject to disagreement (Pacific Salmon Commission, Joint Interceptions Committee 1993). Given the sensitivity and uncertainty of those estimates, the Joint Interceptions Committee of the PSC has not published a report for several years. Obviously, such an accounting process becomes even more difficult when one attempts to include also indirect and nonmarket societal aspects of value in the fishery. It's not surprising that the separate estimates tended to reflect the respective national interests, so the accounting process led not to the desired consensus view but rather to a further polarization of already divergent positions.

Actually, it is not at all necessary to achieve consensus among the parties in the valuation of the assets that are being exchanged, nor indeed is consensus to be expected. What is necessary for an exchange transaction to go through is that each party should value the assets that it receives at least as highly as it values the assets that it relinquishes.

However, this necessary condition is not sufficient. Agreement on exchange depends also on developing an appropriate institutional arrangement for negotiating the terms of exchange. It is here, especially, that the current salmon treaty falls short.

Game theory (especially the theory of auctions and of principle/agent contracting) offers certain insights into how bargaining for exchange should be structured in a situation where there are only two (or only a very few) parties to the negotiations (Fudenberg and Tirole 1991; and, specifically for fisheries, McKelvey 1997). The game models show the nature of the difficulties of negotiation in the face of uncertainty and the importance, therefore, of maximizing information transparency. Indeed, much of the analysis is focused on the creation of incentives for the negotiating parties to reveal the true valuations that they place on the assets.

In our view, it is most important to separate out the two distinct roles of the parties to these negotiations. The first of these is their role as joint owners of the fish stocks and as the custodians of their habitat. The second is their role, or that of their nationals, as resource exploiters gaining benefits from the utilization of this regional natural resource system.

As we have emphasized, the notions of effective use rights and of ownership equity often have clashed, and we believe they must be brought into consonance to deal rationally with future disputes. We have suggested that one way to separate these roles is to conceive of them as operating on different time scales.

Conclusion

The tug-of-war that has characterized the Pacific salmon treaty process is widely perceived as damaging to the interests of both nations. Efforts to balance the value of interceptions have proved unsuccessful. They are currently being frustrated by natural fluctuations in salmon abundance and by the irrationality, from Alaska's standpoint, of making major concessions on its interceptions of salmon spawned elsewhere without commensurate compensation.

The search for sustainable cooperation in the management of North America's Pacific salmon resources should be guided by several overriding general principles. First, it is critically important to respect individual rationality. No party can be expected to adhere to an agreement that leaves this party worse off than it would have been under a purely competitive fishing regime. Recognition of this requires each party to be sensitive to the expectations and competitive strengths of the others. It includes a realization that the policies of each will be driven both by their notions of what is fair and also by how they will judge their relative competitive positions.

Second, because it is the gains from cooperation that provide the parties with the incentive to cooperate, harvest management should be structured to allow full exploitation of those potential gains. This argues for abandoning the efforts to assign fisheries values on a stock-by-stock basis and instead pursuing policies that will optimize binational societal returns to the entire transboundary Pacific salmon fishery.

We have characterized this approach as a binational, multiregional joint venture. Though the concept may seem novel, it has, in fact, a number of precursors in marine fisheries management elsewhere. A recent parallel example is in the proposed management of the southern bluefin tuna fishery, which is being harvested jointly by Australia, New Zealand, and Japan. This would maximize total return by eliminating the harvest of juveniles in the Australian fishing zone and compensating for this by a side payment from Japan (see Kennedy and Pasternak 1991).

We have not spelled out the details of how such a joint venture might function, but the central concept is to separate the role of each nation or jurisdictional entity as owner of a share in the fishery from its role as harvester of the stocks. This extends the usual notion of fisheries joint venture because now each nation and jurisdictional region is simultaneously a principal, or owner, and an agent serving the interests of all of the others.

In this context, new options open up, some not possible under the usual strict jurisdictional constraints. For example, one may conceive of ameliorating an imbalance of Alaskan-British Columbian interceptions by assigning B.C. fishermen a quota in Alaskan waters. This could be paid for by a side payment to Alaska from the Pacific Northwest, which in turn would be compensated for by a reduction in Canadian interceptions of endangered Columbia River stocks.

Third, it is necessary to address the dynamic nature of the fishery and the ubiquitous presence of change—often unanticipated change—that will upset and require revision of settled arrangements. This requirement is served, in particular, by the decoupling of the short-term harvest allocation decision process from the more slowly evolving determination of joint venture profit shares. Game theoretic analysis of institutional design for auctions, joint ventures, and the regulation of public utilities can provide insights here. In particular, because measurement and enforcement costs diminish the net gains available for distribution among the set of cooperating players, it will be important to devise joint venture institutional arrangements that avoid complexity and achieve informational transparency.

Finally, it will be important to strengthen binational institutions, giving them real authority to promulgate and enforce regulations. The advantage of true joint management comes from the fact that it avoids conditions of divided authority under which each party may attempt to use asymmetrically held information to capture short-term gains or may attempt to introduce extraneous or secondary objectives that can destabilize cooperation.

We recognize that nation states do not easily surrender authority to international bodies. But in this instance a new joint salmon commission would be their own creature—an international public utility that would be wholly owned and jointly controlled by policies of the two nations that created it.

These nations need to find the political will to create an agency that is permitted to function independently, shielded from the cross-currents of direct parochial influence. If they succeed, their achievement can be a model for other transnational joint ventures worldwide, not merely in the fisheries sector.

Acknowledgement

The authors would like to thank Gordon Munro for his critical comments, which helped us to sharpen our analysis. We, of course, remain responsible for any errors of fact or interpretation.

References

ADFG (Alaska Department of Fish and Game), 1994. Alaska's District 104 Purse Seine Fishery. ADFG, Juneau.

Crutchfield, J. A., and G. Pontecorvo. 1969. The Pacific salmon fisheries: a study of irrational conservation. Johns Hopkins Press, Baltimore, Maryland.

Culbert, L., and J. Beatty. 1999. Salmon pact "disappointing." Vancouver Sun, June 4, Final edition, A1.

DFO (Department of Fisheries and Oceans, Government of Canada). 1999. Canada and U.S. reach a comprehensive agreement under the Pacific salmon treaty. Press release and backgrounders, June 3. http://www.ncr.dfo.ca/

Downton, M. W., and K. A. Miller. 1998. Relationships between salmon catch and North Pacific climate on interannual and decadal time scales. Canadian Journal of Fisheries and Aquatic Sciences 55:2255–2265.

Francis, R. C., and S. R. Hare. 1994. Decadal-scale regime shifts in the large marine ecosystems of the Northeast Pacific: A case for historical science. Fisheries Oceanography 3:279–291.

Francis, R. C., and N. J. Mantua. 1996. Climatic influences on salmon populations in the Northeast Pacific, report for the National Marine Fisheries Service workshop on "Assessing extinction risk for West Coast salmon." Joint Institute for the Study of the Atmosphere and Oceans, University of Washington, Seattle.

Fudenberg, D., and J. Tirole. 1991. Game theory. MIT Press, Cambridge, Massachusetts.

Gargett, A. E. 1997. The optimal stability "window": a mechanism underlying decadal fluctuations in North Pacific salmon stocks? Fisheries Oceanography 6(2):109–117.

Hare, S. R. 1996. Low frequency climate variability and salmon production. Ph.D. thesis, University of Washington, Seattle.

Higgs, R. 1982. Legally-induced technical regress in the Washington salmon fishery. Research in Economic History 7:55–86.

Jaeger, W. K. 1997. Saving salmon with fishwheels: a bioeconomic analysis. Natural Resources Journal 37(4):785–808.

Kennedy, J. O. S., and H. Pasternak. 1991. Optimal Australian and Japanese harvesting of Southern bluefin tuna. Natural Resource Modeling 5(2):213–238.

Mantua, N. J., S. R. Hare, Y. Zhang, J. M. Wallace, and R. C. Francis. 1997. A Pacific interdecadal climate oscillation with impacts on salmon production. Bulletin of the American Meteorological Society 78(6):1069–1079.

McKelvey, R. 1997. Game theoretic insights into the international management of fisheries. Natural Resource Modeling 10(2):129–171.

Munro, G. R., and R. L. Stokes. 1989. The Canada-U.S. Pacific salmon treaty. Pages 17–35 *in* D. McRae and G. Munro, editors. Canadian oceans policy: national strategies and the new law of the sea. University of British Columbia Press, Vancouver.

Munro, G. R., T. McDorman, and R. McKelvey. 1998. Transboundary fishery resources and the Canada-U.S. Pacific salmon treaty. Occasional papers: Canadian - American public policy, Canadian - American Center, Number 33. University of Maine, Orono, Maine.

Nickelson, T. E. 1986. Influences of upwelling, ocean temperature, and smolt abundance on marine survival of coho salmon (*Oncorhynchus kisutch*) in the Oregon production area. Canadian Journal of Fisheries and Aquatic Sciences 43:527–535.

O'Neil, P. 1999a. Taking stock of the salmon treaty. Vancouver Sun, July 5, Final edition, A1.

O'Neil, P. 1999b. Conservation card played key role in salmon deal. Vancouver Sun, June 5, Final edition, A1.

Osborne, M. J., and A. Rubinstein. 1994. A course on game theory. MIT Press, Cambridge, Massachusetts.

Owen, G. 1982. Game theory, 2nd edition chapter 7: two-person general-sum games. Academic Press, New York.

Pacific Salmon Commission, Joint Interceptions Committee. 1993. Third report on the parties' estimates of salmon interceptions - 1980-1991, Report JIC (93)-1.

Pearcy, W. G. 1992. Ocean ecology of North Pacific salmonids. University of Washington Press, Seattle.

Sands, N. J., and J. L. Hartman. 2000. A simulation model to assess management, and allocation alternatives in multi-stock Pacific salmon fisheries. Pages 435–450 *in*

Knudsen, E. E., C. R. Steward, D. D. MacDonald, J. E. Williams, and D. W. Reiser, editors. Sustainable fisheries management: Pacific salmon. CRC - Lewis Publishers, Boca Raton, Florida.

Shepard, M. P., and A. W. Argue. 1998. Ocean pasturage in the Pacific salmon treaty fact or fiction? Canadian Industry Report of Fisheries and Aquatic Sciences 242. Department of Fisheries and Oceans Canada, Vancouver.

Stevens, T. 1986. United States-Canada salmon treaty negotiations: The Alaskan Perspective. Environmental Law 26:423–430.

Strangway, D. W., and W. D. Ruckelshaus. 1998. Pacific salmon: report to the Prime Minister of Canada and the President of the United States. January 12.

Treaty between the Government of Canada and the Government of the United States of America Concerning Pacific Salmon, March. 1985.

U.S. Department of State. 1999. Diplomatic Note No. 0225 from Canada to the United States; reply; attached agreement, June 30. <*http://www.state.gov*>.

U.S. Senate. 1985. Pacific salmon treaty act of 1985. Public Law No. 99-5, 99 Stat. 7.

Waldeck, D. A., and E. H. Buck. 1999. The Pacific salmon treaty: the 1999 agreement in historical perspective. U.S. Congressional Research Service report, June 18, 1999. Order Code RL30234, Washington D.C.

Yanagida, J. A. 1987. The Pacific salmon treaty. The American Journal of International Law 81:577–592.

Sustaining North American Salmon: Perspectives Across Regions and Disciplines, 369–376

Chapter 17

Challenges and Opportunities for the Future of Pacific Salmon Management

Jim Martin

Introduction

This volume offers an excellent opportunity to contrast the implementation of cooperative management along the Pacific coast under the Pacific Salmon Treaty (PST) with management in the Great Lakes and along the Atlantic coast. It is most important to learn from what is working and what is not so we can improve cooperative management in the future.

The Pacific Salmon Treaty

I served as negotiating assistant for Director Jack Donaldson of the Oregon Department of Fish and Wildlife during the Pacific Salmon Treaty negotiations in 1985. After 20 years of fruitless negotiation, we all experienced near euphoria as we finalized the treaty. There had been a "near-miss" in 1982, when Mike Hunter and Lee Alverson put together the founding principles of the treaty and hit a couple of speed bumps. These reversals were caused by politically influential fishing groups, such as the Southeast Alaska Salmon Trollers and the Puget Sound Purse Seiners, who vetoed the agreements in 1982. Without support of these groups, the U.S. government was unwilling to go forward.

In 1985, the federal government sent the message that there was going to be a treaty whether all the fishing groups signed on or not. Eventually, we agreed to manage this great salmonid resource for mutual benefits and for long-term productivity. We believed that the treaty principles were so strong that we would work out the details, and we were wrong. The period since 1985 has been frustrating for all involved—small problems have been allowed to become big problems, and gridlock has been the general default.

There is no question that the relationship between Canada and the United States in managing salmonid fisheries is worse on the Pacific coast than in the Atlantic and the Great Lakes regions. Why is that? Why do our countries have so much trouble working together on the Pacific coast when we work well together in the Great Lakes and with much less difficulty on the Atlantic coast?

I believe it is because many fisheries-related situations are so different between regions. In the Great Lakes, the status of the fisheries, the overwhelming challenge of sea lamprey predation, and the management of introduced species are all challenges shared commonly by the United States and Canada. In the Great Lakes region, the United States, all the states, Canada, and the provinces are working shoulder-to-shoulder on problems rather than working against one another. I believe this is also basically true in the Atlantic region. But it is not so in the Pacific region.

Productivity Swings in the Pacific

Consider the geographic situation between the Gulf of Alaska and Washington/Oregon. In Alaska is the last bastion of industrial salmonid fisheries and local communities built around fisheries. Alaska still has diverse wild fish stocks and large expanses of unaltered, highly productive freshwater habitat. Alaska and northern British Columbia have experienced 25 years of high ocean productivity. There is a dramatic contrast between the abundance and productivity of the stocks and fisheries of northern British Columbia/Alaska and the depressed stocks, depleted habitat, low marine survival, and endangered species listings in the southern Pacific region. No wonder we have trouble finding consistent and flexible management systems. We might as well be on different planets.

The basic principles that were agreed to in the Pacific Salmon Treaty of 1985 were to take mutual action to conserve the salmon stocks of the region (the conservation principle), to share in the harvest proportionate to the production of each country (the equity principle), and to avoid excessive disruption of traditional fisheries. The basic problem has been that the restrictions to conserve the weak stocks required each country to forgo harvest of stronger stocks in the mix of stocks in the ocean. If each country was to shift its fisheries to meet the conservation principle, where was the incentive to pay the "equity bill" and how could this be done in patterns that minimized disruption to fisheries that were developed specifically to intercept the other country's fish? Examples include the southeast-Alaska troll fishery, which intercepts 90% non-Alaskan chinook, or the Puget Sound sockeye fishery, which intercepts Fraser River sockeye *Oncorhynchus nerka*, or the coho troll fishery off the west coast of Vancouver Island, which catches more than 50% U.S. coho salmon *O. kisutch* stocks.

The approach we took in 1985 was to focus first on the conservation principle and allow the equity principle to be worked out later. Our approach to conservation of chinook, in particular, was that we would constrain harvest to not exceed conservative ceilings, and that would provide stability to fisheries while allowing the stocks to rebuild. We anticipated that the stocks would rebuild by 1988, and then the pie would be larger to split among all the interests. We were wrong in our fundamental assumptions. We failed to anticipate the swings in marine survival, and we underestimated the imbalance in fisheries interceptions (equity) that would result from these swings. The divergence of the productivity of the northern and southern portion of the Pacific region created an equity and conservation imbalance that could not be reconciled without politically impossible restrictions on Alaskan, northern B.C., and tribal fisheries. We have been struggling to find a solution ever since.

In addition to the biological/ecological differences between the northern and southern Pacific regions, the political and economic structures of the communities are completely different. Fisheries restrictions and buyouts that are possible and even reasonable in a diversified community become impossibly onerous in small fisheries-dependent communities of Alaska and northern British Columbia.

Most of us believe that the basic principles of conservation and equity continue to be sound. However, the decision-making structure and the incentives for agreement are fatally flawed. We must develop a mechanism to reconcile these principles with the highly diverse and fluctuating biological setting of the northern Pacific region.

Incentives for Agreement?

All the incentives in the 1985 Pacific Salmon Treaty are working in the wrong direction. There was high incentive to negotiate endlessly without agreement. There was also strong incentive not to come to agreements. The lack of a forcing mechanism or a strong financial incentive allowed the parties to settle on status quo rather than cooperative progress. In the long term, it was bound to cause the collapse of the Pacific Salmon Treaty, which could create a severe loss of value to the region and our future. The answer is to create an incentive for agreement and a forcing mechanism that makes gridlock unacceptable.

I believe McKelvey and Miller (Chapter 16) are correct in describing the need for a "side payment" to correct the equity imbalance. We have to figure out equity payments that have currency in those remote, small communities in southeastern Alaska and northern British Columbia. Simply writing everybody a check is not going to preserve their culture, and that should be important to all of us. It is possible to meet the equity principle and work with people, but we are going to have to develop a much more respectful relationship. We are going to have to develop a forcing mechanism for dispute resolution that has iron-sharp teeth in it.

Back in 1985, one of our greatest difficulties was finding agreement between the Alaskans and the delegation from Washington/Oregon. This is still very difficult, even today. In 1985 we had an Indian allocation lawsuit (the "All Citizens" allocation lawsuit), which was a major motivator for agreement. We anticipated that the Pacific Salmon Treaty would allow us to rebuild the stocks to capacity by 1998 (we were dreaming!), and then the debate would be on how to share the surplus. We agreed that if the post-1998 sharing negotiations resulted in gridlock, then nobody would fish that year. Now there is a forcing mechanism with some teeth in it! This agreement related to sharing between the northern and southern interests on the American side. It did not relate to the Canadians. Some have suggested, however, that such a forcing mechanism could also apply to the U.S. and Canadian interests in the treaty.

As a side note, when 1998 finally came along, all sides agreed that the stocks had not been rebuilt, and all agreed to extend the All Citizens allocation lawsuit agreement for more time to get the stocks rebuilt. They are not rebuilt yet.

We might explore a dispute resolution similar to that included in the All Citizens settlement, where we would come to agreement or nobody fishes. Perhaps we need to restrict the negotiations between the regional interest groups to a very short period of time. Following that, we could switch to mediation. If mediation fails, we could shift to arbitration. If we fail to accept the arbiter's solution, nobody fishes that year. I think that is the kind of forcing mechanism that would have moved the 1985 treaty forward. The key question is how to get such a mechanism in the system over the political opposition of those who value their current veto power.

We must change the dynamic where the regional governments protect their fishermen at the expense of the resource. This kind of brinksmanship does the greatest disservice to the resource, fishermen, and communities in the long term. However, it seems to work politically in the short term. Fishermen who gain the greatest value from the resource should be paying the equity bill, not the general taxpayers. Fishermen who gain the most from the resource should feel the sacrifice if gridlock is reached, as with the no-fishing-without-agreement provision. The current gridlock is created by the expectation that the equity obligation can be avoided without serious costs.

This focus on divisive allocation disputes and the dramatic variability in stock status across the Pacific region is fundamentally different from the situation in the Great Lakes and the Atlantic. It is no wonder that the management relationship is different as well.

In the past few years, there has been a major breakthrough in renegotiating the fishing annexes of the Pacific Salmon Treaty. In June 1999, the United States and Canada signed a new agreement that amends the 1985 treaty. Much credit goes to the federal, state, and tribal negotiators on both the Canadian and U.S. sides. As anticipated, the agreements involve major fishing constraints among the southern fisheries off Canada and Puget Sound without much change in the Alaskan or tribal fisheries. Substantial money from the United States is required to fund the enhancement accounts in the northern and southern regions (see MacDonald et al., Chapter 15; and McKelvey and Miller, Chapter 16, for more information on these endowment funds). This may turn out to be a lasting way of responding to the equity and conservation problems. Time will tell whether we move forward or revert to fingerpointing and gridlock. Even this level of progress would have been deemed unlikely a few short years ago. It is unlikely that the long-term problems of decision making under the Pacific Salmon Treaty are over as long as all the key players wield a veto.

Exploring Common Themes

I have been attending the American Fisheries Society (AFS) meetings for a long time, and I rarely see the kind of synthesis and cross-communication between the three North American salmonid management systems as explored in this volume: the North Atlantic, the Great Lakes, and the North Pacific. AFS should establish a systematic mechanism for such collaboration.

We have common themes such as international structures and cooperation across boundaries. In each salmon region of North America, we have degraded streams where whole watershed restoration is going to be a key component. We share these big, fluctuating marine systems or, in the case of the Great Lakes, this big, highly unstable freshwater ecosystem. We recognize that these systems are oscillating in natural productivity, and we must design flexibility to respond to the variability. Of course, in the Great Lakes, we recognize that some of the system variability is due to human-caused effects, such as pollution and exotic species impacts. However, the fluctuations must be planned for, regardless of cause.

We are struggling with highly altered fish community structures and asking how to make these altered systems function. We share a heavy reliance on hatchery production as a common element of each of these three management systems. We are struggling to meet longstanding social and economic goals and Indian treaty obligations at the same time that the attitude of the public is rapidly shifting from a commodity perspective to a focus on conservation. We have endangered species and depressed stock situations popping up all over the place. At the same time, fish have moved from the sports page to the front page. Fish are no longer just the commodity—they are also the indicator of ecosystem health. This dramatic shift has happened over the short span of our careers, and we are struggling to get out in front of the changing demands for conservation and management in each of the three regions of North America.

This is a great time for AFS to take the lead in collaboration between salmon managers in the Great Lakes, Atlantic, and Pacific regions. We should be learning collaboratively from one another rather than independently struggling with the same issues in parallel. As we design a mechanism for on-going collaboration, we should not forget the fourth great salmonid ecosystem on the planet, the western Pacific region. The Russians and Japanese are also dealing with the same questions. In eastern Russia, one of the last unaltered salmonid ecosystems in the world exists, but the professionals there have too few resources to effectively study the system. We are out of their communication loop and they are out of ours, though some progress has been made recently.

Suggestion: A Role for the American Fisheries Society

Perhaps we could institutionalize a binational, cross-regional symposium as a standing part of the AFS annual meeting schedule every four years. The primary themes (genetics, habitat restoration, hatcheries, fisheries exotics, etc.) could vary, but the basic idea would be to update our collective thinking on salmonid science and management issues across the breadth of these regions on a systematic basis. I have been invited to speak in the Great Lakes region several times and have found professionals struggling with the same problems that are on the "front burner" in the Pacific region. Friends from the Atlantic region discuss similar problems. A regular forum for collective learning and a system of visitations for detailed follow-up would create value. Our current mechanism is to refer to the literature, but this falls short of our need. We need to have the academics and the managers from these regions collaborating on a systematic basis.

As an example, I would offer my experience working with Brian Riddel, of the Canadian Department of Fisheries and Oceans. When Brian entered the Pacific Salmon Treaty process, he had just come from work on the Atlantic salmon problem on the East Coast. He was able to transfer much of the experience to help us. Brian went on to become the leader of the Bilateral Chinook Science Team for the Canadian side. His professionalism and broad understanding helped keep the nationalism of the delegations and the divisiveness of the arguments from penetrating the science team for the most difficult and contentious issues managed under the PST. This is the kind of professionalism that must be rewarded and supported. For his efforts, Brian was broadly respected by Canadian and U.S. delegations alike.

We need to think more creatively about long-term solutions to problems we share. It is important to distinguish the urgent from the important. As an example, we plan habitat restoration a project at a time, while instinctively knowing that the long-term development pattern will drive the health of the watershed in 50 years. So how do we take action that will help shape the long-term picture while doing the short-term habitat restoration projects? Can we help society and politicians see the value of healthy watersheds and fish populations in the future? In the current political climate in the Columbia basin, it seems probable that we will make progress in the next decade. The listing of so many stocks in the Columbia under the Endangered Species Act has clearly changed the balance of politics. Many development projects that would have enjoyed automatic approval in the past (e.g., filling, dredging, irrigating, unscreened water diversions, power production strategies) are now seriously questioned on scientific and legal adequacy with respect to fish impacts. Why, we are even thinking of breaching major hydropower dams and losing 140 mi of barging in the Snake River. Such things were unthinkable two decades ago.

The question is whether we have the staying power to institutionalize the incentives and disincentives so that we make progress during the next four decades. What a great exercise it would be to discuss the major societal factors driving long-term success in the Asian, Pacific, Great Lakes, and Atlantic regions.

The broad thesis of this paper is how interconnected we are. Fish conservation and management have moved from the state as the dominant venue to the region. The next great leap is to plan at the regional level and to collaborate between regions to facilitate quick learning. Let me offer an example based on an emerging global problem.

The Challenge of Global Climate Change

Consider the challenge that global climate change may create for natural resource managers. If the climate changes rapidly over the next 50 years, as some have suggested, then many stocks might be unable to adapt new environmental tolerances to keep up. Therefore, many stocks may go extinct, only to be recolonized by stocks from southern latitudes that are already adapted to the new environmental conditions. It is quite possible that the stocks of northern California may be best adapted to the temperatures and environmental patterns of the Columbia basin in 50 years. The Columbia stocks may well be the founding stocks for the British Columbia rivers, and so on.

If this is so or even could be, then it is in all our interests to preserve the whole stock complex with maximum variety now. We cannot afford to conclude that the preservation of the stocks of each state is simply up to that state. It may well be that we will depend on those stocks for our future. We have never experienced the pace of environmental change that could occur during the next 50 years, and we will depend on one another to have choices available. We are more interdependent on one another today in the Pacific region than is generally recognized. That is why working the bugs out of the Pacific Salmon Treaty is so important. That is why within-region and between-region collaboration is so important to shorten the learning time and quicken the response.

The Best of Times

After 30 years of working on these problems, it is important to reflect on what a wonderful time this is. For most of our history as biologists, we have had lots more ideas than receptive audiences. The political decision-making system was not very interested in our issues. As a society, we were busy taming the earth. Development was our agenda. Because of the Clean Water Act, the Endangered Species Act, and environmental concerns around the globe, we are suddenly in demand to offer good advice. We are speaking to whole new audiences, advising political leaders and pushing the limits of our own thinking. Governors are even inviting fish chiefs to be their salmon advisors! We are actually studying whether huge hydropower dams should come out to save salmon. What a time!

We cannot allow short-term thinking, parochial viewpoints, or lack of collaboration to stand in the way of our potential. Consider the generations of biologists who struggled against huge odds to preserve as many choices as possible for us today. They fought the battles against the development juggernaut in obscurity, in understaffed agencies, long before "ecosystems" and "biodiversity" were household words. Our society built the dams, channelized the rivers, degraded the streams, and paved over the wetlands without so much as a second thought, except for the heroic resistance of our "tribal elders" among the biologists. We can best honor them by taking full advantage of the opportunities of these times. We have the attention of the political system right now. We have a massive opportunity to learn from one another in these large regional ecosystems. Let's learn together collaboratively, and let's have a good time doing it.

Summary

Sustaining North American Salmon: Perspectives Across Regions and Disciplines, 379–395

Chapter 18

Sustaining Salmon Fisheries in Canada and the United States: Recommendations for Policy and Management

K. D. Lynch, M. L. Jones, and W. W. Taylor

Introduction

A great deal has been written about the need for sustainability of North American salmon fisheries (e.g., Knudsen et al. 2000) and about the challenges impeding progress toward this goal. We believe the chapters assembled herein make valuable contributions to the review and analysis of salmon sustainability in the United States and Canada. By compiling and comparing information from these chapters about salmon management across regions and disciplines, we offer further insight into the factors that hinder as well as promote salmon sustainability. From this, we develop recommendations for progressing toward sustainability in the areas of policy, management, communications, and research.

Contrasting Themes Across Regions

The preceding chapters illustrate vast differences in the origins of today's tribal, commercial, and recreational salmon fisheries across North America. Historically, the North Atlantic wild salmon fishery, as described by Kocik and Brown (Chapter 1) and Beland and Bielak (Chapter 3), made important yet—when compared with the North Pacific salmon fisheries—relatively modest contributions to the region's early tribal and commercial harvests. Many other fishes and food sources were available to early communities, and early colonial settlements were quick to utilize and block rivers for industrial purposes. For the Pacific Northwest (including British Columbia and Alaska), Glavin (Chapter 13) described the salmon's central role in aboriginal history and culture, as well as the significance of these fish to the commercial fisheries of the past two centuries. In this region, salmon resources helped provide the foundation on which cultures, economies, and communities were built, and they continue to figure prominently in defining the heritage and identity of the Pacific Northwest. In contrast to these two maritime regions, the Great Lakes salmon fishery was established only a few decades ago. Aboriginal and early European communities in this region depended upon other fish species and resources. Tanner and Tody (Chapter 7) described how the introduction of salmon signaled a shift in management focus away from restoration of Great Lakes commercial fisheries (largely based on lake trout and various coregonines) toward a salmon-based recreational fishing industry that continues to thrive in and around the lakes. Recreational angling is also of considerable economic importance in the Pacific Northwest, but it does not dominate the commercial and aboriginal fisheries the way it does in the Great Lakes and the Atlantic regions.

Differences in the origins of the salmon fisheries are not surprising, considering the range of social and ecological conditions across the continent that set the stage for fisheries development. The salmon fisheries that have developed in each region now display considerable variations in scale and scope. For example, salmon in the North Atlantic are still caught in some international commercial fisheries, but this salmon fishery is nearly nonexistent in the United States and Canada (Department of Fisheries and Oceans 2001). In the Canadian sport fishery, however, Whoriskey and Glebe (Chapter 4) described how anglers spend CAN$74.1 million to pursue Atlantic salmon. The recreational fishery is even larger in the Great Lakes, where U.S. and Canadian sport fishers spend more than US$1.3 billion (not including river angling), largely in pursuit of salmonids (Bence and Smith 1999). Great Lakes commercial fisheries, in contrast, generated US$52 million dockside value in 1991, and salmon make a negligible contribution to this fishery (Brown et al. 1999). The Great Lakes salmon fishery is dwarfed by the size of the Pacific salmon fishery. Glavin (Chapter 13) described a coast-wide aboriginal salmon fishery that was estimated to provide as much as 130 million pounds of salmon. Several million pounds of salmon continue to be harvested annually, and the exvessel value of commercial landings across the entire region was estimated near US$600 million in the early 1990s (Huppert 1995; Alaska Department of Fish and Game 2001). Despite recent declines in the salmon fisheries of the Pacific Northwest, commercial, tribal, and recreational fisheries are still both regionally significant and collectively much larger than those fisheries in other regions of North America.

Although Lupi and Jester (Chapter 10) argued that the value of salmon is far greater than can be estimated using consumption-based economic indicators, the values reported here do reflect the relative regional significance of these fisheries. Differing values lead to differences in regional priorities for salmon. Rideout and Ritter (Chapter 5) and Whoriskey and Glebe (Chapter 4) explained how the closure of the Atlantic salmon recreational fishery in the United States has reduced angler interest in salmon, and even the relatively small commercial Atlantic salmon fishery receives little federal support. Tanner and Tody (Chapter 7) and Knuth (Chapter 9) both indicated that salmon are more of a public and political priority in the Great Lakes, where sport fisheries stakeholders increasingly recognize and seek to maintain this important regional fishery. Given the large human population of the region and the associated diversity of issues and industries throughout the basin, however, this nonnative, hatchery-dependent salmon fishery does not always top the wider list of regional priorities. Our Pacific Northwest salmon chapters clearly showed that salmon are a much greater priority in this region. According to Glavin (Chapter 13), Knudsen (Chapter 14), MacDonald et al. (Chapter 15), McKelvey and Miller (Chapter 16), and Martin (Chapter 17), wider sectors of society—even those outside the traditional stakeholder set of commercial and recreational fishers—seem to be aware of the ecological, economic, and cultural significance of salmon in the region, and as a result, the public may be more likely to consider or support various social and political measures for salmon conservation and restoration.

Having different origins, histories of use, and current priorities in their respective regions, salmon fisheries across North America also demonstrate different stages of collapse and restoration. Atlantic salmon stocks are in critical condition, especially in the United States and the Inner Bay of Fundy. Habitat loss, a history of excessive commercial fishing, and, more recently, poor marine survival have resulted in the near elimination of wild Atlantic salmon production. In the Great Lakes, Hansen and Holey (Chapter 8) described a drastic reduction in chinook abundance and catch rates in Lake Michigan in the late 1980s, but this collapse was different from that of Atlantic salmon, which is now listed under the Endangered Species Act (ESA). In the Great Lakes, salmon productivity is more a question of stocking strategies, food web dynamics, hatchery techniques, and diseases (although habitat conditions may also affect wild salmon production), and fishery managers are not yet dealing with large-scale crisis conditions. If one imagines that the Atlantic salmon and Great Lakes salmon fisheries represent two ends of a stock status continuum, then the North Pacific salmon fisheries could be described as being between these two ends. Numerous Pacific salmon stocks have diminished in size and diversity, and many stocks in the southern part of the range are now threatened or endangered (Knudsen et al. 2000). Regionally, however, salmon production still supports significant commercial, tribal, and recreational fisheries. Overfishing, ocean conditions, and a range of watershed-based habitat issues, as described by all the authors addressing Pacific salmon, have combined to diminish—though not yet eliminate—the productive capacity of Pacific salmon. Many stocks could still be restored or expanded through aggressive habitat- and stock-conserving measures, although it is likely that certain genetic stocks have been lost over the years.

Not only are salmon fisheries in each region at different stages of collapse and restoration, but also, each region faces a different set of issues that contribute toward stock status. On the eastern coast of North America, Kocik and Brown (Chapter 1), Rideout and Ritter (Chapter 5), and Friedland (Chapter 6) described how Atlantic salmon managers are becoming increasingly concerned about the impacts of aquaculture operations, which present challenges in maintaining genetic diversity and reducing the risk of diseases. Aquaculture is also expanding on the west coast, with similar concerns beginning to be expressed about the potential for impacts on wild salmon (Auditor General of Canada 2000). Cage-based aquaculture is in the early stages of development in the Canadian portion of the Great Lakes and has not yet attracted much attention. Still, the hatchery-based salmon fishery of the Great Lakes also has to deal with issues of relatively low genetic diversity and increased risk of disease outbreaks. Great Lakes salmon managers must also increasingly deal with predator-prey balance, contaminants, and introduction of ecosystem-altering exotic species (e.g., zebra mussels). Other factors at play on the east and west coasts—including habitat loss, overfishing, and stock interceptions—are still often viewed as impediments to recovery, yet despite significant harvest reductions, especially on the east coast, stocks continue to decline for reasons that are poorly understood.

Because each region's salmon fisheries and their related conservation issues are so different, we find that stakeholders in each region also appear to have different expectations for and types of dependence on the fisheries. In the eastern United States and Canada, commercial fishers now depend on nonsalmon-based fisheries to make their living, and recreational salmon fishers have either switched to other species or have the means to enjoy the elite fishing camps of Canada, as described by Beland and Bielak (Chapter 3). Whoriskey and Glebe (Chapter 4) also described some areas in Canada where local economies are largely dependent on salmon angling, despite the relatively low regional priority placed on salmon. Because of limited access to other Atlantic salmon rivers and restrictions on fishing, as in the eastern United States, other recreational anglers appear to be lowering their expectations for salmon angling opportunities. In the Great Lakes, however, recreational salmon fishers have developed great expectations for the salmon fishery (Gale 1987), and many coastal communities emphasize this recreational potential in their tourism promotion. Large capital investments in the supporting facilities (e.g., marinas) are based on the assumption of a sustainable and highly productive fishery. Because it is a relatively new fishery, however, stakeholders in the Great Lakes do not share the multigenerational expectations and salmon-dependent lifestyles built around salmon fisheries, as in the Pacific Northwest. In the West, many fishing-dependent families and communities may have more difficulty in adjusting to reductions in Pacific salmon harvests because they may have expectations based on the productive potential that they experienced in past years, and the economic alternatives for remote communities are limited. More subsistence and recreational fishing also occur there, and salmon habitat issues are more deeply intertwined with other aspects of the Pacific Northwest lifestyle (e.g., dam dependence, cultural identity, etc.). Therefore, stakeholders throughout the Pacific Northwest region tend to be more aware of and affected by salmon conservation measures such as fishery closures or dam removal than those in other regions of North America.

All these differences among regions likely contribute to differences in the origins of and purposes for binational institutions whose concerns include salmon conservation. For example, the North Atlantic Salmon Conservation Organization (NASCO), created in 1983 to address allocation issues, was motivated by concern over interception, especially in Greenland fisheries. For the Great Lakes region, Tanner and Tody (Chapter 7), Goddard (Chapter 12), and Dochoda and Jones (Chapter 11) discussed the original and continuing role of the Great Lakes Fishery Commission (GLFC), which was initially established to control sea lamprey, a common threat to many Great Lakes fisheries. The GLFC does not have authority to address fishery allocation issues, but it has provided an interjurisdictional forum for discussing other shared fisheries issues and concerns. On the Pacific coast, the Pacific Salmon Commission was established principally to resolve debates arising from the Pacific Salmon Treaty about responsibility for and allocation of commonly harvested stocks. Dochoda and Jones (Chapter 11) argued that the GLFC has played a key role in fostering cooperative fishery management in the Great Lakes region, while the Pacific and Atlantic institutions have tended to operate in a medium of conflict, primarily over allocation. Dochoda and Jones (Chapter 11) suggest that the creation of a binational institution to address common concerns rather than to provide a forum for resolving conflict helps engender a greater spirit of cooperation among the parties before conflicts arise.

This book may not provide an exhaustive list of the differences between the salmon fisheries of the three regions, but the chapters collectively point to different outlooks or prospects for the future of salmon conservation in each region. According to Friedland (Chapter 6), Kocik and Brown (Chapter 1), and Rideout and Ritter (Chapter 5), the future of wild Atlantic salmon lies in our ability to determine and address the causes of continued low marine survival as well as the effects of the aquaculture industry. River habitat quality and availability will also need to be improved for the abundance of Atlantic salmon to increase. It is likely that restoration will hinge on the answer to the question of how much the Atlantic salmon is worth. The Great Lakes salmon fishery appears to face a more certain future, at least in the short run, assuming managers can understand changes in food web balance, maintain coordinated hatchery production, and adapt to other ecosystem changes. Despite some lingering concerns about salmon's status as an introduced species, salmon receive a great deal of public, economic, and political support throughout the basin. Salmon management efforts will continue, and the sustainability of this fishery may lie in balancing stakeholder expectations with fluctuations in salmon productivity.

Great Lakes salmon managers experience varying degrees of certainty over the future of their fisheries, but this is not the case in the Pacific Northwest. In this region, the future of salmon fisheries depends on potentially significant changes in both mixed-stock fishery allocation and watershed-based habitat conditions. Because of persistent stock interception issues, Glavin said that commercial fisheries should shift from mixed-stock (open-water) to terminal (in-river) harvests; McKelvey and Miller (Chapter 16) described how side payments to compensate for interceptions might help. In addition, Knudsen (Chapter 14) and MacDonald et al. (Chapter 15) emphasized that salmon habitats affected by numerous watershed and water use issues need to be addressed in a more holistic, integrative manner. Clearly, the future of Pacific salmon fisheries depends on our ability to create equitable and sustainable fishing patterns as well as conservation and restoration of many upstream habitats; *both* fronts must be addressed if any degree of sustainability is to be achieved.

Common Themes Across Regions

Despite substantial differences among regions, these salmon fisheries also share many common features. Perhaps foremost, ecological and behavioral similarities between Pacific and Atlantic salmon lead to similar concerns about the impacts of environmental stresses on these species, particularly in the two maritime regions. The anadromous, migratory nature of salmon is the reason that interception fisheries have been such a concern in both the Atlantic and the Pacific. Their migratory nature also leads salmon stocked into the Great Lakes by one jurisdiction to be harvested by others (e.g., Ontario anglers frequently harvest chinook salmon stocked into Lake

Ontario by the state of New York); yet in contrast to the marine salmon fisheries, there has never been any controversy surrounding this fact. Great Lakes anglers do not seem to view this type of interception as a problem, despite voicing strong concerns about incidental catch of salmon in commercial fisheries.

Anadromy also means that wild salmon depend on productive and accessible habitats in both freshwater and marine environments. In both the Pacific and Atlantic oceans, there is now good evidence that salmon productivity is greatly influenced by large-scale climatic phenomena and that the effects of climate shifts are felt in common by different stocks originating from the same region. For example, in the Pacific, we note declines in productivity of southern stocks and increases in northern stocks during the 1980s and 1990s, as discussed in several of the Pacific salmon chapters. These so-called marine regime shifts have had an enormous impact on the interplay among the parties to negotiations about commercial harvest allocations, as discussed eloquently by McKelvey and Miller (Chapter 16). In the Atlantic, Friedland (Chapter 6) presented evidence that changing marine conditions have affected growth and survival of Atlantic salmon stocks. Kocik and Brown (Chapter 1) and Rideout and Ritter (Chapter 5) view understanding effects on marine survival as a key to the future recovery and sustainability of Atlantic salmon.

Do the salmon stocks of the Great Lakes share this connection between climate and ocean survival? For Great Lakes salmon, the "oceans" are large lakes, so the conditions they experience differ considerably from those experienced by marine anadromous stocks. Nevertheless, there is emerging evidence that climate is important to salmon productivity in the Great Lakes, at least for chinook salmon. It has been hypothesized that the collapse of chinook stocks in Lake Michigan, described by Hansen and Holey (Chapter 8), may have been triggered by an interaction between reduced consumption of alewife and the thermal stresses experienced in the Great Lakes by overwintering salmon. Salmon in the Great Lakes must endure winter temperatures of 4°C or less, much colder than the temperatures normally faced during winter by the (southern) salmon stocks from which the Great Lakes stocks were derived. These cold temperatures may lead to more rapid depletion of winter lipid reserves in Great Lakes salmon and, thus, to elevated stress in late winter and early spring.

In the Great Lakes, much ecological research has focused on the effects of changes in the food web on salmon productivity, as discussed by Hansen and Holey (Chapter 8). In the Pacific and Atlantic, greater emphasis has been placed on correlating ocean physical conditions with growth and survival indices. We speculate that these might, in fact, be two ways to look at much the same problem, with the different approaches being motivated mainly by practical issues of observability. Conducting food web studies in vast marine systems is extremely difficult, but it seems plausible that the causal linkages between climate changes and salmon productivity include a food web pathway. Perhaps the trophodynamic research that has contributed to our understanding of Great Lakes salmonine systems offers a basis for developing hypotheses about the changes in salmon productivity that have been observed in marine systems. Conversely, as Great Lakes salmon abundance time series grow longer, it might be fruitful to examine the linkages between climate and productivity in these systems.

Ecological similarities among regions are even more evident when one considers freshwater riverine habitats. During their migratory periods, salmon need to bypass dams and other obstructions in their search for suitable habitats. Evidence abounds from both the Pacific and the Atlantic that barriers to migration have profoundly affected salmon stocks, particularly in the United States. In the Great Lakes, current levels of production of wild salmon are similarly limited by access to spawning and rearing habitats. Concerns about other watershed-based effects on salmon productivity, perhaps most strongly emphasized by Knudsen (Chapter 14) and MacDonald et al. (Chapter 15) for the Pacific Northwest, are certainly relevant to all three regions. For example, the effects of land cover changes (e.g., timber removal, urbanization) on stream hydrology are equally important to the freshwater survival of Pacific and Atlantic salmon.

Pacific and Atlantic salmon also have in common a remarkable diversity of life history patterns throughout their range. Hutchings (Chapter 2) made this abundantly clear for Atlantic salmon, but the members of the genus *Oncorhynchus* that make up Pacific salmon exhibit at least as much variability in life history characteristics. Even Great Lakes stocks of salmon, especially steelhead, show considerable variation. Hutchings (Chapter 2) argued that this variability reflects local adaptations to environmental conditions faced by individual populations. Both Knudsen (Chapter 14) and Hutchings (Chapter 2) emphasized that the success of conservation and, especially, restoration efforts will hinge on the preservation of genotypic and phenotypic diversity. In the Great Lakes, the extirpation of lake trout stocks from most of the basin has led to a loss of genetic variation that is believed to pose a serious impediment to the restoration of these stocks (Burnham-Curtis et al. 1995; Krueger and Ihssen 1995).

Phenotypic diversity will likely become increasingly important in all regions as global ecosystem change occurs. Salmon are sensitive to temperature, and changes in ocean temperature may affect their range of available habitat, cause shifts in productivity, and disrupt their seasonal migration timing. Similarly, warming of rearing streams may alter the suitability of these habitats for juvenile salmon. Managers and other stakeholders can try to anticipate and adjust stocking and harvest patterns to respond to these changes, but the long-term sustainability of Atlantic and Pacific salmon populations may depend on their ability to adapt.

Another shared trait among regions is some degree of dependence on artificial enhancement. Hatchery-based production and stocking are used for salmon management across the continent to supplement natural production, introduce new strains or stocks, reestablish stocks in their historical range, or simply increase salmon population sizes. Hatchery managers face many challenges related to genetic diversity, disease, and negative interactions between hatchery and wild stocks—especially on the east and west coasts. Whether being used as a temporary stock-enhancement measure (i.e., meant to be phased out as natural reproduction expands) or as an on-going source of production for fisheries, hatchery production will very likely continue to play an important role in all regions for the foreseeable future, especially if wild stocks continue to decrease and ecosystem changes increase.

Other common traits across regions can be found among the social aspects of salmon fisheries. Earlier, we explored the regional differences in stakeholders' expectations and types of dependence on salmon, but in all cases, regardless of the fisheries' size or scale, the stakeholder population is complex and diverse. Salmon managers and policy-makers in each region must deal with a range of stakeholder interests, whether for commercial, recreational, tribal, industrial, cultural, or other reasons. Knuth (Chapter 9), Beland and Bielak (Chapter 3), and MacDonald et al. (Chapter 15) illustrated the diversity of values, attitudes, and goals held within and between various types of salmon stakeholders, and these differences contribute to complexity in the social landscape. In addition, salmon managers in each region need to deal with nontraditional salmon stakeholders, such as dam operators and regulators, land use and urban planners, irrigators, and others who have competing uses for and impacts on salmon habitat. Salmon managers across North America cannot afford to consider stakeholder concerns as an afterthought or add-on to their management strategies; by misunderstanding or oversimplifying multistakeholder concerns and impacts, they risk their salmon management being incomplete and ineffective.

The social landscape is further complicated by the mix of political jurisdictions and institutions with roles or responsibilities related to salmon. In the eastern United States and Canada, the Great Lakes Basin, and the western United States and Canada, complex patterns of political boundaries are superimposed on the watersheds and coastlines of salmon habitat. Multiple nations (including tribal nations), states and provinces, counties, townships, and municipalities have various types of agencies or offices that directly or indirectly affect salmon or their habitat. Many of these political units are now joining together, through agreements or treaties, to form interjurisdictional decision-making or advisory bodies. For example, Dochoda and Jones (Chapter 11), Rideout and Ritter (Chapter 5), and MacDonald et al. (Chapter 15) described how the United States and Canada are coordinating some management activities through these types of groups. With the federal-provincial-First Nations relationship in Canada and the federal-state-tribal relationship in the United States differing in their respective roles and responsibilities, and with the variations in other institutions' direct and indirect roles related to salmon management, there is a great deal of institutional complexity that must be understood and sorted out in the process of building effective and efficient interjurisdictional partnerships.

Progress Toward Sustainability

Given these differences and similarities between regions, what lessons can be learned by comparing the experiences of salmon managers across the continent? Can any of these regions provide relevant contributions to management approaches in other regions? Despite some obvious regional differences, we believe this book provides evidence of general factors that either hinder or promote sustainable salmon production. By identifying and understanding these factors, managers may be better able to remove the hindrances and encourage those activities that promote sustainability. There is enough common ground between regions to believe that some factors described in one region may be adapted in other regional management strategies, thereby increasing the likelihood of making progress toward sustainability.

Factors that Hinder Sustainability

Perhaps the most important impediment to sustainable salmon production is habitat loss. Knudsen (Chapter 14), Rideout and Ritter (Chapter 5), Kocik and Brown (Chapter 1), and several others emphasized the ecological effects of dams, landscape change, and climate change in reducing available freshwater and marine salmon habitat. Many of these habitat threats are a consequence of the higher values (real and perceived) that society places on alternative uses of water resources and river systems. In the past, many decisions to develop hydroelectric power, barging routes, irrigation, and other water allocations were not made in the context of salmon habitat considerations. Habitat conservation and restoration is consistently listed as a top priority in salmon management.

The future is likely to bring even greater challenges. Demand will almost certainly continue to grow for freshwater resources—for water to drink, to irrigate crops, and to produce energy. At the same time, it is likely that freshwater supplies will become increasingly scarce in many regions because of climate changes or continued overallocation of water. As a result, the waters of the Great Lakes and salmon-producing rivers across North America will be the focus of numerous competing social, economic, and political demands. Salmon habitat concerns will likely increase—in some cases, the preservation of salmon stocks may become a luxury for which society may not be willing to pay.

Many harvesting practices also limit salmon sustainability. Overharvests and stock interceptions were and are especially evident on the east and west coasts, where mixed-stock fisheries are thought to lack sufficient conservation incentives. Commercial fishers managing businesses tend to operate within a short-term time horizon and try to maximize profits, which in turn are influenced by global market pressures. Salmon imports, often from foreign fish farms, tend to keep worldwide market prices low; consequently, North American operations need to harvest more fish to generate a profit. Harvest regulations alone (e.g., gear, open seasons, and harvest restrictions) have, so far, failed to protect many stocks from becoming threatened or endangered. With the exception of Atlantic salmon in the United States, overharvesting in recreational or subsistence fisheries is not generally thought to be an issue, but it is not inconceivable that further restrictions may be necessary if stocks continue to decline.

Closely related to habitat loss and overharvesting is a factor that deserves additional consideration: lack of information. A great deal of scientific uncertainty surrounds habitat and overharvesting issues (e.g., how and why is marine habitat changing? how should stock units be defined? how many fish are really out there?). In fact, disagreements about U.S. and Canadian salmon population and harvesting data have been significant roadblocks in Pacific salmon conservation and allocation. Scientific uncertainty affects other aspects of salmon conservation as well. Much more needs to be learned about long-term climate change, marine survival, stock genetics, and disease resistance, for example. Such information could help managers conserve habitat and adjust stocking and harvesting rates, but some degree of uncertainty will always remain. Salmon ecosystems are far too large and complex to be fully and completely understood given our current assessment technology; therefore, ecosystem management needs to be adaptive and accommodate some level of uncertainty (Walters 1986).

Salmon ecosystems also cross political and jurisdictional boundaries, and any lack of coordination or communication about salmon management between agencies within these units further hinders efforts at sustainability. With the exception of salmon raised in private aquaculture facilities, salmon are common, public resources, and various sectors of the public, such as the communities of place and communities of interest described by MacDonald et al. (Chapter 15), often develop competing expectations for salmon use or conservation. Agencies with public trust responsibilities not only need to understand the expectations held by stakeholders within their jurisdiction, but they also need to consider how the patchwork of expectations across the landscape affects their ability to conserve habitats and allocate salmon harvest. Consider Michigan's Pere Marquette River and its tributaries. This blue-ribbon salmon and trout river system spans only 153 miles, yet its watershed includes national forest land, state forest land, parts of four counties, and several cities and villages. Stakeholders have strong opinions about harvesting rates, sea lamprey control, riparian land use, access site development, and competing uses for this river (Cassuto 1994). Therefore, agencies with different types of responsibility for the river need to understand how this social landscape (a relatively simple one, compared with the Columbia or Fraser River systems) affects their ability to develop and achieve their goals. MacDonald et al. (Chapter 15), Knuth (Chapter 9), and Rideout and Ritter (Chapter 5) provided further evidence that salmon management institutions, as well as other stakeholder organizations, need to establish and maintain channels of communication and strategies for coordinating their activities that affect salmon and their habitats. Without such channels and strategies, the potential for costly conflicts increases, along with the potential to offset any progress toward sustainability.

Factors that Promote Sustainability

Whether one looks at ecological or social systems surrounding salmon, it is clear that recognition of the interconnectedness of components within *and* between these systems promotes sustainable management systems. Decisions in one agency can have ramifications for communities and ecosystems hundreds of miles away. Conversely, changes in one ecosystem component can have far-reaching impacts for salmon populations in multiple jurisdictions. This latter type of impact is clearly illustrated by the case of sea lamprey in the Great Lakes. On the heels of the catastrophic declines in lake trout throughout the Great Lakes, Dochoda and Jones (Chapter 11) and Goddard (Chapter 12) described how the sea lamprey was recognized as a common problem that required joint U.S.-Canadian solutions and, thus, how it served as an impetus for the formation of the binationally funded Great Lakes Fishery Commission. The interconnectedness between sea lamprey and multiple agencies responsible for Great Lakes fishery management is manifest in this international institution, which continues to provide a forum for addressing other common management issues (e.g., via the Joint Strategic Plan described by Dochoda and Jones, Chapter 11). Interestingly, the sea lamprey also forced managers to address the interconnectedness between the Great Lakes and its tributaries, as efforts to control this pest concentrate on the riverine life stages of the lamprey.

Through the GLFC, management agencies have established relationships for jointly addressing both upstream and open-water issues; these relationships have further facilitated joint action on salmon habitat and population dynamics. Furthermore, recognition of interconnectedness is also important in creating appropriate regulatory tools and conservation incentives. McKelvey and Miller's analysis (Chapter 16), leading to a proposal for side payments, is an excellent example. Their proposal is based on recognition of the potential interconnection between salmon conservation and restoration in one region and the allocation of harvest benefits in another. To the extent that salmon harvests can be regulated and habitats protected via complementary political and economic conservation tools, progress toward sustainability is much more likely.

Even when appropriate conservation regulations and incentives have been developed and implemented, support of stakeholders is needed to promote sustainability. Laws or other regulatory tools that lack support from the stakeholder community are less likely to be followed, potentially leading to costly enforcement and litigation and to negative impacts on salmon and their ecosystems. Communities that support and follow regulations and other management measures are more likely to contribute toward sustainability. This support is more likely when regulated communities understand and have some degree of input in forming regulations, as described by MacDonald et al. (Chapter 15). It is also facilitated when stakeholders have long-term expectations for the fishery that are in line with ecological limits on salmon productivity—this can lead to economic and social stability in salmon-dependent communities. For example, in eastern Canada, as described by Whoriskey and Glebe (Chapter 4) and Beland and Bielak (Chapter 3), many local communities rely on salmon-related tourism, and they may be more willing to support conservation measures (e.g., catch-and-release angling) that maintain expected levels of production. In the Great Lakes, salmon anglers developed expectations—based on the extraordinary growth and survival rates salmon enjoyed in the early years of this fishery—for catch rates that may be unrealistic in the long run (Gale 1987). More recently, successful efforts to inform stakeholders about the longer term ecological limits to Great Lakes salmon production have facilitated agreements to reduce salmon stocking levels in Lakes Ontario and Michigan. Finally, support is further facilitated by the extent to which conservation-oriented attitudes and values are held by a community, as described by Knuth (Chapter 9), and such values may spread through the processes of education or social learning. In short, the combination of supporting knowledge, expectations, and values among all types of stakeholders can be a powerful facilitator of sustainability.

Realistically, however, support of the stakeholder community is likely to be incomplete or tenuous. Therefore, managers need to consider other factors that promote sustainability: alternatives and adaptability. When stock declines necessitate reductions or closures of a fishery, communities that have other economic or recreational options may fare better than those that have greater dependence on salmon. If communities have the option to shift to other species or activities for income or pleasure, the potential for long-term conflict or economic hardship is reduced. At the same time, conservation and restoration of salmon may proceed with less interference, thereby enhancing the prospects for stock recovery. Though the importance of identifying and adapting to alternatives is clear for these communities, the same principles are equally applicable to the institutions that undertake scientific research or international cooperation. Researchers need flexibility to pursue alternative theories, models, and applications; access alternative funding sources; and adapt to changing stock and ecosystem conditions. International forums need to be flexible in response to changing social, economic, political, and ecological realities; they need to avoid the bureaucratic inertia that hinders information sharing, joint decision making, equity evaluation, and coordination of activities. For all types of salmon stakeholders, just like the salmon themselves, the ability to access alternatives and adapt to changing conditions will likely be the key to sustainability.

Recommendations

Given the inherent complexity of the ecological and social systems surrounding salmon fisheries, we are aware that the above descriptions of factors that hinder and promote sustainability may be oversimplified or incomplete. Nevertheless, we believe that these factors can help us form recommendations for enhancing the prospects of sustainable salmon production across North America.

In all salmon-producing regions, policy-makers and managers should consider several general guidelines. First, in light of the uncertainty surrounding the future of salmon stocks and their habitats, they should use a precautionary approach in deciding how to regulate the interactions between humans and fish. Friedland (Chapter 6) made an excellent argument for this approach, urging that lack of information should not limit conservation measures. Loss of salmon diversity has irreversible and probably undesirable consequences, so caution is warranted in the interest of preserving long-term conservation options. Second, though a precautionary approach points to limits on harvesting activity, creative combinations of harvesting opportunities and conservation incentives should continue to be explored. For example, the side payments proposal of McKelvey and Miller (Chapter 16) could play a role as part of a comprehensive conservation initiative in places where allocation and equity are at issue. To the extent that regionally appropriate conservation incentives can be developed, fishing limits under a precautionary approach may meet less stakeholder resistance. Third, those institutions that make decisions on regulations or incentives need to be flexible, adaptive, and transparent. Institutions generally need stakeholder support to meet their goals, especially as conditions surrounding the fishery change. As stakeholders have opportunities to access and observe the development of changes in management strategies, they may be more likely to understand and support (or at least not hinder) conservation-oriented decisions. Open, clear decision-making processes may help build trust between and among institutions and stakeholders. Fourth, institutions with roles and responsibilities related to salmon need to consider how well their current structures and processes conform to ecosystem function. MacDonald et al. (Chapter 15) emphasized the importance of organizing decision making around watersheds because landscape-level coordination can result in more complementary and effective forms of river-based habitat and stock conservation. Even in the marine mixed-stock fishery, those nations, states, and provinces that provide home waters for salmon need to work together in determining harvesting and allocation and in understanding large-scale habitat changes. Just as sea lamprey provided a "common enemy" around which Great Lakes institutions were organized, other large-scale ecosystem threats (e.g., climate change) may further facilitate the development of ecosystem-oriented institutional collaboration.

The importance of communication is evident in many of the above recommendations, but we offer additional guidelines for enhancing the exchange of information and ideas among salmon stakeholders. In an editorial in *Fisheries*, Robert Lackey (2001) asked readers to consider the drastic and pervasive lifestyle changes that would be needed to restore and sustain wild Pacific salmon (e.g., reduced consumption and standard of living), and he challenged fishery managers to assess the likelihood of achieving sustainability in light of the public's apparent unwillingness to make these changes. We agree that individuals and communities across the continent need more practical information about their direct and indirect impacts on salmon, and this underlies our first recommendation for communications. Public information and education campaigns are badly needed to increase awareness of and support for salmon conservation. Long-term impacts on public attitudes and values toward salmon require sustained and consistent messages about diverse issues related to their conservation; therefore, encouraging on-going public dialogue about salmon should be a primary communication goal. Second, on-going dialogues should also be fostered among all types of salmon fisheries professionals. Even within a single nation, state, or province, salmon professionals work in different disciplines, industries, and agencies, and for other special interests. If they can share information and learn from one another, the likelihood of developing comprehensive approaches to salmon conservation will increase. Third, communication between these professionals and other stakeholders can be facilitated by examining their patterns or networks of interaction. By mapping and analyzing the network of stakeholder relationships surrounding an issue such as salmon conservation, one can gain insight into how and why influence and power shape policy and management decisions and outcomes (see Knoke 1990; Wasserman and Faust 1994). We recommend such analyses be conducted for stakeholders in salmon issue-based networks—these can reveal how to utilize and possibly expand existing channels of communication among a representative and comprehensive set of stakeholders.

Our third area of recommendations builds on the need for further research and analyses in both ecological and social sciences. From an ecological standpoint, Knudsen (Chapter 14) stated it well: salmon "may arguably be the most studied group of exploited fishes," yet there remains a great deal that we do not know about their ecology—knowledge that would greatly aid our efforts to ensure sustainability. Some key research topics are effects of global change, causes and importance of phenotypic diversity, habitat requirements for conservation and restoration, and food web effects on salmon production. If we look at fisheries as social systems, on the other hand, further research is also needed about the people and communities that affect and are affected by salmon. Such research may reveal the roots of conflicts (and their resolution), identify and create appropriate conservation incentives, reveal the dynamics of influence and power in decision making, and enhance the effectiveness of communication and collaboration strategies. Economists, political scientists, anthropologists, and other social scientists can help identify and, if needed, ameliorate the social impacts of aggressive conservation measures. Both ecological and social scientists studying salmon bear an equal burden to share their understanding of their respective systems. These ecological and social systems are interrelated, and they both contain factors that may limit the achievement of salmon sustainability.

The future of salmon across North America depends on decisions made today—decisions made in context of historical shortcomings, current uncertainties, and future expectations. Salmon professionals in each region of North America and from various academic disciplines can and should offer their own experiences and insights into the factors that hinder and promote sustainability. Then our collective understanding of these factors can facilitate progress toward the goal of a sustainable future for salmon.

References

Alaska Department of Fish and Game. 2001. Alaska commercial salmon harvest data. <*www.cf.adfg.state.ak.us/geninfo/finfish/salmon/catchval/blusheet/94exvesl.htm*>.

Auditor General of Canada. 2000. The effects of salmon farming in British Columbia on the management of wild salmon stocks. Chapter 30 *in* 2000 Report of the Auditor General of Canada. Government Services Canada, Ottawa.

Bence, J. R., and K. D. Smith. 1999. An overview of recreational fisheries in the Great Lakes. Chapter 10 *in* W. W. Taylor and C. P. Ferreri, editors. Great Lakes fisheries policy and management. Michigan State University Press, East Lansing, Michigan.

Brown, R. W., M. Ebener, and T. Gorenflo. 1999. Great Lakes commercial fisheries: historical overview and prognosis for the future. Chapter 11 *in* W. W. Taylor and C. P. Ferreri, editors. Great Lakes fisheries policy and management. Michigan State University Press, East Lansing, Michigan.

Burnham-Curtis, M. K., C. C. Krueger, D. R. Schreiner, J. E. Johnson, T. J. Stewart, R. M. Horrall, W. R. MacCallum, R. Kenyon, and R. E. Lange. 1995. Genetic strategies for lake trout rehabilitation: a synthesis. Journal of Great Lakes Research 21 (Supplement 1):477–486.

Cassuto, D. N. 1994. Cold running river. The University of Michigan Press, Ann Arbor, Michigan.

Department of Fisheries and Oceans. 2001. Fisheries and Oceans Statistical Services, Canadian Landings Information. <*www.dfo-mpo.gc.ca/communic/statistics/landings/land_e.htm*>.

Gale, R. P. 1987. Resource miracles and rising expectations: a challenge to fishery managers. Fisheries 12(5):8–13.

Huppert, D. D. 1995. Understanding the U.S.–Canada salmon wars: why the Pacific salmon treaty has not brought peace. University of Washington, School of Marine Affairs.

Knoke, D. 1990. Political networks: the structural perspective. Cambridge University Press, Cambridge, UK.

Knudsen, E. E., C. R. Steward, D. D. MacDonald, J. E. Williams, and D. W. Reiser. editors. 2000. Sustainable fisheries management: Pacific salmon. Lewis Publishers, Boca Raton, Florida.

Krueger, C. C., and P. E. Ihssen. 1995. Review of genetics of lake trout in the Great Lakes: history, molecular ecology, physiology, strain comparisons, and restoration management. Journal of Great Lakes Research 21 (Supplement 1):348–363.

Lackey, R. T. 2001. Defending reality. Fisheries 26(6):26–27.

Walters, C. 1986. Adaptive management of renewable resources. McMillan, New York.

Wasserman, S., and K. Faust. 1994. Social network analysis: methods and applications. Cambridge University Press, Cambridge, UK.